The Data Literacy Cookbook

edited by Kelly Getz and Meryl Brodsky

Association of College and Research Libraries
A division of the American Library Association
Chicago • 2022

The paper used in this publication meets the minimum requirements of American National Standard for Information Sciences–Permanence of Paper for Printed Library Materials, ANSI Z39.48-1992. ∞

The ACRL Cookbook series was conceived of and designed by Ryan Sittler and Doug Cook.

Other books in this series:
The Scholarly Communications Cookbook, edited by Brianna Buljung and Emily Bongiovanni
The Teaching with Primary Sources Cookbook, edited by Julie M. Porterfield
The Library Outreach Cookbook, edited by Ryan L. Sittler and Terra J. Rogerson
The Critical Thinking about Sources Cookbook, edited by Sarah E. Morris
The Sustainable Library's Cookbook, edited by Raymond Pun and Gary L. Shaffer
The Library Assessment Cookbook, edited by Aaron W. Dobbs
The First-Year Experience Cookbook, edited by Raymond Pun and Meggan Houlihan
The Discovery Tool Cookbook, edited by Nancy Fawley and Nikki Krysak
The Embedded Librarian's Cookbook, edited by Kaijsa Calkins and Cassandra Kvenild

Library of Congress Control Number: 2022946511

©2022 by the Association of College and Research Libraries, a division of the American Library Association. All rights reserved except those which may be granted by Sections 107 and 108 of the Copyright Revision Act of 1976. Printed in the United States of America.

26 25 24 23 22   5 4 3 2 1

TABLE OF CONTENTS

vii   Introduction

SECTION 1. INTERPRETING POLLS AND SURVEYS

3   [[Ch1]]Survey Literacy: A Skills-Based Approach to Teaching Survey Research
Jesse Klein

6   [[Ch2]]Setting the Scene with Surveys: Using Polling Software to Demonstrate Primary and Secondary Data
Wendy G. Pothier

9   [[Ch3]]The Mini-study: A Three-Part Assignment for Original Data Creation, Summation, and Visualization
William Cuthbertson, Lyda Fontes McCartin, and Sara O'Donnell

SECTION 2. FINDING AND EVALUATING DATA

17   [[Ch4]]Three-Step Data Searching
Annelise Sklar

21   [[Ch5]]Transforming Research Questions into Variables: A Recipe for Finding Secondary Data
Alicia Kubas and Jenny McBurney

25   [[Ch6]]Sweeten the Search: Discover Data for Reuse with a Tool That Links Publications to the Underlying Data
Elizabeth Moss

30   [[Ch7]]The Most Vital Statistics: Finding and Analyzing Historical Mortality Rates
Alisa Beth Rod and Jennie Correia

34   [[Ch8]]Understanding the Enumerated World: Making Sense of Data as an Information Source
Alexandra Cooper, Elizabeth Hill, and Kristi Thompson

38   [[Ch9]]Looking at Data
Kay K. Bjornen

43   [[Ch10]]Interrogating the Data: What Data Sets Can and Cannot Tell Us
Kristin Fontichiaro

46   [[Ch11]]Data Zines: A Hands-On Approach to Community Curiosities
Tess Wilson

49   [[Ch12]]On the Hunt: Understanding and Analyzing GSS Data Extraction for Incorporation within Sociological Research Projects
Amy Dye-Reeves

52   [[Ch13]]Using Statistics to Define the Problem: Data and Service Learning
Amy Harris Houk and Jenny Dale

56   [[Ch14]]Data and Statistics in the News and Media
Kaetlyn Phillips

SECTION 3. DATA MANIPULATION AND TRANSFORMATION

61   [[Ch15]]A Kinesthetic Approach to Data: Moving to Understand Nominal, Ordinal, Interval, and Ratio Relationship in Data
Wendy Stephens

64   [[Ch16]]Text Mining Charcuterie Board
Yun Dai and Fan Luo

67   [[Ch17]]Anyone Can Cook (R)! Open Data with R, a Five-Week Mini-mester
Jay Forrest and Ameet Doshi

70   [[Ch18]]Software Carpentry Al Dente: Rendering Tech Training for Online Artisans
Peace Ossom-Williamson, Shiloh Williams, and Hammad Rauf Khan

73   [[Ch19]]A Recipe for Improving Online Instruction for the Carpentries
Kay K. Bjornen and Clarke Iakovakis

SECTION 4. DATA VISUALIZATION

79   [[Ch20]]Correlation Does Not Equal Causality: Introducing Data Literacy through Infographics and Statistics in the Media
Nick Ruhs

83   [[Ch21]]Pies, Bars, Charts, and Graphs, Oh My! A Data Visualization Appetizer
Haley L. Lott

86   [[Ch22]]Data Visualizations: The Good, the Bad, and the Ugly
Kaetlyn Phillips

89   [[Ch23]]Seasonal Visual Literacy: Using Current Events to Teach Data and Spatial Literacy Skills with Adaptable LibGuides
Jacqueline Fleming and Theresa Quill

93   [[Ch24]]To Visualize Is to Experience Data
Chelsea H. Barrett and Gerard Shea

97   [[Ch25]]Upping the Baseline for Data Literacy Instruction
Jessica Vanderhoff

Table of Contents

101 [[Ch26]]A Literacy-Based Approach to Learning Visualization with R's ggplot2 Package
Angela M. Zoss

104 [[Ch27]]Build Your Own Data Viz Pizza: A Modular Approach to Data Visualization Instruction
Rachel Starry

108 [[Ch28]]Veggie Pizza: Choosing a Data Visualization Tool
Rachel Starry

111 [[Ch29]]Four-Cheese Pizza: Color and Accessible Design
Rachel Starry

114 [[Ch30]]Data Visualization using Web Apps in a Rainbow Layer Cake
Yun Dai and Fan Luo

117 [[Ch31]]Graphical Abstracts: Creating Appetizing Infographics for Your Research Article
Aleshia Huber

SECTION 5. DATA MANAGEMENT AND SHARING

123 [[Ch32]]Making File Names for Digital Exhibits
Kate Thornhill and Gabriele Hayden

126 [[Ch33]]Data Management Failures: Teaching the Importance of DMPs through Cautionary Examples
Richard M. Mikulski

131 [[Ch34]]Low-Fat Research Data Management
Elizabeth Blackwood

134 [[Ch35]]Managing Qualitative Social Science Data: An Open, Self-Guided Course
Sebastian Karcher and Diana Kapiszewski

136 [[Ch36]]Seven Weeks, Seven DMPs: Iterative Learning around Data Management Plan Creation
Emma Slayton and Hannah C. Gunderman

140 [[Ch37]]Equitable from the Beginning: Incorporating Critical Data Perspectives into Your Research Design
Jodi Coalter, David Durden, and Leigh Amadi Dunewood

SECTION 6. GEOSPATIAL DATA

147 [[Ch38]]Challenge Accepted: Introducing Geospatial Data Literacy through an Online Learning Path
Joshua Sadvari and Katie Phillips

151 [[Ch39]]GIS for Success Series: Learning the Basics of QGIS Workshop
Kelly Grove

154 [[Ch40]]GIS for Success Series: Let's Make a Map in QGIS Workshop
Kelly Grove

157 [[Ch41]]Statistical and Geospatial Literacy for Integrative Genetics
Jay Forrest and Chrissy Spencer

161 [[Ch42]]Web Map Layer Cake: Teaching Web Mapping Skills with Leaflet for R
Sarah Zhang and Julie Jones

SECTION 7. DATA IN THE DISCIPLINES

167 [[Ch43]]Data in Context: How Data Fit into the Scholarly Conversation
Theresa Burress

171 [[Ch44]]Let the Dough Rise! Integrating Library Instruction in a Digital Humanities Course
René Duplain and Chantal Ripp

175 [[Ch45]]Ethics and Biodiversity Data
Rebecca Hill Renirie

179 [[Ch46]]Data Decisions and the Research Process in the Sciences and Social Sciences
Nicole Helregel

182 [[Ch47]]Financial Data for Economics Students
Jennifer Yao Weinraub

183 [[Ch48]]Stuffed Shiny App with Business Intelligence
Yun Dai and Fan Luo

189 [[Ch49]]Fast Casual Marketing Strategies
Juliann Couture, Halley Todd, and Natalia Tingle Dolan

192 [[Ch50]]When and Where: A Framework for Finding and Evaluating Social Science Data for Reuse
Ari Gofman

197 [[Ch51]]Data Literacy Layered Lasagna for Preservice Teachers
Brad Dennis and Allison Hart-Young

SECTION 8. DATA LITERACY OUTREACH AND ENGAGEMENT

203 [[Ch52]]Data Visualization Day: Promoting Data Literacy with Campus Partners
Wenli Gao

206 [[Ch53]]Getting Messy Ourselves: An Experiential Learning Curriculum for Subject Librarians to Engage with Data Literacy
Adrienne Canino

211 [[Ch54]]Research Data Management Stone Soup: Gauging Team Competencies
Michelle Armstrong, Megan Davis, Ellie Dworak, Yitzhak "Yitzy" Paul, and Elisabeth Shook

214 [[Ch55]]**Data Literacy Family Style: Full-Day Professional Development**
Molly Ledermann, Emilia Marcyk, Terence O'Neill, and Dianna E. Sachs

217 [[Ch56]]**Everyone Is Welcome at the Table: Outreach for Data Management and Data Literacy in Research Assignment Design**
Shannon Sheridan and Hilary Baribeau

220 [[Ch57]]**Seasoning and Simmering: Cultivating Data Literacy Skills through an Open Data Hackathon**
Peace Ossom-Williamson

223 [[Ch58]]**From Soup to Nuts: Finding Your Way around the Data Services Buffet**
Jane Fry and Chantal Ripp

226 [[Ch59]]**Teaching Data Literacy and Computational Thinking in Educational Technology**
Lesley S. J. Farmer

SECTION 9. DATA LITERACY PROGRAMS AND CURRICULA

231 [[Ch60]]**Cooking Up a Data Literacy Course**
Claire Nickerson

238 [[Ch61]]**Baking a Data Layer Cake: Scaffolding Data Skills through Video Vignettes**
Shannon Sheridan

241 [[Ch62]]**Building Data Literacy through Scaffolded Workshops: Experiences and Challenges**
Jiebei Luo and Yaqing (Allison) Xu

244 [[Ch63]]**Data Literacy Appetizers: LibGuide Data Instruction Modules for Undergraduates**
Beth Hillemann and Aaron Albertson

247 [[Ch64]]**Data as Curation: Framing Data Creation as a Critical Practice through Collections-Based Research Inquiry**
Gesina A. Phillips, Tyrica Terry Kapral, Matthew J. Lavin, and Aaron Brenner

253 [[Ch65]]**Quantitative Data Skills for Undergraduates: A Seminar Series for Social Science Students**
Whitney Kramer and Amelia Kallaher

Introduction

No one would dispute that today we are awash in data in ways that we weren't in the past.[1] From today's smart devices to customer reviews and social media posts, students create and are confronted with data in many formats. Data literacy enables individuals to access, interpret, critically assess, manage, handle, and ethically use data.[2] Data literacy falls under the umbrella of information literacy. As students apply critical thinking skills to navigate the data ecosystem and enter into the scholarly conversation through collection, analysis, interpretation, and evaluation of data, they engage in the core concepts of ACRL's *Framework for Information Literacy for Higher Education*.

Librarians have published about data literacy instruction for almost twenty years,[3] sometimes in regard to what data librarians should teach,[4] sometimes as part of information literacy,[5] and sometimes as an expanded role in helping researchers through the research data life cycle.[6] Technological limitations previously hindered students and researchers from easily conducting some of these data processes, but advances in computational power and interface design that occurred during the twenty-first century have enabled students and researchers to participate in the data life cycle with fewer barriers.

Although some libraries have designated data librarians and entire departments devoted to research data services, many rely on reference librarians, subject specialists, or information literacy librarians to provide and promote data literacy instruction. With few courses offered in library schools that directly address data literacy instruction,[7] a brigade of academic librarians have acted as the Julia Childs of their era—cooking up data literacy lessons from scratch. In this cookbook, these librarians offer well-tested recipes to help hungry readers create their own classic dishes.

This volume of the Cookbook series includes a variety of approaches to teaching data literacy, from self-paced learning modules to for-credit courses. There are recipes for those new to data literacy instruction looking to taste test and recipes for more specialized chefs, including some discipline-specific and project-based recipes. The chapters range from simple activities to one-shot sessions to full semester-long courses. We aim to provide librarians with recipes that guide students as both consumers and producers in the data life cycle. We've categorized most chapters into sections based on the learning outcomes and an additional section, Data Literacy Programs and Curricula, which offers entire meal plans with large-scale learning objectives. The sections are

- *Interpreting Polls and Surveys:* recipes designed to help students understand the data we commonly gather from surveys and polls
- *Finding and Evaluating Data:* chapters on how to search for, locate, and evaluate data collected by others
- *Data Manipulation and Transformation:* chapters that describe how to categorize, clean, manipulate, and transform data and prepare them for analysis
- *Data Visualization:* lessons on how to create and evaluate information visualizations
- *Data Management and Sharing:* lessons on data management planning considerations as well as the ethical responsibilities of those who collect and share data
- *Geospatial Data:* recipes including ways to *visualize* and tell stories using geospatial data and GIS software
- *Data in the Disciplines:* recipes crafted specifically for the areas of business, ecology, education, social sciences and the humanities
- *Data Literacy Outreach and Engagement:* workshops and activities about marketing data literacy and data services to your colleagues and the campus community
- *Data Literacy Programs and Curricula:* sequenced and scaffolded multipart instruction that takes place over a series of weeks or workshops and spans large learning objectives

Many sections have overlapping learning outcomes. We invite the reader to sample recipes from multiple sections and combine them to whip up a scaffolded curriculum!

Introduction

Our chefs have been very generous in sharing their lesson plans, their advice, and their teaching materials. Many chefs included websites with more information and offered CC-BY licenses for their images. We are very grateful for their creativity and their expertise in bringing data literacy instruction to a broad audience. Please try these recipes, experiment, write in the margins, add your own spices, and let us know how they came out.

While we've enjoyed collecting these recipes, our aim is to share them and improve everyone's data literacy instruction skills. Bon appétit!

Kelly Getz and Meryl Brodsky

NOTES

1. Nicole Lazar, "Data, Data, Everywhere…," *Harvard Data Science Review 2, no. 2 (May 27, 2020)*, https://doi.org/10.1162/99608f92.a6e7a24e.
2. Javier Calzada Prado and Miguel Ángel Marzal, "Incorporating Data Literacy into Information Literacy Programs: Core Competencies and Contents," *Libri 63, no. 2 (May 31, 2013)*, https://doi.org/10.1515/libri-2013-0010.
3. Karen Hunt, "The Challenge of Incorporating Data Literacy into the Curriculum in an Undergraduate Institution," *IASSIST Quarterly* 282, no. 3 (Summer/Fall 2004): 12–16, https://iassistquarterly.com/public/pdfs/iqvol282_3hunt.pdf.
4. Jake R. Carlson, Michael Fosmire, Chris Miller, and Megan R. Sapp Nelson, "Determining Data Information Literacy Needs: A Study of Students and Research Faculty," preprint, Libraries Faculty and Staff Scholarship and Research, paper 23 (2011), https://docs.lib.purdue.edu/lib_fsdocs/23/.
5. ACRL Research Planning and Review Committee, "Top Trends in Academic Libraries: A Review of the Trends and Issues Affecting Academic Libraries in Higher Education," *College and Research Libraries News 75, no 6 (2014)*, https://doi.org/10.5860/Crln.75.6.9137.
6. Carol Tenopir, Robert J. Sandusky, Suzie Allard, and Ben Birch, "Research Data Management Services in Academic Research Libraries and Perceptions of Librarians," *Library and Information Science Research 36, no. 2 (April 2014): 84–90*, https://doi.org/10.1016/j.lisr.2013.11.003.
7. Lisa Federer, "Defining Data Librarianship: A Survey of Competencies, Skills, and Training," *Journal of the Medical Library Association 106, no. 3 (July 2018): 294–303*, https://doi.org/10.5195/jmla.2018.306.

Section 1.
Interpreting Polls and Surveys

3   [[Ch1]]**Survey Literacy: A Skills-Based Approach to Teaching Survey Research**
Jesse Klein

6   [[Ch2]]**Setting the Scene with Surveys: Using Polling Software to Demonstrate Primary and Secondary Data**
Wendy G. Pothier

9   [[Ch3]]**The Mini-study: A Three-Part Assignment for Original Data Creation, Summation, and Visualization**
William Cuthbertson, Lyda Fontes McCartin, and Sara O'Donnell

Survey Literacy
A Skills-Based Approach to Teaching Survey Research

Jesse Klein, PhD, Assistant Research Scientist, Social Data Science Center, University of Maryland, jrklein@umd.edu

NUTRITION INFORMATION

Surveys are powerful tools used in many disciplines and organizations to gain valuable insight on the attitudes, behaviors, and beliefs of any population. Despite their popularity and versatility, however, surveys tend to be overused and misunderstood. The proliferation of online easy-to-use survey platforms, such as Qualtrics, Google Forms, and SurveyMonkey, has contributed to an abundance of surveys in many areas of research. However, although these tools ease the task of survey creation, they also contribute to mistakes in survey design, such as too many questions, inappropriate question types, lack of foresight about the data being collected, spelling and grammatical errors and so on. Therefore, the responsibility to engage with the principles of effective survey research—or survey literacy—is more important today than ever before.

TARGET AUDIENCE AND NUMBER SERVED

We all consume information supported with statistics that are informed by survey data, making the concepts of survey literacy ubiquitous and critical to becoming information literate. While this lesson can be easily adapted for most audiences, a group of 20 to 30 undergraduate students or first- and second-year graduate students are the ideal audience for a high-impact, interactive session.

LEARNING OBJECTIVES

As a result of engaging in these activities, participants will be able to

- navigate secondary data repositories for examples of survey questions
- translate a topic or question into measurable concepts
- clearly define concepts
- describe several ways to measure a concept
- operationalize concepts into single- or multiple-item survey questions
- think critically about how survey design relates to survey data
- consider ways that survey bias can be introduced through survey design
- become a good steward for effective and ethical practices in survey research

COOKING TIME

60 to 90 minutes

DIETARY GUIDELINES

The Research as Inquiry frame from ACRL's *Framework for Information Literacy for Higher Education* is easier to teach through practice, which provides a good foundation for integrating survey literacy into broader information literacy instruction. With the focus on research questions and methods, navigating this frame with specific methods and examples make the content more relatable for participants.

INGREDIENTS

- Presentation tool (e.g., PowerPoint, Post-it Easel Pad, markers, etc.)
- Example infographic, report, or fact sheet that provides statistics
- Secondary data repository to search for survey questions (e.g., Inter-university Consortium for Political and Social Research [ICPSR], General Social Survey [GSS] Data Explorer, etc.)
- Online survey platform (e.g., Qualtrics, Google Forms, SurveyMonkey)

PREPARATION

To get students engaged with survey literacy, the session starts by asking them about their experiences with surveys and interpreting statistics from survey data. As a bridge between these introductory questions and the exercise, you will use an example infographic, report, or fact sheet that provides statistics and ask them to walk you through the data and concepts being communicated. Prior to this initial step as described in the

Section 1. Interpreting Polls and Surveys

instructions, you will want to engage with the example first and adjust its theme and sophistication depending on the discipline and skill of the learners. To prepare for the exercise, based on the audience and available time, you should also take the time to select between three and five concepts varying in complexity, such as academic success, happiness, poverty, health, stress, job satisfaction, and so on. Then gather some examples of real survey questions that measure those concepts for reference.

INSTRUCTIONS

1. Discuss surveys as a research method and relate to students' experiences taking surveys or interpreting statistics in the media based on survey data (e.g., polls, market research, etc.). For example, ask students if they have ever taken a survey, if they were aware of the survey's purpose, whether it was clear what the survey was measuring, if they have ever noticed biased language in a survey, and/or whether they interpreted survey results presented as statistics or graphs in the media.
2. Present an infographic, report, or fact sheet that provides statistics. For example, many government agencies and nonprofit organizations have fact sheets or infographics to easily communicate data to the public.
3. To get students thinking critically about survey data, ask them, "What type of survey question(s) would you need to gather the data that support this statistic?" and "What is the concept behind this statistic?"
4. Based on the discipline and audience, present your preselected three to five concepts to the group. Provide enough space for you to write or type around the concepts, including space for "Define" and "Measure" under each concept (preferably side-by-side).
5. Ask, "How would you define this concept?" Then, one by one, go through the concepts with participants and write down their definitions of the concepts under "Define." There will be several definitions. This is an important part of the exercise.
6. Next to each definition, in the space titled "Measure," go through each definition and ask, "Now, how would you measure the concept defined this way?"

Note: Students may suggest open-ended measurements and, while it is worth noting that there is value to giving survey participants space for open-ended responses, for the purpose of this exercise you want to move students toward closed-ended questions with distinct response categories. Translating complex concepts into precise measurements with specific response categories is central to survey literacy as explored in this exercise.

7. Facilitate a discussion where students refine the suggested measurements into survey questions with clearly outlined response categories, such as "Strongly agree," "Agree," "Neither agree nor disagree," "Disagree," and "Strongly disagree."
8. If there is time, this is a great opportunity to discuss how bias gets introduced to surveys through words and phrases. Within the context of the concepts, discuss bias-free and inclusive language in the wording of the survey questions you constructed together.

Note: Chapter 5 in the 7th edition of the *Publication Manual of the American Psychological Association* provides "Bias-Free Language Guidelines," which you can use to help facilitate this discussion and encourage students to think critically about bias in their writing in general.

9. Demonstrate how to locate secondary data repositories, specifically ICPSR and the GSS Data Explorer because those have survey questions and variables for all disciplines. For example, on the ICPSR main page, go to Find Data, and then select Search/Compare Variables. Explain to students that survey questions turn into variables after data collection and that these terms can often be used interchangeably.
10. In the Search/Compare Variables search box, type any of the concepts you just discussed. For example, type in "health," then scroll through the results to show participants the bevy of survey questions that exist to help explore the ways in which to measure health. This is just one example; instructors may choose surveys from other disciplines as well.
11. This can be a good stopping point if the session is time-restricted.
12. If there is time for a longer instruction session, the next step would be to take the

concepts that the class created measurements for and build part of a survey with a few of the questions. Although this would require demonstrating some of the more technical aspects of creating a survey on a platform such as Qualtrics, you can demonstrate and have students watch or follow along as a section is built that can then be previewed. This extra step brings the process full circle for participants and puts the lesson into a visual survey they can relate to, and they now know how to articulate the process on their own.

Note: This is a good time to introduce students to the concept of pilot testing, where the survey goes through edits for spelling, grammar, and inclusivity and tests for measurements, platform navigation and accessibility, and proper data collection. Many mistakes we see In contemporary surveys can be prevented through proper pilot testing.

REVIEWS/ASSESSMENT STRATEGY

After going through the preselected concept definitions and measurements, give students (alone or in groups) a few minutes to select their own concept, define it, and develop two possible ways to measure it. Share as a group, and focus on discussing further the concepts that posed the greatest difficulty for students. For an embedded course with the opportunity for a formal assignment, ask students with preselected topics to use GSS Data Explorer to choose three demographic questions, three questions with any response scale, and three questions with the same response scale. Using these GSS questions, students can then create a Qualtrics or Google Forms survey to share with the instructor or class to get experience with the tool and to think critically about whether the questions measure what they are intended to measure.

ADAPTING THE RECIPE

This activity can be applied to any discipline because surveys are used to gather attitudes, behaviors, and beliefs on every aspect of society. Data repositories house survey data on any topic representing concepts from STEM (science, technology, engineering, and math) to humanities and everything in between.

ADDITIONAL RESOURCES

American Psychological Association. *Publication Manual of the American Psychological Association*, 7th ed. Washington, DC: American Psychological Association, 2020.

GSS (General Social Survey) Data Explorer, https://gssdataexplorer.norc.org.

Inter-university Consortium for Political and Social Research (ICPSR), https://www.icpsr.umich.edu.

Setting the Scene with Surveys
Using Polling Software to Demonstrate Primary and Secondary Data

Wendy G. Pothier, MSLS, MS, Associate Professor and Business & Economics Librarian, Dimond Library, University of New Hampshire, wendy.pothier@unh.edu

NUTRITION INFORMATION
This recipe bakes up an icebreaker activity that provides students with an introduction to the differences between primary and secondary data and how both serve a purpose in telling a research story. Through active learning, this short activity is done at the start of an instruction session to demonstrate the definitions of primary and secondary data and how both are used in research. Though this recipe is available in many flavors and packages, this chef usually serves it in the cuisine of business market research.

TARGET AUDIENCE AND NUMBER SERVED
This recipe serves a wide range of audiences, but the chef has served it to undergraduate business students doing market research in class sizes of 10 or more.

LEARNING OUTCOMES
This icebreaker activity helps set the stage for the rest of the library research session on using secondary data resources.

Students will
- utilize secondary data in their marketing research project
- discern the different roles that primary and secondary data play in telling the research story

COOKING TIME
This recipe usually takes between 5 and 7 minutes at the start of an hour-long instruction session.

DIETARY GUIDELINES
This recipe references the frame Information Has Value from ACRL's *Framework for Information Literacy for Higher Education*. It demonstrates how secondary data add context and benchmarking to the primary data students collect for their marketing plans. It also introduces the concept that data have a fiscal value, as students learn how businesses acquire data to make decisions.

INGREDIENTS
- Students
- Polling software (This chef uses Mentimeter—https://www.mentimeter.com.)
- A related chart or data graphic from a secondary research resource (This chef uses graphs from Statista.)

PREPARATION
Create a polling slide or survey question in advance of the session. This question should relate to students. In this case, the librarian asks about student preferences for local pizza restaurants.

Find a statistic that relates to the student polling or survey question. For example, the librarian asks students about their pizza restaurant preferences and then later shares a statistic relating to national trends in pizza restaurant preferences. The instructor frequently uses the Statista database for finding topical data.

INSTRUCTIONS
To start the session, the librarian uses polling software to engage students with a hands-on activity. This will serve as the introduction to a session on finding secondary data sources. The librarian asks the students a question (see figure 1) regarding which pizza restaurant they prefer. Once students respond to the poll, which captures the data and displays them in real time to the class, the class discusses what we can learn from the results.

FIGURE 1
Example of primary data collection using polling software (conducted live in class session).

FIGURE 2
Example of primary data collection with another audience (previous class session).

Next, the librarian shares data from other classes' responses to this question and together the class discusses how the data compare (see figure 2).

Finally, the librarian shares screenshots of data from a library database (see figure 3) that looks at the pizza restaurant industry and provides additional context for how we might interpret the local data versus the industry or business trends. Does the industry data survey a demographic sample that is different from the class demographics? The librarian asks students how location/geography or respondent demographics, survey method, and sample size affect the survey results. This leads to a brief discussion of which factors (also called variables) might make secondary data relevant to their research story and how it will inform their collecting of primary data.

This demonstration helps students think about how survey questions vary between primary data collection and using secondary data sources. It encourages students to think about what is already known while developing their primary research study. The activity demonstrates the definitions of primary and secondary data through active learning. This activity also introduces students to the value of utilizing both primary and secondary data to create their narrative, as well as what secondary data can show about a topic to help businesses make data-driven decisions. It serves as an introduction to a full session on secondary research data resources related to their market research projects.

Section 1. Interpreting Polls and Surveys

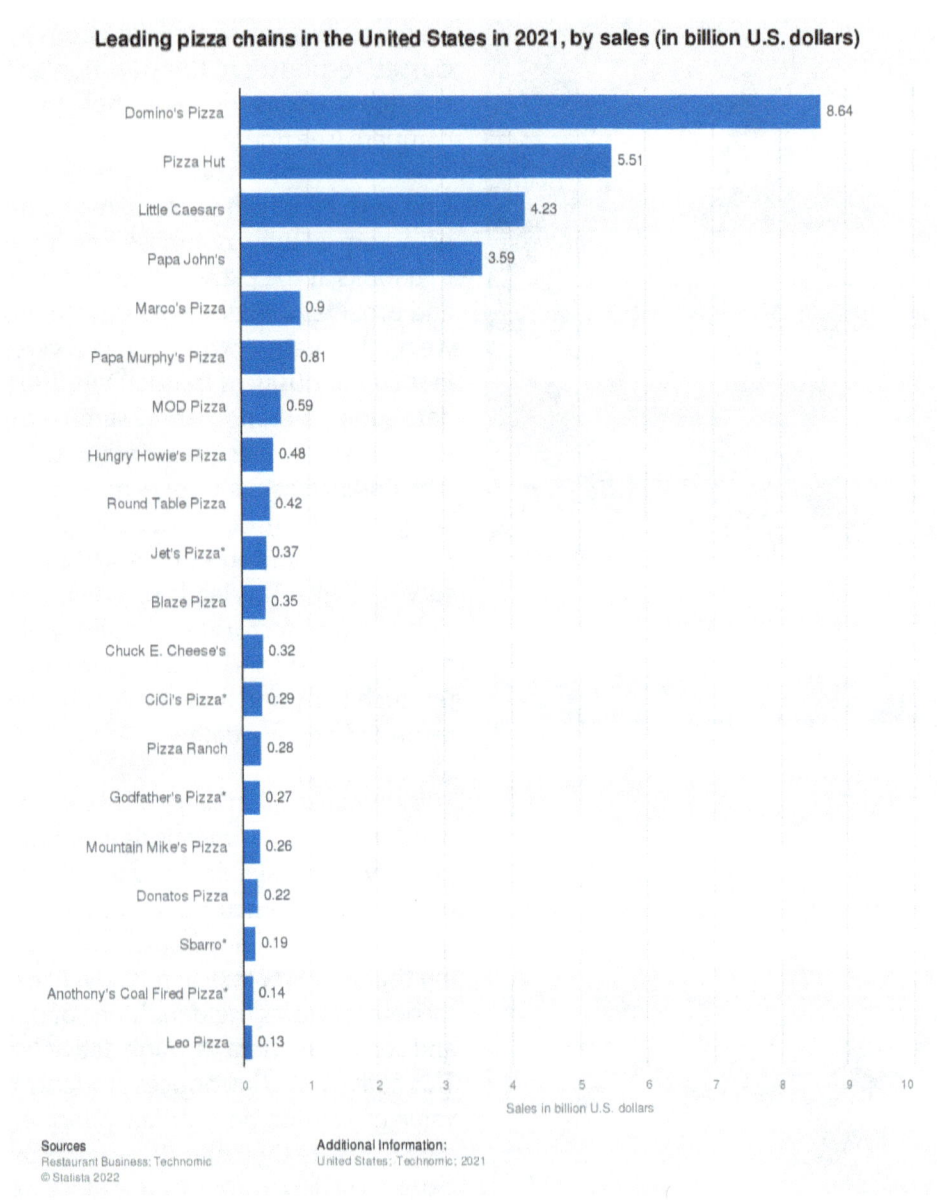

FIGURE 3
Restaurant Business. (August 3, 2022). Leading pizza chains in the United States in 2021, by sales (in billion U.S. dollars) [Graph]. In Statista. Retrieved September 04, 2022, from https://www.statista.com/statistics/261888/ranking-of-pizza-chains-based-on-us-sales/.

REVIEWS/ASSESSMENT STRATEGY
For assessment, the librarian uses the polling software for a follow-up quiz question at the end of the session to make sure the icebreaker concepts were understood by the class.

ALLERGY WARNING
Students will need to have a laptop or other device to participate in using the polling software.

CHEF'S NOTES
The chef has conducted the icebreaker in a variety of settings—in person, virtual, and hybrid. The live polling software enables participation in all three scenarios with a real-time display of results.

The Mini-study
A Three-Part Assignment for Original Data Creation, Summation, and Visualization

William Cuthbertson, Instruction Coordinator and Undergraduate Engagement Librarian, Meriam Library, California State University, Chico, wcuthbertson@csuchico.edu; Dr. Lyda Fontes McCartin, Director, Center for the Enhancement of Teaching and Learning, University of Northern Colorado, lyda.mccartin@unco.edu; Sara O'Donnell, Children's Librarian, Ledding Library of Milwaukie, Oregon, odonnells@milwaukieoregon.gov

NUTRITION INFORMATION

This series of assignments is designed to be the culminating project of a credit-bearing information literacy course, incorporating the knowledge and skills that students have practiced over the duration of a semester. While elements of the assignments can be integrated into a one-shot class, students will be most successful if they already have a background in basic research skills. Before beginning this assignment, students will need to have competency identifying and writing a research question, finding supporting literature, and reading and understanding scholarly articles. In short, they must be capable consumers of academic research. This assignment will introduce them to content creation—developing their own data collection tool, writing and formatting a brief research article using their findings, and finally presenting their data in a visual format.

This recipe consists of a sequence of three assignments that introduce students to data creation, visualization, and reporting. It provides students with hands-on experience in creating data in order to format and present the results of a research study.

TARGET AUDIENCE AND NUMBER SERVED

This recipe is designed for first-year students new to college-level academic research in any discipline. It was designed as a sequence of assignments for a library credit course but can be modified for one-shot instruction. This recipe can be used for small or large classrooms and with students working individually or in groups.

LEARNING OUTCOMES

Students will
- design and execute a small research study
- collect, analyze, and report original research data
- write up their results as an academic research study
- present their findings in a visual format, such as a presentation or infographic

COOKING TIME

The mini-study is designed as a sequence of three assignments for a library credit course where students are engaged in a long-term research project. Through the assignments, students conduct a brief study relating to their research question. They may conduct their study using interviews, a questionnaire, or a combination of these methods. Students then analyze their data and present them in a visual format that illustrates their study's findings.

When used in a credit course, the assignment takes students multiple weeks to complete. Approximately 3–4 days should be dedicated to the assignment and presentations. For this assignment to be successful, students must be introduced to, and have practiced skills in, reading and understanding scholarly research *before* beginning the mini-study. Students should also be working on a research project and have a research question.

If modifying the full mini-study for one-shot instruction, a librarian must have at least three separate instruction sessions of 60–75 minutes. A librarian may need to provide supplemental materials such as videos for students to watch to cover some topics. The assignment works best when integrated into a discipline-specific course. See the Chef's Notes for ideas on how to break down the assignment for use in one-shot instruction without an embedded research assignment.

Section 1. Interpreting Polls and Surveys

DIETARY GUIDELINES
This sequence of assignments provides undergraduate students the opportunity to create and present data. By presenting their own data, students engage in real research and learn how academic literature connects to original research. Through this sequence of assignments students apply these frames from ACRL's *Framework for Information Literacy for Higher Education*:

- *Research as Inquiry:* Students engage in real-life research and connect abstract research topics to real data.
- *Information Creation as a Process:* By going through the research process and incorporating their own research into a paper or project, students learn how to create information.

INGREDIENTS
- Suggested lesson plan on reading scholarly research: https://libguides.unco.edu/bridgethegap
- Mini-study assignments (see Instructions)
- Readings and videos
  - Krum, Randy. "The Science of Infographics." Chapter 1 in *Cool Infographics: Effective Communication with Data Visualization and Design*, 1–51. Indianapolis, IN: Wiley, 2013.
  - Tufte, Edward. *Visual Explanations: Images and Quantities, Evidence and Narrative*. Cheshire, CT: Graphics Press, 1997.
  - Time. "A Look at Income Inequality in the United States." February 20, 2020. YouTube video, 3:36. https://youtu.be/qc7g6Uhi1i4.
  - Politizane. "Wealth Inequality in America." November 20, 2012. YouTube video, 6:23. https://youtu.be/QPKKQnijnsM.
- Infographic creation tools. The following tools are free and will help students create an infographic. Verify that the selected tool allows students to download or present the graphic in a suitable format. Confirm that these tools are accessible with a screen reader for students needing accommodation. Hand-drawn infographics are also acceptable and often preferred by students.
  - Canva, https://www.canva.com/create/infographics/
  - Easelly, https://www.easel.ly
  - Piktochart, https://piktochart.com
  - Infogram, https://infogram.com
  - Venngage, https://venngage.com
  - PowerPoint, Microsoft Office software

PREPARATION
Student Preparation
- Students must have a research question selected for a library course or discipline-specific course research project.
- Students should be introduced to reading and understanding scholarly research articles. We recommend Brianne Markowski's lesson plan for teaching students how to read academic articles (and how to synthesize sources in student writing) found at https://libguides.unco.edu/bridgethegap.

Librarian Preparation
- Review the three assignments and determine which you will use.
- Review the materials in the Ingredients section and determine which resources to use in your instruction sessions.
- If using the mini-study with one-shot instruction sessions, consult with the faculty member about the assignment and make adjustments as needed for the course.
- Determine how many sessions you are able to schedule with the students and adjust your instruction as needed.
- Determine, in consultation with the faculty member, how students will learn to read a research study and when they can develop their research question.
- Edit the assignments for your context. Print or post the assignment in your learning management system.

INSTRUCTIONS
The following provides the detailed sequence of three assignments. At this point in a credit course, students should have a research topic selected and have experience reading and understanding scholarly literature.

Note that the sequence below culminates in a five-to-seven-minute lightning talk presentation. The sequence can also be modified to work as an end-of-semester academic poster presentation. In this case, the write-up of the mini-study can be included on the poster itself, while students talk through their findings and infographics with the audience. Both the lightning talk and poster presentation provide students the experience of verbally sharing their experience in creating, interpreting, and visualizing their data sets.

The mini-study provides students with practice in presenting research findings and discussing scholarly research. You may adjust the assignment details for your context. For example, in assignment 1 you may want to lower the sample size. Additionally, if using these assignments in one-shot instruction sessions, you may choose, in consultation with the faculty member, to use only assignment 1. Given the time you are allotted and how collaborative your faculty member wants to be, you may combine assignment 1 with 2 or 3.

Assignment 1: Mini-study
Assignment 1 asks students to conduct a mini research study. They use the results of the study to learn how to format an academic article and discuss scholarly research. Students must have a sample size of at least 6 people and may use interviews, surveys, or a combination.

Instructions for Students
For your mini-study you will conduct a study to submit now and to incorporate into your lightning talk at the end of the course. You must conduct your study with a sample size of *at least six* students. You may choose to conduct your study using interviews, a survey or questionnaire, or a combination of the two methods.

Part 1: Find Participants and Collect Demographic Data
- Sex
- Age
- Major
- Year in college

Part 2: Conduct the Study
You are required to ask at least three questions relating to your research question. You may add more questions if you are so inclined. *For example,* if you were writing a study to discover whether students had plagiarized, you might ask the following questions:
1. Did you plagiarize written assignments in high school?
2. Have you ever plagiarized a written assignment for a class in college?
3. Why did you plagiarize? Select all that apply. (If you are interviewing, just write the answers.)
    » I did not care about the assignment.
    » I did not start the assignment on time.
    » I had other assignments to complete at the same time.
    » Other (please explain).

Ask the questions and record the answers in a consistent manner. Feel free to ask the instructor for help in writing your questions.

Part 3: Analyze the Results
Once you have collected your data, analyze the results. The goal is to meaningfully connect the response information from your sample to your research question. You can also use the demographic breakdowns of your participants and connect those to your research question.

For example, the study could answer the following questions:
- How many students in your study plagiarized?
- How many students in your study plagiarized in both high school and college?
- What is the most common reason given for why students plagiarize?
- Do women plagiarize more than men?
- Is there an age group which plagiarizes more than others?

Part 4: Write Up Your Study
You will write up your findings in the format of a small research study. You *must* include the following sections:
- *Abstract:* Write 2–3 sentences. One should explain your research question and why it's important. The other(s) summarize your findings overall.
- *Introduction:* Relate your survey to your research question here. How do they connect? How does the information you gather matter and relate to your research question?
- *Methods*
    - *Participants:* What are their demographics? How did you select them?
    - *Data collection method:* Did you do an interview? A survey? Something else?
- *Results and Analysis:* See part 3.
- *Discussion*

In addition, you must use proper American Psychological Association (APA) formatting (http://apastyle.apa.org/style-grammar-guidelines/references/basic-principles) and include
- Title page
- Running header on each page
- Page numbers

Assignment 2: Infographic

In this assignment students take raw data from their own research and present them clearly for their peers.

Instructions for Students

Based on the principles of information presentation you learned from the Krum and Tufte readings, your assignment is to take *at least two points of data* from the data you gathered in your mini-study, statistical information you found on your topic or within the publications you found through your research, and represent that information in visual form.

Creating Your Infographic

The following software programs will help you to create an infographic. Before you commit to a program, however, ensure that it allows you to download the end product in a format you are able to submit for class.

Microsoft PowerPoint and *Google Slides* are viable options for creating infographics. There are other recommended infographic creation tools in the Ingredients list.

Hand-drawn infographics, so long as they meet the grading criteria, are a perfectly acceptable option.

Note: Icons and pictures can be downloaded to be integrated into your infographic from online searches or via the Noun Project: http://thenounproject.com. These graphics must be attributed on your infographic as per the artist's or creator's wishes.

As you create your infographic, consider the following questions:
- As it relates to my topic, what are the most relevant pieces of information to convey from my selected sources?
- How would visualizing this information allow for better comprehension by readers?
- In what ways can I present this information visually so that it
  - simplifies the complexities of the information being conveyed?
  - brings to light correlations that may have been obscured before?
  - engages readers in the overall topic and my approach to the research?

Questions to consider as you review and explore other infographics and prepare your own:
- What specific information is this graphic asking me to consider?
- How do the visual elements illuminate the data presented?
- How do the visual elements distract from the data presented?
- Is the graphic effective in presenting its data overall? How so?
- If it is not effective, what would you suggest to make the data—and the information conveyed through those data—clearer to readers?

As you present your infographic to the class, you may want to hand out copies of the infographic to help illustrate some of the information that you uncovered in your research. Note that infographics that you share must be created by *you*. They *cannot* be printouts of graphics that you found online. The goal is to take *raw data* from the research you have found and present that information in a clearer, more visual manner.

Assignment 3: Lightning Talk

The lightning talk provides students the opportunity to combine visual and verbal presentation skills. During the lightning talk, students provide an overview of their infographic. If using only assignments 1 and 3, you will need to adjust the assignment instructions.

Instructions for Students

You will have five to seven minutes to orally present a summary of your research topic, the conclusions within the research you found, and your infographic to the class.

In your talk, the following objectives must be met:
- Present your research question. You must present the final *effective research question* you selected.

- Discuss your mini-study by walking the class through your infographic. Discuss:
  - Who did you poll?
  - What did you put on your infographic?
  - How did what you learned in your mini-study influence your approach to your topic?
- Answer your research question based on the evidence you found.
  - Based on the evidence you found, what is the answer to your research question?
  - Don't give opinions; give facts.

Pro Tips!

- Practice ahead of time so that you know what you want to say and how to say it.
- Dress the part of a researcher. You'll give better presentations when you appear prepared.
- Talk slowly. Take your time to give a professional presentation.
- You are among classmates who went through the same process—find the things that excited or frightened you about your research and share those with the class.
- Be the kind of audience you want to have in your presentations. People recognize respect.

REVIEWS/ASSESSMENT STRATEGY

The grading criteria are shared with the student prior to the assignment and completed by the teaching librarian. Grading criteria for assignments 1, 2, and 3 are available for download at https://digscholarship.unco.edu/infolit/26/.

CHEF'S NOTES

The assignment sequence can be done through a series of instruction sessions. If students are not working on their own research project and do not have their own data, the assignment can be adapted. For example, to focus on the analyzing step, provide students with preexisting data. Data or statistics for visualization can come from academic journal articles or from many national and international organizations, such as the US Census Bureau, ProQuest's Statistical Abstract of the United States, the National Center for Education Statistics, or the World Health Organization. For a data collection experience, give students 15 minutes to survey each other, allowing time to discuss the results.

Section 2.
Finding and Evaluating Data

17 [[Ch4]]Three-Step Data Searching
Annelise Sklar

21 [[Ch5]]Transforming Research Questions into Variables: A Recipe for Finding Secondary Data
Alicia Kubas and Jenny McBurney

25 [[Ch6]]Sweeten the Search: Discover Data for Reuse with a Tool That Links Publications to the Underlying Data
Elizabeth Moss

30 [[Ch7]]The Most Vital Statistics: Finding and Analyzing Historical Mortality Rates
Alisa Beth Rod and Jennie Correia

34 [[Ch8]]Understanding the Enumerated World: Making Sense of Data as an Information Source
Alexandra Cooper, Elizabeth Hill, and Kristi Thompson

38 [[Ch9]]Looking at Data
Kay K. Bjornen

43 [[Ch10]]Interrogating the Data: What Data Sets Can and Cannot Tell Us
Kristin Fontichiaro

46 [[Ch11]]Data Zines: A Hands-On Approach to Community Curiosities
Tess Wilson

49 [[Ch12]]On the Hunt: Understanding and Analyzing GSS Data Extraction for Incorporation within Sociological Research Projects
Amy Dye-Reeves

52 [[Ch13]]Using Statistics to Define the Problem: Data and Service Learning
Amy Harris Houk and Jenny Dale

56 [[Ch14]]Data and Statistics in the News and Media
Kaetlyn Phillips

Three-Step Data Searching

Annelise Sklar, Assistant Director, Scholarship Tools and Methods Program, University of California San Diego Library, asklar@ucsd.edu

NUTRITION INFORMATION
This lesson is designed for upper division undergraduate students or early stage graduate students who are required to find and analyze social sciences data for original research, such as for an honors or capstone project. It is meant to follow and build on a literature searching instruction session. The purpose of this lesson is to introduce students to the concepts they need to find reusable data. After a brief introductory lecture, the students will conduct a "reference interview" with themselves to identify avenues for finding research data.

TARGET AUDIENCE AND NUMBER SERVED
This lesson is most interactive in discussion-sized courses, but it can be scaled up to a lecture.

LEARNING OUTCOMES
By the end of the session, students will be able to
- explain how data are collected and created by different types of researchers or organizations and described in different types of publications
- identify and use search tools to find data based on discipline, format, or creator's affiliated organization type

COOKING TIME
This lesson is meant to follow and build on a library instruction session on literature searching. Walking through the points on this worksheet with students requires about 15–30 minutes. More time can be added for students to apply the questions to their own research topics, for a librarian to demonstrate databases, or for students to explore recommended databases. For the whole session, 45 minutes to an hour is generally recommended.

DIETARY GUIDELINES
This lesson touches on the following frames from ACRL's *Framework for Information Literacy for Higher Education*:
- *Information Creation as a Process:* Students will be prompted to think about why data are collected and by whom.
- *Information Has Value:* The instructor will introduce the concepts of open data versus licensed data and explain that students simply may not have access to some data due to restricted use agreements and cost-prohibitive products.
- *Research as Inquiry:* The instructor will reiterate the importance of the literature review and building on existing disciplinary knowledge. Students will be reminded to review publications' methodology and data sections for references to data that have been used in previous studies as well as known limitations of available data.
- *Searching as Strategic Exploration:* Students will understand that data can be described and published in a variety of sources and that they will likely have to search multiple resources using a variety of search terms and strategies.

INGREDIENTS
- Handout/worksheet with definitions of key vocabulary relating to data and questions for students to consider (see appendix)
- Examples of library-licensed or open data sets that match student topics; for example, ICPSR, the Dataverse Project, DataPlanet, Social Explorer, Data.census.gov, ProQuest Statistical Insight, and so on.
- Presentation slides or other visual aid with concepts from handout (optional)
- Computers with internet connections for students to practice searching individually or in small groups (optional)

PREPARATION
This recipe works best when supplemented with a few targeted examples that demonstrate different types of data and sources. For example, political science students who are comparing countries might be introduced to panel data from international organizations

Section 2. Finding and Evaluating Data

such as the World Bank or OECD, the Correlates of War data set, or the Europa World Plus database. Urban studies students researching their local community might be taught about census data, local government data, or databases such as Social Explorer and SimplyAnalytics. Ideally, the librarian will request and receive students' topics ahead of time in order to choose sources that match individual students' research questions.

The librarian should customize the handout and slides to promote available data sources that match course topics.

INSTRUCTIONS

Pass out the handout to students at the beginning of class so that they may follow along.

Begin by sharing brief definitions of the word *data*; descriptions of different types and formats; a differentiation between microdata, aggregate data, and statistics; and explanations of the terms *data set* and *repository*.

Following the handout and lecture:
1. Prompt students to think about the kinds of data that would be useful to their projects and to jot down ideas about who might have collected or created these data (e.g., government agencies, nonprofit organizations, private business or industry, or academics). This part is often tricky for novice researchers, so instructors may want to phrase the question in such a way that the students consider each potential creator one by one.
   » For example, the instructor may ask, "Based on what you know, do you think the government would have collected these data?" Depending on the nature of the data, the instructor may follow up by suggesting students think about government forms they've filled out and whether their question was asked, such as whether their voter registration card asked about ethnicity.
   » After a pause, the instructor may follow with, "Is there a nonprofit [or nongovernmental] organization that might be tracking these data? This might be particularly likely for data that are too controversial for a government to collect—or be trusted to collect impartially."
   » Then, "Is this information a business might want to collect, or an industry group on behalf of businesses, maybe for marketing or monitoring the competition? Or are these data someone might collect and sell, like a professional public opinion poll?"
   » And finally, "Is this something academic researchers might collect, maybe in a survey or experiment?"
2. Explain that different kinds of researchers publish their results in different types of publications (scholarly articles, reports, etc.) and recommend students pay attention to how data are described or cited in these publications. Searching literature is one way of finding data. If students have already found relevant articles, prompt them to look at the methodology section to see what they can learn about the data. Connecting to the first question, point out that academics typically publish in scholarly articles and books, but experts who work for the government, think tanks, nonprofits or nongovernmental organizations, and intergovernmental organizations often communicate through reports and websites published by their organizations (this is called *gray literature*). Likewise, industry and commercial research may be distributed through trade publications. No matter the publication type, students should find out where the data originated. They should try to determine whether the authors created their own data via original research, or whether they are using data from another source and there is a data citation, or whether the article merely gives the names of the individuals or organizations that collected the data.
3. Finally, explain that, depending on their discipline or even their institutional culture, creators of data have different distribution norms. Data may be available in institutional repositories, in licensed databases, or on an organization's or researcher's web page. There also may be accessibility differences. The data may be available only upon request, only for a fee (possibly with a restrictive license), or sometimes not at all, depending on privacy issues, proprietary information, or a researcher's individual inclination or grant specifications.

If time permits and students have developed their topics, the librarian may demonstrate

select databases and search tools and give time for students to explore these databases and tools on their own.

REVIEWS/ASSESSMENT STRATEGY

The handout/worksheet associated with this lesson is intentionally not designed for library assessment purposes. It is instead intended to guide students to apply the concepts to their own research topic during class time (to "workshop" their topic) and then serve as reference notes as they pursue further inquiry on their own.

Ideally, students should have time and equipment during the workshop to search for data sets using recommended strategies and tools, with the librarian available for individual follow-up questions. Librarians can evaluate the effectiveness of their teaching based on these follow-up questions. As students learn to identify and access data in order to analyze them for a project such as an honors thesis or capstone paper, the true measure of their learning is their own successful completion of their work. No matter how successful this instruction session is, it is likely that students will individually consult with librarians and faculty throughout the life of their research project.

Appendix A. Library Help for Data Research

Research Guide: link to research guide
Librarian: name and email

DATA—DEFINED

da·ta noun plural but singular or plural in construction, often attributive \'dā-tə, 'da- also 'dä-\
1. factual information (as measurements or statistics) used as a basis for reasoning, discussion, or calculation
2. information output by a sensing device or organ that includes both useful and irrelevant or redundant information and must be processed to be meaningful
3. information in numerical form that can be digitally transmitted or processed
(From Merriam-Webster, http://www.merriam-webster.com/dictionary/data)

TYPES OF DATA
- **Observational:** Captured in real-time, typically outside the lab
  Examples: Sensor readings, survey results, images, audio, video
- **Experimental:** Typically generated in the lab or under controlled conditions
  Examples: test results
- **Simulation:** Machine generated from test models
  Examples: climate models, economic models
- **Derived / Compiled:** Generated from existing datasets
  Examples: text and data mining, compiled database, 3D models

COMMON FORMATS
- **Text:** field or laboratory notes, survey responses
- **Numeric:** tables, counts, measurements
- **Audiovisual:** images, sound recordings, video
- **Models, computer code, geospatial data**
- **Discipline-specific:** FITS in astronomy, CIF in chemistry
- **Instrument-specific:** equipment outputs

KEY TERMS
- **Microdata:** Data directly observed or collected from a specific unit of observation
  – Examples
    - Census: the unit of observation is probably an individual, a household, or a family
    - Survey or poll: the responses of a single respondent
- **Aggregate Data:** Is higher-level data that have been compiled from smaller units of data
  – Examples: inflation rate, consumer price index, demographic data for city or state

Section 2. Finding and Evaluating Data

- **Statistics:** Numerical data that has been organized and interpreted, usually displayed in tables
- **Datasets:** A dataset or study is made up of the raw data file and any related files, usually the *codebook* and *setup files*.
  - Most datasets require at least basic *statistical analysis* (Stata, SPSS, R, etc.) or *spreadsheet* programs (Excel) to use.
- **Repositories:** A *data repository* is a collection of datasets that have been deposited for storage and findability.
  - They are often discipline specific and/or affiliated with a research institution
  - Examples
    - ICPSR
    - Harvard Dataverse Network
    - UC San Diego Digital Collections

Step by Step Plan for Finding Datasets and Statistics

1. Think about who might collect the data.
- Could it have been collected by a government agency?
- A nonprofit/nongovernmental organization?
- A private business or industry group?
- Academic researchers?

Who? _____

_____

_____

2. Look for publications that use the kind of data you're looking for and cite the dataset
- e.g. scholarly articles or government reports.
- **Quick tip:** use your article search strategies for Google Scholar and add the word "dataset"

Sample search: _____

3. Once you know that what you want exists, it's time to hunt it down.
- Is it freely available on the web?
  - Google Dataset Search: https://toolbox.google.com/dataset-search
  - IQSS (Harvard) Dataverse
  - Quality of Government Institute (free)
  - MacroData guide
  - data.census.gov
  - UNdata
  - World Bank World Development Indicators
  - **Tip:** Check regular Google—you never know!

- Or part of a package to which the library already subscribes?

• ICPSR (requires reg w/ UCSD email) • UC San Diego Dataverse • Cross-National Time Series	• CQ Voting & Elections Collection • Roper iPoll • Proquest Statistical Insight • Data-Planet • OECD iLibrary

- Can it be requested directly from the researcher? There's a reason articles usually include author contact information…

Transforming Research Questions into Variables
A Recipe for Finding Secondary Data

Alicia Kubas, Government Publications and Data Librarian and Regional Depository Coordinator, University of Minnesota, Twin Cities, akubas@umn.edu; Jenny McBurney, Social Science Librarian, University of Minnesota, Twin Cities, jmcburne@umn.edu

NUTRITION INFORMATION
We created this course-integrated workshop to address the increasing need for social sciences students to find and use secondary data in their coursework. This workshop was specifically designed for an undergraduate honors business class, but the model can be adapted for other subject areas that have similar needs: for students to better understand where data come from, how to access data, and how to use data in their own projects.

TARGET AUDIENCE AND NUMBER SERVED
This session is targeted toward upper level undergraduates in courses with a data component. There is no limit on number of students or participants, but for classes over thirty students we recommend a co-instructor to assist with discussion and class facilitation.

LEARNING OUTCOMES
At the end of the session, students will be able to
- translate their research questions into variables in order to determine the scope of their data needs
- identify the organization or type of organization that would collect or produce those data in order to choose appropriate search tools
- articulate how data resources are organized in order to find relevant data using appropriate search strategies

COOKING TIME
One to three hours of preparation time and 45 to 60 minutes of instruction time

DIETARY GUIDELINES
This recipe addresses the frame Searching as Strategic Exploration in ACRL's *Framework for Information Literacy for Higher Education*. Increasingly, we see that students must find secondary data and do more of their own data analysis for their coursework. The foundational data literacy skills students begin to develop in this type of workshop can help prepare them for future courses, as well as for graduate school or their future careers. In this workshop, students learn that they have to be flexible when determining variables from their research questions, creative in their thinking to identify where to find the data they want, and iterative in their overall process as they work through their questions and discover new solutions.

INGREDIENTS
- Slide deck
- LibGuide for data and statistical resources
- Worksheet/assessment tool
- Instructor computer with projector
- Computers for students
- Whiteboards or other brainstorming tools (optional)

PREPARATION
- Create a slide deck that contains examples of research topics, questions, or theses that require a major data component.
- Create a LibGuide or research guide of data and statistical resources, both freely available online and in library subscription databases. Some recommended examples include the US Census website and resources, DataPlanet, and IPUMS from the Minnesota Population Center.
- Create a flowchart worksheet (see figure 1) that visually guides students through the process of breaking down their research question into variables, data sets, and sources. The worksheet can also include a link to the LibGuide and your contact information.

Section 2. Finding and Evaluating Data

Research Question:

Why do craft brewery clusters develop in certain regions of the US and not in others

Variables	Datasets	Possible Sources
Voter participation (proxy variable for sense of community)	% of voting age population that voted in the last election in Minnesota	- MN Secretary of State - Data-Planet (library database)
Arts, entertainment, and recreation companies per capita (proxy variable for openness to experience)	# of companies classified with NAICS code 71 per capita in Minnesota	- Bureau of Labor Statistics - Reference Solutions (formerly ReferenceUSA)

FIGURE 1
Sample flowchart

INSTRUCTIONS

1. *Introduction:* Explain what secondary data are and why data are important in making or supporting an argument. Also provide some grounding expectations when searching for data such as being flexible and adaptable as students may not find the perfect data they want, being prepared to use multiple data sets, understanding that they will likely need to search in many places and resources to find their data, and understanding that persistence and patience are key. Lastly, briefly outline the objectives and what you will go over during the instruction session, and note that they'll have time at the end of the session to apply the exercises outlined in the presentation to their own topic and time for their own data searching.

2. *Activity/Instruction:* Introduce the concept of translating a research question into variables and then specific data sets.

   Example 1: Start with a simple example such as "price behavior and structure of the organic produce market in the United States." Ask students what variables they can identify from this research topic. In this case, the variables to guide them to are conventional and organic produce prices for markets in the US.

   Next, help them think through what specific data sets will look like for these variables. For this simple case, the variables are easily translated into the data points of specific prices for produce, which can be found in many sources, both freely online and in library databases. Remind students that the data they use may depend on what's available and realistic (e.g., geographic regions and types of produce).

   Next, ask students to brainstorm who might collect and disseminate the data— for this example, the US government and specifically the US Department of Agriculture. Then, demonstrate searching Google, as this is a good place to start looking for government data. Don't forget to mention tips for searching online for government data: keep your search terms broad; use the site:.gov filter; and if you find references to the data, try to trace back to the original source.

   Example 2: Now move on to a more complex example that requires proxy variables, such as "why craft brewery clusters develop in certain regions of the US and not in others." As with the first example, ask students to identify variables from this research question. This time the variables are not obvious, so let students

brainstorm for a bit. Explain that if you don't know how to identify variables from a topic, you may need to do some market research to learn what factors impact your topic and could be used to identify variables. In this example, we talk about sense of community, openness to experience, and well-being among a region's residents as potential factors. Since these factors are not measurable, ask students what proxy variables they might use. The example list below from a previous student thesis project (see Additional Resources) illustrates some potential proxy measurements for this topic.

Key factor: Sense of community

Potential proxy measurements: Voter participation; volunteerism; nonprofit public charity activity per capita; clubs, sports, and other social capital organizations per capita

Key factor: Openness to experience

Potential proxy measurements: Recreational drug use; arts, entertainment, and recreation companies per capita

Key factor: Well-being

Potential proxy measurements: Well-being index score

Now, students are ready to identify specific data sets for those variables. For example, talk students through finding voter participation data by starting with potential sources (government, research organizations, educational institutions, etc.). This is a good time to show students the data and statistical resources LibGuide that contains library subscription databases as well as free online sources.

Last, demonstrate some searches in a few commonly used resources for data and statistics, such as DataPlanet, IPUMS, and Data.census.gov.

3. *Assessment:* Once the instruction portion concludes, students go through this process on their own in a hands-on activity. They choose their own research question and use the flowchart worksheet to demonstrate their understanding of the process, which doubles as an assessment tool.

REVIEWS/ASSESSMENT STRATEGY

We use two strategies for assessing student learning. First, throughout the demonstration portion of the class, students are encouraged to participate by answering questions orally or brainstorming their own ideas relating to the topics on whiteboards. This helps us gauge student learning throughout the session so that we can slow down or speed up the content accordingly. Second, during work time, students fill out a worksheet relating to their own topic. This allows us to quickly scan their sheet for understanding and progress and also start individual conversations with each student about their questions or problems that arise.

ALLERGY WARNING

Student allergies: Searching for data can be extra challenging for students. Since there is no one-stop shop like a library catalog or go-to database, students typically have to search many different resources and often use data sets from multiple sources in their projects. Remind them that finding data is an iterative process!

Instructor allergies: Data websites and resources are always changing and being updated, especially if you use government sources. Make sure to run practice searches ahead of time so that you know what kind of results to expect!

ADAPTING THE RECIPE

Online instruction: If you will be teaching this workshop online, we recommend finding a colleague who can be an extra set of hands during work time. Organize virtual breakout rooms for students with similar topics so that they can discuss what they are finding and so that you and your colleague can each help multiple students at once. The paper worksheets could be swapped for individual or small-group Google Docs. If you want to have students brainstorm ideas visually, you could use Jamboard or a similar tool instead of a whiteboard.

Large classes: For larger classes, we recommend inviting a colleague along to help during work time, whether you will be teaching in person or online. This way you can each have more one-on-one and small-group interactions with students, and students who are shyer or falling behind will have more opportunities to ask questions and get caught up. Plus, co-teaching with a colleague can help your co-teacher learn from you and prepare them for teaching data literacy themselves in the future.

ADDITIONAL RESOURCES
Example LibGuide
Kubas, Alicia. "Comprehensive Guide for Data and Statistics." LibGuide, University of Minnesota Libraries. Accessed January 28, 2021. https://libguides.umn.edu/dataandstatistics2.

Example Student Paper
Rosas, Michelle. "A Recipe for Success: Exploring Cluster Formation in the American Craft Brewing Industry." Undergraduate honors thesis, University of Minnesota, 2013. University of Minnesota Digital Conservancy. https://hdl.handle.net/11299/155308.

Sweeten the Search
Discover Data for Reuse with a Tool That Links Publications to the Underlying Data

Elizabeth Moss, Librarian, Inter-university Consortium for Political and Social Research (ICPSR), Institute for Social Research, University of Michigan, *eammoss@umich.edu*

NUTRITION INFORMATION
It can be challenging to find shared social and behavioral health data with the appropriate ingredients for reuse in class assignments or academic research. Help sweeten the data discovery process with a tool that is especially appealing for students who are hesitant about using quantitative data. In a three-part activity, students start by creating a user story to assess needs and determine the best terms for an effective search. Then they take an active tour of a freely available collection of scholarly literature linked to the wide array of social and behavioral research data available for reuse at a large research data archive. While on the tour, students perform searches to discover relevant data. After the tour, they perform a final self-assessment of their data discovery process.

TARGET AUDIENCE
Undergraduate and graduate students

LEARNING OBJECTIVES
Students will
- construct a user story to inform a search strategy
- tour and conduct searches in a research data archive's data-related publications database to find and recognize relevant content when evaluating data for reuse
- assess their data discovery process

COOKING TIME
45 minutes

DIETARY GUIDELINES
ACRL's *Framework for Information Literacy for Higher Education* frames: Research as Inquiry; Searching as Strategic Exploration

INGREDIENTS
- Web browser access for students and librarians
- 1 copy per student of User Story Template and Examples (appendix A)
- 1 copy per student of Tour of the ICPSR Bibliography of Data-Related Literature (appendix B)
- 1 copy per student of Discovery Process Self-Assessment Template and Example (appendix C)

PREPARATION
Before holding the class
- Adapt the User Story Template (appendix A) by customizing the examples to align with the likely subject area of the class members.
- Create an example user story to use when instructing the students during Activity 1.
- Go through the Tour document (appendix B) yourself so you are familiar with a search strategy and outcome you want to use as an example for the class during Activity 2.
- Prepare ahead for Activity 3 by anticipating how you would evaluate your example discovery experience when demonstrating how to fill out the short, six-step Self-Assessment document (appendix C).
- Provide an online location for students to access the three documents (appendices A, B, and C) that they will use in the class, or print them out and provide one copy of each of the three documents to each student.

INSTRUCTIONS
Explain to the class that they will be learning how to find research data of interest to them by searching a large database of publications that have been linked to the data underlying the findings, called the "ICPSR Bibliography of Data-Related Literature." The data being analyzed in the publications are located at

Section 2. Finding and Evaluating Data

ICPSR, an archive of shared social and behavioral research data. Stress that before you begin searching any database, it is best to get clarity about what you are searching for, and why. This leads you to Activity 1.

Activity 1
- Ask students to open a browser window or tab to access the user story examples provided via the link in the document.
- Once they have reviewed the examples, guide class members through the User Story Template in appendix A by using the example you prepared.
- Then ask them to create their own user story statements and to note specific data needs and search terms to use.
- Encourage them to think about alternative or synonymous terms to use when searching the database.
- When class members have finished, move to Activity 2.

Activity 2
- Together with the students, read the introductory paragraph of the Tour document in appendix B.
- Important: Point out to the students the information in this recipe's Chef's Notes.
- Together with the students, open a new browser window or tab and navigate to the links provided in the document in Step 1 and Step 2.
- Answer any questions they may have about the Search Tips (from Step 2).
- With the class, go to Step 3 in the Tour document and demonstrate entering the search terms you prepared. Show how you can try different terms if the first combination is unsuccessful.
- Briefly demonstrate sorting and filtering in Step 4.
- For Step 5, demonstrate choosing a search result that interests you based on the needs identified in your user story. Show how to click out to the full text of the publication, as well as how to link to the study or studies associated with that publication. Tips:
  - *When in a publication:* Point out why any aspect discussed catches your eye; for example, the authors discussed an interesting approach they used to analyze the data; the authors pointed out specific variables that you would want to analyze; the authors mention that they used the public-use or restricted-use version of the data for a particular reason that is relevant to you; the authors were able to use one data set in combination with another to achieve an interesting finding; unique aspects of sample population match your needs; the authors discuss in-depth the methodology used to collect the data that makes you want to access the data for your own use; you are reading a report with summary statistics—and that is what you want to paste in your term paper and cite, and so on.
  - *When in the study description:* Point out what may be of interest to you; for example, the sample size, population, or location is appropriate for your needs; you want to download the codebook to become more familiar with the data; the data are available for online analysis; the data files you need are publicly available; there are other data-related publications associated with the study that you want to look at more closely, and so on.
- Once you have finished your demonstration, invite the students to type their terms in the search box and begin the process that you just demonstrated, as outlined in Steps 3–5 in the Tour document.

Activity 3
- Once students have completed the Tour, at least ten minutes prior to the end of the class, tell the students it is time to document their experience by filling out the Self-Assessment document (appendix C).
- Using the six assessment aspects listed in the document, briefly state out loud your self-assessment for the example discovery experience you provided during the class.
- Ask the class to take five minutes to fill out their own self-assessment document based on their own discovery experiences.
- End the class by encouraging the students to continue exploring this resource for current or future data needs.

REVIEWS/ASSESSMENT STRATEGY
See appendix C, "Discovery Process Self-Assessment Template and Example"

CHEF'S NOTES

- Be sure students understand ICPSR's definition of a study. At ICPSR, a study is a collection of one or more data files and the documentation needed to understand how to use those files. Each study has its own description page on the ICPSR website.
- Be sure students understand that they can search ICPSR's study catalog directly for data. Point out that the Bibliography of Data-Related Literature is an additional resource allowing them to find data via the publications using the studies in the ICPSR catalog. That resource is the focus of this class.
- The ICPSR Bibliography does not index the full text of publications. While searching the bibliography yields fewer results than searching the study catalog, those results are very relevant. If a word is important enough to go into the citation, then the publication often will have significant content using that word or concept.

ADDITIONAL RESOURCES

Association of College and Research Libraries. *Framework for Information Literacy for Higher Education*. Chicago: Association of College and Research Libraries, 2016. https://www.ala.org/acrl/standards/ilframework.

ICPSR. "ICPSR Bibliography of Data-Related Literature." 2021. https://www.icpsr.umich.edu/web/pages/ICPSR/citations/.

ICPSR. "User Stories." 2021. https://www.icpsr.umich.edu/web/pages/ICPSR/citations/user-stories.html.

Appendix A: User Story Template and Examples

Create a user story to clarify search goals and data needs.

Start by examining the existing user stories for ICPSR Bibliography users, found on the ICPSR website:

> https://www.icpsr.umich.edu/web/pages/ICPSR/citations/user-stories.html

Create a user story, incorporating this statement to state your goals:

As a <user role>, I want <goal> so that <benefit>.

Example:

As a senior in a capstone course, **I want** to find data I can use to write a paper **so that** I can generate statistics online and use them as part of the assigned work.

In addition, note specific data needs to consider when examining your search results.

Example:

I need data containing variables about high school students and attitudes about race and equality.

Based on the identified goals and data needs, list likely search terms to use when conducting the database search.

Example:

"race relations" "high school students"

Appendix B: Tour of the ICPSR Bibliography of Data-Related Literature

With your user story and search terms, you are ready to find social science or behavioral health research data that will meet your goals. You will tour the ICPSR Bibliography of Data-Related Literature and search for scholarly works of interest as you go. Those works are all linked directly to the full text of the publications (where possible), *and* they are linked to the data used in the publication. Those data are described and archived at the Inter-university Consortium for Political and Social Research (ICPSR), a large repository of curated, digitized social and behavioral health science data archived for people like you to access and reuse.

Step 1: Start your tour at the ICPSR Bibliography's search portal: https://www.icpsr.umich.edu/web/pages/ICPSR/citations/

Step 2: Review the search tips located just below the main search box.

Step 3: Enter in the search box the term or terms defined in your user story.

Example: "race relations" "high school"

- Look for search results to be returned in the tab called "Data-related Publications."
- This search yields no results in the "Data-related Publications" tab. A modified (in this case, broader) query, e.g., *race* and *"high school"*, will yield many publications related to many data collections, which we will refer to as studies. ICPSR studies contain one or more data files together with documentation files needed to understand how to use the raw data. Each study can be found on a study home page, which contains links to these files, along with an extensive description of the study and download links.

Step 4: Once you have a set of search results, you can change how the results are sorted, and filter by:

- *Publication year:* This works well for students interested in only the most recent publications.
- *Publication type:* There are 11 different types, and you may want to see only journal articles. If you do not intend to download data and you want to use the database to find publications containing summarized statistics that you can quote, you may want to limit your results to reports.
- *Journal title:* Specific journals are seen as significant in some fields of research, so you can limit your search results to see only what is in the database published in a particular journal.
- *Author:* You may want to see publications by a particular author whose work may be connected to available data. Keep in mind that the names in the database are not authority controlled, so they may appear in slightly varying forms, e.g., Smith, John A., could be the same as Smith, John.
- *Study:* Try clicking "view all" in this filter. It gives you a consolidated list of all the studies in the ICPSR collection that are associated with publications that are retrieved by your query.

Step 5: Now that you have sorted and filtered your results, you have reached a point in the tour where you can explore any search results of interest. You can choose to

- link to the full text, where possible, and read one or more publications to learn how the authors made use of the data
- select one or more citations to export into your bibliographic software
- click on the link to the underlying study or studies used in any publication to access the study's description page to
    - explore other publications listed in the Data-related Publications tab associated with that study so you can read more ways the data have been used.
    - read the study's description for information, e.g., to see if its data files are available for online analysis, to examine its variables, to read the methodology used to collect the data, to see the sample used, to read its online codebook, or to learn if there are restricted files in the study
    - click on the series link (it will display at the top of the study page if the study is part of a larger grouping of studies) to find more publications associated with the studies in the series

Appendix C: Discovery Process Self-Assessment Template and Example

Once your searching is done, complete a self-assessment and briefly document your
1. *Original goal:* (Restate your original user story statement.)
2. *Search terms:* (What terms worked?)
3. *Challenges:* (Were there any stumbling blocks?)
4. *Results:* (What goal was reached?)
5. *Surprises:* (Did you find something unexpected during the process?)
6. *Next steps:* (What direction can you pursue in future explorations of the database?)

Example

1. *Original goal:* As a senior in a capstone course, I wanted to find data I could use to write a paper so that I could generate statistics online and use them in the paper as part of the assigned work.
2. *Search terms:* race and "high school"
3. *Challenges:* I could not find any publications using my chosen terms, so I modified "race relations" to race so they were not so specific.
4. *Results:* I found a very relevant 2020 article in a peer-reviewed journal linked to a data set that I can analyze online. The article clearly used variables in the data that captured whether a respondent will be enrolled in college, as well as parental attitudes about college, and other race/ethnicity variables captured in the data. So I have confirmation that I can get the data I need to create statistics for my paper.
5. *Surprises:* I thought I would have to know more about a statistical analysis package to be able to generate statistics for my project. But I read the data description, which tells me I can use online analysis without having to know SPSS, SAS, or R!
6. *Next steps:* The paper I read gave me some great ideas for analysis, so not only can I use this data set for my assignment this weekend, using online analysis, but I also might want to try something more complex with the data in the future as I become more proficient in stats software.

The Most Vital Statistics
Finding and Analyzing Historical Mortality Rates

Alisa Beth Rod, PhD, Research Data Management Specialist, McGill University Library, alisa.rod@mcgill.ca; Jennie Correia, Personal Librarian for the Social Sciences and Associate Director of Teaching, Learning, and Research Services, Barnard College, jcorreia@barnard.edu

NUTRITION INFORMATION

This recipe describes a hands-on data literacy instruction session designed for a US economic history undergraduate course. The session aims to teach students to find, critically analyze, and contextualize secondary data related to historical urban mortality rates in the United States. Students learn how to use a spreadsheet application (e.g., Excel) to interrogate historical trends in several major US cities and how to use library tools to discover supplementary sources to help explain trends that arise.

This session is optimized when co-taught by two information professionals: a data librarian (or other data-related specialist) and a subject librarian. Combining the skills of functional and subject experts introduces or deepens students' understanding of the research-related services that are offered by many contemporary academic libraries.

TARGET AUDIENCE AND NUMBER SERVED

Social science undergraduate junior and senior students. This session may serve up to one large lecture course but is ideal for a class of around 50 students or fewer.

LEARNING OUTCOMES

- Conduct secondary and primary research to answer or elaborate on a well-defined empirical research question.
- Recognize how tabular data may be transformed from print secondary sources to a digital format.
- Create a visualization of numeric data using a spreadsheet application (Excel).
- Describe the process for analyzing and interpreting secondary quantitative empirical evidence.
- Articulate the relationship between primary and secondary sources of information and evidence.
- Gain confidence using library and other research tools to discover and identify relevant primary and secondary sources.

COOKING TIME
75–90 minutes

DIETARY GUIDELINES

This session incorporates several frames from ACRL's *Framework for Information Literacy for Higher Education*:

- *Information Has Value:* The data for this workshop were collected and transformed from primary US Census sources by Michael R. Haines. The data are accessible to member institutions of the ICPSR data repository for educational purposes, reflecting one dimension of value. This allows for a discussion about knowledge practices and dispositions, including citing creators of information/giving credit, intellectual property rights and respecting Haines's effort in creating these data and making them available for reuse, and access to licensed data in the context of secondary information sources.
- *Research as Inquiry:* The creation of the visualization of these data points to further potential research questions. For example, trends in mortality rates drop inconsistently across different urban areas in the middle of the nineteenth century. We ask students to hypothesize possible explanations and then find additional sources to support their contextualization and understanding of the trends.
- *Searching as Strategic Exploration:* Students learn how to search and interpret library catalog and journal database results to identify relevant primary and secondary sources.

INGREDIENTS

- Data set (cleaned version: see Additional Resources)
- Excel (or another spreadsheet application)
- Library research guide
- Library catalog
- Computers
- Internet access
- Projector or online learning environment with option to share instructor's screen.

PREPARATION

Create or adapt an existing library research guide with recommendations for relevant economic history databases and discovery tools, including signal words for identifying primary sources in the library's catalog, scholarly tertiary sources for background information, common sources for economic secondary data sets, digitized primary source compilations, and ways to get individualized help from the library. See the Additional Resources section for an example of one of our guides.

INSTRUCTIONS

The data librarian (or functional data specialist) instructs students to download the data set from GitHub and demonstrates the process, then opens and displays the Excel spreadsheet containing the cleaned data set. Prior to 2018, the data set for the chart exercise portion of this session required significant advance preparation: retrieving the information from a PDF of a scanned print document (e.g., manually transferring data from an appendix table to an Excel worksheet).[1] In 2018, a digital copy of the data was deposited in the ICPSR repository, though access is restricted to member institutions.[2] In the ICPSR data set, the data for each of the five cities are contained in separate tabular files along with multiple variables (including our variable of interest, crude death rates). We have further cleaned the ICPSR data set and created a single spreadsheet containing crude death rates for all five cities. This cleaned data set is stored on GitHub and licensed under an MIT license, which allows for reuse without attribution (see Additional Resources). This process and the timeline are worth discussing with students, including showing the original publication and tables.

1. Students may follow along on their own computers or by watching the demonstration. Ask students to describe how the data are organized (with a column for years on the left and several additional columns representing five urban areas).
   a. Ask students how crude death rates are calculated by historians and to identify any limitations with this indicator.
   b. Demonstrate how to create a chart in Excel to visualize the mortality rates through the following steps:
      i. Go to the Insert tab at the top of the Excel worksheet containing the data. Direct students to the chart icons in the middle of the tab. Ask students which chart type would be useful for visualizing trends or patterns over time. This is a nonintuitive step as many students will suggest the line chart. However, the line chart in Excel treats variables as categorical, not numerical, and since the two variables (year and mortality rate) are numerical, the line chart is not appropriate. Instead, direct students to click on an empty cell in the worksheet and then insert a scatter plot with straight lines.
      ii. Although it is possible and more efficient to preselect data before inserting a chart, students may benefit from learning how to build a chart from scratch. In the Chart Design menu bar, direct students to click on the Select Data button.
      iii. In the Select Data dialog box, direct students to click Add listed under "Legend Entries (Series)" (left-hand side of the dialog box).
      iv. In the Edit Series dialog box, enter the first city name (Baltimore) for "Series name." For "Series X-Values," explain that "X" represents the horizontal axis, or the independent variable. Since the data are time series (e.g., plotted over years), and time is almost always inde-

pendent, direct students to select the full range of years (1802–1920) as the X series. Students should select the mortality rates in the Baltimore column as the "Series Y Values." (But first, critically, they must delete the ={1} that appears as a default. Otherwise, they will receive an error message.)

v. Click OK and return to the Select Data dialog box, then repeat step iv for each of the four other cities. Et voilà! A time series chart!

vi. The final step in creating a chart is to edit chart elements to improve readability and reduce distortion of the data. Click on the green plus sign at the top right-hand corner of the chart. Check the boxes for Axis Titles, Chart Title, and Legend. Right-click on the horizontal axis and select Format Axis… from the menu that appears. In Axis Options, set the minimum bound to 1802 and the maximum bound to 1920.

c. After completing the chart (see figure 1), guide students to discuss an analysis of the visualized trends. How do mortality rates shift over time across the cities? What patterns do they observe? Are there specific years or periods when the mortality rates increase or decrease substantially? What is going on in New Orleans in the mid-nineteenth century? Students can employ additional options in Excel to help in their analysis (e.g., fitting a trend line, charting each year individually, adjusting the horizontal axis to focus on a specific set of years, etc.).

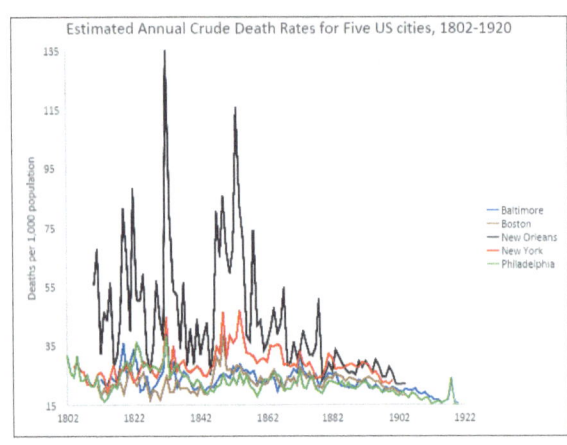

FIGURE 1
Chart comparing historical US urban mortality rates. Figure created by the author, Alisa B. Rod, in Excel. See Additional Resources for full data set citation.

2. The subject librarian walks students through possible steps to locate information sources related to the questions that arose during the data analysis portion.
   a. Transition with a brief discussion of the relationship between primary, secondary, and tertiary sources in economic history, using the idea of analysis as the point of distinction.[3]
   b. Use a tertiary source (e.g., *The Oxford Encyclopedia of Economic History*) for background information to contextualize and understand the trends observed. Our search showed entries related to multiple research paths such as public health, epidemics, and the demographic transition.
   c. Pick one topic from the tertiary source suggestions to search for a wider variety of source types in the library catalog. For example, we narrowed the broad topic of public health with catalog filters such as region/geography, time period, authorship, subject, and genre. We discovered primary source records from the New York Board of Health and plans from specific public works projects in our own libraries' archives, as well as book-length secondary treatments of the history of the city's public health infrastructure.
   d. Conduct a similar search in a specialized secondary scholarship tool (e.g., EBSCO's America: History and Life) to see how the advanced search techniques learned in the catalog can translate to similarly structured databases.
   e. Using an example of an article in the group's search results in America: History and Life, review the strategy of mining the footnotes of secondary sources for leads on primary sources of evidence and additional secondary scholarship

to read. The circular nature of this example reinforces the iterative nature of research.

REVIEWS/ASSESSMENT STRATEGY
In-class assessment involved posing frequent questions to students about the concepts and techniques as we demonstrated them in our presentation. After the session, students made appointments for library consultations or dropped in for quantitative analysis help at the Empirical Reasoning Center where we informally observed the students' engagement with the concepts introduced in the workshop.

CHEF'S NOTES
Use texts that the students are already familiar with (e.g., the Haines article[4]) to demonstrate how to mine footnotes and reference lists for additional primary or secondary sources. If the class size and time allow, students can work in groups to identify relevant leads, assess whether the lead would be a primary or secondary source for them, and try to select the right library tools to search for and obtain the full text of those sources. In reporting back to the larger group, the groups can work with the librarian and the whole class to refine their strategies.

Thanks: We adapted this session from an assignment and initial Excel training developed for an Urban Studies Junior Colloquium course at Barnard College. Barnard faculty members Gergely Baics (associate professor of history) and David Weiman (Alena Wels Hirschorn '58 Professor of Economics and faculty director of the Empirical Reasoning Center) developed several iterations of the assignment and provided guidance on the content, scope, and learning outcomes for this session.

ADDITIONAL RESOURCES
Barnard Empirical Reasoning Center. "Mortality in Five US Cities, 1802–1920." Data for US economic history course, Barnard College, Spring 2020. https://github.com/barnarderc/workshops/blob/ba2e1710c27087f21b0a9d-53b4b19c2f96ad21cd/Spring%202020/Economic%20History%20US/Mortality%20in%20Five%20US%20Cities_1802-1920.xlsx.

Correia, Jennie. "ECON 3013: Economic History of the United States." LibGuide, spring 2021. https://guides.library.barnard.edu/ECON-X3013-001.

Haines, Michael R. "Mortality in Five American Cities in the 19th and 20th Centuries, 1800–1930," version 1. ICPSR 37155. Inter-University Consortium for Political and Social Research, November 14, 2018. https://doi.org/10.3886/ICPSR37155.V1.

NOTES
1. Michael R. Haines, "The Urban Mortality Transition in the United States, 1800–1940," *Annales de démographie historique* 101, no. 1 (2001): 33, https://doi.org/10.3917/adh.101.0033.
2. Michael R. Haines, "Mortality in Five American Cities in the 19th and 20th Centuries, 1800–1930," version 1, ICPSR 37155, Inter-University Consortium for Political and Social Research, November 14, 2018, https://doi.org/10.3886/ICPSR37155.V1.
3. Bates College Library, "Primary vs Secondary Sources," LibGuide, accessed January 28, 2021, https://www.bates.edu/library/primary-vs-secondary-sources/ (page discontinued).
4. Haines, "Urban Mortality Transition."

Understanding the Enumerated World
Making Sense of Data as an Information Source

Alexandra Cooper, Data Services Coordinator, Queen's University, coopera@queensu.ca; Elizabeth Hill, Data Librarian, Western University, ethill@uwo.ca; Kristi Thompson, Research Data Management Librarian, Western University, kthom67@uwo.ca

NUTRITION INFORMATION
This recipe is a guide to preparing an instructional session aimed at postsecondary students in the social or health sciences or related disciplines on locating, evaluating, and using secondary data sources as information resources.

Who collects data? Where can you access them? Why are data available on some topics and not others? Why are some statistics available at a detailed level of geography and others only nationally? What are some key limitations of official statistics, and where can information be found to fill in the gaps? This session uses these questions to encourage students to consider how data are used as information sources.

Governments are a primary creator and distributor of statistical information. They collect data to help develop policy and to allow for planning and assessment of services. Thinking about these purposes can help us understand why some topics are covered in depth and others are covered in less detail or not at all. We will draw distinctions between administrative data and survey data, with examples of how both types of data play key roles in generating government statistics. Topics include privacy and confidentiality, data suppression, and the role of sampling with regard to survey data. This session will explore a range of data dissemination formats from reports and summary tables to full data sets.

TARGET AUDIENCE AND NUMBER SERVED
Up to 40 students in a classroom setting. This session has been developed to be palatable to data novices. It best serves students in social science, business, or health science courses, which frequently use data as a source of information: to profile a community, write a business plan, find prevalence of a health condition, determine audience for a program, or investigate a social problem.

LEARNING OUTCOMES
Students will
- describe what data are and their importance in society and research
- identify government and other organizations as sources of secondary data
- evaluate usefulness of secondary data sources for research

COOKING TIME
- *Preparation:* one day.
- *Lecture and workshop:* one class session that is 1½ to 2 hours in duration. Lecture should take no more than one hour. Remaining time should be used to introduce and start the assignment.
- *Assignment:* Assignment should be planned with the faculty member in advance of the session. Students should be given an appropriate length of time outside of class to complete the assignment, working individually or in small groups, depending on the size of the class. The librarian should be available for consultation with the students by appointment over this period.

DIETARY GUIDELINES
This session fits into ACRL's *Framework for Information Literacy for Higher Education*. The focus is specifically tied to the second frame, Information Creation as a Process: "Information in any format is produced to convey a message and is shared via a selected delivery method. The iterative processes of researching, creating, revising, and disseminating information vary, and the resulting product reflects these differences."

The goal of this session is to help students develop into informed and critical consumers of information by teaching them how to locate, access, and evaluate data.

INGREDIENTS

Kitchen Equipment
Computer with an internet connection. If each student has their own computer, they can follow along more easily. Consider taking screenshots of interactive sites in case there are internet issues.

Ingredients
Data resources to highlight based on the subject being taught, the content of the course, and relevant geography. Students are generally more engaged when the instructor shows them data on their hometown.

PREPARATION
The lecture portion highlights examples of the three levels of data collection: international, national, and local. Determine the resources to highlight in consultation with the faculty member leading the course. See the Additional Resources section for a list of suggested sites of data sources.

A key ingredient is the national census (American, Canadian, United Kingdom, or whichever is relevant). You will use part of the lesson to briefly demonstrate how to search, browse, and download data from two or three sites. We suggest selecting the census, an international data source such as the World Development Indicators at the World Bank, and possibly an additional site that is relevant to the course.

The assignment should be developed in conjunction with the faculty member and should be directly relevant to course outcomes. One option that works well is to have students profile a community, group, area, or country on a topic of interest, drawing information from at least two of the three levels of data collection (international, national, local) that will be identified.

INSTRUCTIONS

Introduction: What Are Data?
The introductory part of the lesson is conducted as a traditional lecture with slides, with frequent stops to engage the class in discussion. Start with a simple definition of *data* to pique students' interest and introduce the topic in a humorous and nonintimidating way. Start with the question "What are data?" Encourage students to offer examples.

After some discussion, explain: Numeric data come from people counting things, and most of the data we deal with in the social sciences are either administrative data or survey data. Differentiate the following concepts:
- Administrative data are data that are collected as by-products of administering something (think hospital admissions or tax records).
- Survey data are data that are collected specifically for information or research purposes, by formally asking questions of a population (or sample) of respondents.
- A census is a survey that tries to collect information about every member of some population.

Data are released as microdata or macrodata. Explain the difference.

You may choose to highlight microdata examples for a course with an analysis component, or primarily emphasize macrodata released in tables and reports for courses where the student is not expected to conduct an analysis. Many of the sites we suggest feature both. (See Additional Resources).

Section: Where or Who Do Data Come From?
Explain the differences between public and private data sources:
- Public sources such as municipal, state, and national governments; governmental organizations like the UN and the World Bank. Nonprofit organizations and academic researchers are other sources of public data.
- Private sources such as companies, although these data are generally very expensive or proprietary and not released. Some limited data may be made available in shareholder reports and similar documents.

Publicly funded institutions have a mandate to spend their money toward certain goals and are held accountable to the public. Private institutions and businesses have no such mandate and are not accountable beyond what is required by law. Therefore, most publicly available data come from public institutions.

Instruct the class to begin a data search with these questions:
- Who cares about this topic?
- Who has access to this population?

- Who has funding to collect these data and a mandate to release the data?

Encourage them to think of service providers and advocacy organizations in addition to governmental sources.

Highlight that data on many topics may not be available for reasons including
- Organizations interested in the topic did not have the funding to collect it.
- Data were collected but not released for privacy or ethical reasons.
- Population is difficult to collect data on (e.g., the homeless population).

Use examples to demonstrate techniques like extrapolating from a similar population or a different level of geography when a data search is not producing results.

Section: Surveys, Samples, and Suppression

Explain that data collection methodology influences the level of detail that will be available in the statistics derived from those data. Sample surveys are designed to gather information on particular populations and subpopulations, and geographic precision and other detailed information about individuals is often suppressed or generalized in microdata to preserve respondents' confidentiality. A larger survey will give more reliable estimates for smaller subpopulations. Surveys are often designed to obtain accurate estimates for specific subgroups and geographic regions. National health surveys provide a good example of this.

Section: Government Data

Learning about why data are collected will help students better understand the data they are using and develop strategies to find them. Explain that governments collect data to allow for planning and assessment of policies and services. Specific data collection programs survey as few people as possible to collect enough data for specific government needs. Governments also collect administrative data from sources such as tax forms and birth or death records.

Highlight examples such as a health needs assessment survey. In Canada, community health needs are managed by regional health authorities, so the Canadian Community Health Survey collects and releases data at this geographic level. The US National Health Interview Survey similarly collects and releases data for the four census regions (Northeast, Midwest, South, and West), and within census regions by areas determined by metropolitan and nonmetropolitan status.

Section: Top Down, Bottom Up

Differentiate between the following techniques for locating data on a small geographic area.
- *Top down:* Look at national data and drill down to a local area. Demonstrate searching census data for information on a state or city.
- *Bottom up:* Start with locally collected data, for example from a municipality or a local advocacy organization. Provide locally relevant examples to illustrate the point.

REVIEWS/ASSESSMENT STRATEGY

The session works best if it is tied directly to a research assignment where the student needs to find statistical information; the students' success in doing so can serve as a proxy for evaluating the impact of the session.

CHEF'S NOTES

These types of sessions have been a good fit across multiple disciplines. They can easily be adapted by selecting appropriate data sources; for example, by focusing on sources of environmental, economic, or health data.

ADDITIONAL RESOURCES

Suggested sites for secondary data
- International
  - European Union, Eurostat, https://ec.europa.eu/eurostat/
  - Organisation for Economic Co-operation and Development, OECD Data, https://data.oecd.org
  - United Nations, UNdata, https://data.un.org
  - World Bank Open Data, https://data.worldbank.org
- National
  - Statistics Canada, https://www.statcan.gc.ca/eng/start
  - UK, Office for National Statistics, https://www.ons.gov.uk
  - US, Data.gov, https://www.data.gov
- Census
  - Statistics Canada, Census of Population https://www12.statcan.gc.ca/census-recensement/index-eng.cfm
  - UK, Office for National Statistics, Census, https://www.ons.gov.uk/census

- US Census Bureau, https://www.census.gov
- Local
  - United States
    - Chicago Data Portal, https://data.cityofchicago.org
    - New York City Open Data, https://opendata.cityofnewyork.us/data/
    - Open Government, Data.gov, https://www.data.gov/open-gov/
    - Lists of US city and county open data sites
  - Canada
    - Toronto Open Data, https://open.toronto.ca
    - Vancouver Open Data Portal, https://opendata.vancouver.ca/pages/home/

Looking at Data

Kay K. Bjornen, Research Data Initiatives Librarian and Assistant Professor, Oklahoma State University Library, kay.bjornen@okstate.edu

NUTRITION INFORMATION

Looking at Data introduces undergraduates to data information literacy skills. In the first module, articles about current events are discussed and students are encouraged to use critical thinking skills similar to those used for information literacy. Students are asked to check on data sources and evaluate the credibility of the message conveyed and how well it is supported by the data. An article based on COVID-19 is used as an example. A second module uses data about New Zealand Expeditionary Force World War I casualties to look at the history of data visualizations and their value for conveying a story.

TARGET AUDIENCE AND NUMBER SERVED

This lesson was created for use in a for-credit undergraduate honors class, "They Wouldn't Put It on the Internet if It Wasn't True": Information, Society, Trust and Technology in a "Post-truth" World. The lesson uses the context and concepts taught in information literacy instruction. Students are asked to read and analyze information in class. The exercises were created for a small class (15–20). It can be used for larger classes, though that may reduce the amount of time available for discussion and interaction. The lesson can be modified for any general audience, including secondary students and the general public.

LEARNING OUTCOMES

Following the lesson, students should be able to
- explain what *data literacy* means
- align data literacy with information literacy
- evaluate the use of data in popular media for credibility and objectivity
- apply data evaluation skills to non-text information (data visualizations)

COOKING TIME

This lesson is planned for a 50–75-minute class period.

DIETARY GUIDELINES

Academic librarians are often called upon to provide instruction in information literacy to first-year undergraduate students. ACRL's *Framework for Information Literacy for Higher Education* provides a guide for creating lessons that are interconnected to create a broad foundation for developing lifelong skills. Though data literacy is clearly a subset of information literacy, it is often overlooked with regard to this foundation. Applying the same critical skills to data that they use for media and text-based information means students are able to adapt them for use in a variety of settings, including more advanced classes and the workplace.

INGREDIENTS

- Computers or devices with internet access for each student or team of two to three students.
- Websites, blogs, or online publications about a topic of current interest. Topic should be current so that students can be better grounded in potential bias and have a foundation of knowledge to evaluate the information.
- Data and associated visualization. A historical source such as New Zealand Expeditionary Force World War I casualties works well.

PREPARATION

Prepare for the first module by choosing a current topic as a basis for student discussion. Locate one or more articles, blogs, or websites that use data to communicate a viewpoint or message. The message or viewpoint can be overt or subtle but should cite publicly accessible data sources. If time allows, find sources that communicate opposing viewpoints. Review the data sources cited and determine whether the data have been used in a fair manner. Be prepared to discuss and point out any instances of manipulation or omission that will affect the validity of using the data to support the argument.

For the second module, locate effective data visualizations and the associated data source.

A number of excellent resources are listed for both good and bad visualizations in the Additional Resources section. The visualization of World War I casualty data for the New Zealand Expeditionary Force is used as an example because it is engaging and effective. It also makes the point that, although data visualization has received a lot of attention in recent years, the concept is not new. Be prepared to ask students about their perceptions of the data before they are visualized and whether students have ideas for alternative visualization techniques.

INSTRUCTIONS

This unit is intended to be used as a subset of information literacy instruction. Specifically, you will teach data information literacy from the perspective of a data consumer rather than manipulating data for research or analysis.

The two aspects of data literacy that will be examined are
- reading and understanding data and information based on data
- effectively using data to validate and communicate a viewpoint

Through information literacy instruction, students should have been introduced to and become familiar with confirmation bias and the need for critical evaluation of information. They may have already been introduced to the CRAP test.[1] The acronym is commonly used to evaluate text-based information, especially online. It can be applied equally well to data. The acronym stands for
- *Current:* Is this recent information? When was it published or the website updated?
- *Reliable:* Is the data source linked or cited? Can the data be verified? Is the information presented in a fair and balanced manner?
- *Authoritative:* Is the source reputable? What is its expertise?
- *Purpose:* What story is being told? Does the storyteller have an agenda?

If appropriate, introduce or review the CRAP test and give one or more examples of how to apply it. Show students how to locate the date of publication for an article or website update. If this information is not available, is the source reliable? Discuss criteria for determining if a data source is authoritative and review an example where an incomplete or manipulated data set is used.

An editorial entitled "Five COVID-19 Charts Democrats Can't Let You Know About" is useful for this purpose.[2] The graphs used to support the authors' arguments were generated from two reputable data sources, but the data are used selectively and clearly intended to advance a political objective.

Module 1
1. After discussing the CRAP criteria, ask students to review the articles you have selected either individually or in teams. Ask students to summarize the content and then use the CRAP acronym to identify conclusions or questions generated by each criterion. For example, "Is there reason to believe that the conclusions created from the data may have changed since the article was published?" or "How current is current?" "Do you agree with the conclusions drawn from the data, or can you identify an alternate conclusion?" "Are there reasons to doubt the reliability of the data source or of the author who used the data?"
2. The authors of the article "Five COVID-19 Charts Democrats Can't Let You Know About" included three data visualizations to make their point. This fact provides an excellent segue into the next section, a discussion of evaluating data visualizations. After examining the text and its underlying data source, set this source aside and return to it after introducing effective visualization techniques, if time allows.

Module 2
1. Introduce the second module about data visualizations. Share the data from your chosen example in table form and ask students questions about trends they can see and conclusions that they can draw. Note how long it takes them to arrive at an answer.
2. Ask, "Why do we use data visualization?" Some conclusions that may come from the discussion:
   » It makes large data sets coherent and may clarify trends not easily identified when data are in tables or unstructured form.
   » It reveals levels of detail not immediately available through inspection of the data.
   » It makes comparing multiple data sources easier.

Section 2. Finding and Evaluating Data

3. Show table 1, a data table of casualties suffered by the New Zealand Expeditionary Force, and compare it to figure 1, a hand-drawn visualization of the same data.

NEW ZEALAND EXPEDITIONARY FORCE.

Return showing Number of Casualties (Deaths only) by Units of N.Z.E.F. while on Active Service.

Unit.	Killed in Action.		Died of Wounds.		Died of Sickness.		Other Causes.		Totals.		Grand Totals.
	Officers.	Other Ranks.	Officers.	Other Ranks.	Officers.	Other Ranks.	Officers.	Other Ranks.	Officers.	Other Ranks.	All Ranks.
Divisional Headquarters	..	1	1	2	..	5	..	..	1	8	9
Mounted Brigade Headquarters	..	1	..	1	..	..	..	..	..	2	2
Auckland Mounted Rifles*	15	194	8	68	..	42	..	6	23	310	333
Canterbury Mounted Rifles	13	179	5	70	4	60	1	1	23	310	333
Wellington Mounted Rifles	12	187	9	66	3	44	..	3	24	300	324
Otago Mounted Rifles	4	88	2	35	1	22	..	3	7	148	155
N.Z. Company Imperial Camel Corps	2	21	4	13	..	3	..	1	6	38	44
N.Z. Cyclists	1	34	1	20	..	4	..	1	2	59	61
N.Z. Mounted Rifles (Unattached)	..	..	..	..	..	13	..	3	..	16	16
N.Z. Field Artillery	33	379	19	243	5	146	1	15	58	783	841
N.Z. Engineers (Field Companies)	2	85	1	71	..	34	1	5	4	195	199
N.Z. Engineers (Field Troop)	..	1	..	2	..	7	..	1	..	11	11
N.Z. Engineers (Signal Service)	1	24	1	24	..	15	..	1	2	64	66
N.Z. Wireless Troop	..	..	..	..	1	4	..	..	1	4	5
N.Z. Tunnelling Company	1	24	1	16	..	19	..	1	2	60	62
Light Railways Operating Company	..	3	1	..	..	4	..	1	1	8	9
Headquarters, Infantry Brigades	1	3	2	1	..	1	..	..	3	5	8
Auckland Infantry Regiment	75	1,576	30	495	3	96	..	19	108	2,186	2,294
Canterbury Infantry Regiment	59	1,601	22	548	4	125	2	17	87	2,291	2,378
Otago Infantry Regiment	73	1,718	27	579	1	115	4	23	105	2,435	2,540
Wellington Infantry Regiment	57	1,530	15	463	4	97	6	21	82	2,111	2,193
N.Z. Rifle Brigade	107	2,198	28	759	..	178	2	30	137	3,165	3,302
Infantry unattached	..	..	..	..	1	214	1	15	2	229	231
N.Z. Maori (Pioneer) Battalion	8	122	4	62	2	130	1	7	15	321	336
N.Z. Machine Gun Corps (Mounted)	..	19	1	18	1	12	..	5	2	54	56
N.Z. Machine Gun Corps (Infantry)	10	235	10	109	7	20	..	4	27	368	395
N.Z. Army Service Corps (Mounted)	..	..	..	..	..	2	..	..	..	2	2
N.Z. Army Service Corps (Divisional Train)†	..	11	1	8	1	33	1	1	3	53	56
N.Z. Entrenching Battalion	..	1	..	86	2	26	..	25	1	2	4	139	143
N.Z. Divisional Employment Company	..	..	..	9	..	2	..	8	..	1	..	20	20
N.Z. Area Employment Company	..	..	..	..	..	2	..	8	..	1	..	11	11
N.Z. Medical Corps	..	9	..	53	3	49	8	47	1	27	21	176	197
N.Z. Army Nursing Service	..	..	..	..	..	..	2	..	10‡	..	12	..	12
N.Z. Dental Corps	..	..	..	..	..	..	..	..	1	..	1	..	1
N.Z. Veterinary Corps	..	..	..	..	..	..	..	6	3	1	3	7	10
N.Z. Chaplains Department	..	4	..	..	..	..	1	..	1	..	6	..	6
N.Z. Army Ordnance Corps	..	..	..	..	..	..	..	2	..	1	..	3	3
N.Z. Postal Service	..	..	..	..	..	..	..	3	..	..	..	3	3
N.Z. Army Pay Corps	..	..	..	..	..	..	1	2	..	1	1	3	4
Rarotongan Company	..	..	..	..	..	..	..	11	..	..	..	11	11
N.Z. Samoan Forces	..	..	..	..	..	..	..	13	..	2	..	15	15
Totals	488	10,382	198	3,752	52	1,570	35	220	773	15,924	16,697

* Includes one officer killed while serving with Royal Air Force. † Including A.S.C. Field Bakery and Butchery and Depot Units of Supply. ‡ Ten nurses drowned from "Marquette," torpedoed in the Mediterranean Sea.

TABLE 1
Casualties of the World War I New Zealand Expeditionary Force

Note that the tabular data do not easily convey trends or the magnitude of the data. Ask whether students can identify specific events and trends. Note how much more quickly the answers can be found in the graphic form. The graph shows New Zealand Expeditionary Force casualties from May 1915 until April 1919. Beginning with the top line: total casualties, followed by the categories of casualties in descending order: Wounded, Killed, Died of wounds, Died from disease, Died—cause unknown and drowned, Prisoners. Key battles are written above the total casualties line. Ask students to talk about the impact of the casualty numbers when they see them on the visualization. Also point out that the graph (visualization) was drawn by hand between 1914 and 1919.

4. Discuss how a good data visualization tells a story. Time may preclude an in-depth discussion of rules for effective data visualizations. Some guiding questions for basic evaluation of visualizations are these:
   » What story does the visualization tell?
   » Are there elements that clutter or confuse the story?
   » Is the information complete and well labeled?
   » Are the data shown clearly?
   » Are any elements (axes, scale, key) manipulated?

Mention Edward Tufte as one authoritative source of rules for good data visualizations. He created the term *chart*

junk to refer to unneeded elements that obscure rather than enhance the data story.³

5. Present examples of bad visualizations from a blog, such as "The Nine Worst Data Visualizations Ever Created" listed in the Additional Resources section. Using the CRAP test and visualization guidelines, ask students to identify what story each example is trying to tell and what makes it a bad visualization. This is a good way to start because the problems are easy to identify with the worst data visualizations.

6. Examine some good visualizations for comparison. The site *Information Is Beautiful* has a number of good choices that display data about current topics. "Coronavirus Riskiest Activities," from this site, is an example of an infographic, a different type of data visualization (https://informationisbeautiful.net/visualizations/covid-19-coronavirus-infographic-datapack/#activities). Return to the question "What story is being told, and is it effective?"

7. Finally, return to the editorial used for the first exercise ("Five COVID-19 Charts Democrats Can't Let You Know About") and examine the graphs published with the article. Use the same criteria and see if students can determine what story is being told and whether the message communicated in the article is supported by the data used to create the visualizations. Ask students to find the original data used to create the graphs and evaluate whether they support the graphs as shown. This exercise can be modified to accommodate available class time since it may require significant time to be thorough.

REVIEWS/ASSESSMENT STRATEGY

Some possible assessment options include these:

- Provide students with a web page or publication that addresses a topic of current interest, and ask them to summarize the content. Evaluate the credibility and objectivity of the supporting data.
- Align data literacy with information literacy by asking students to evaluate the authority of different data sources used

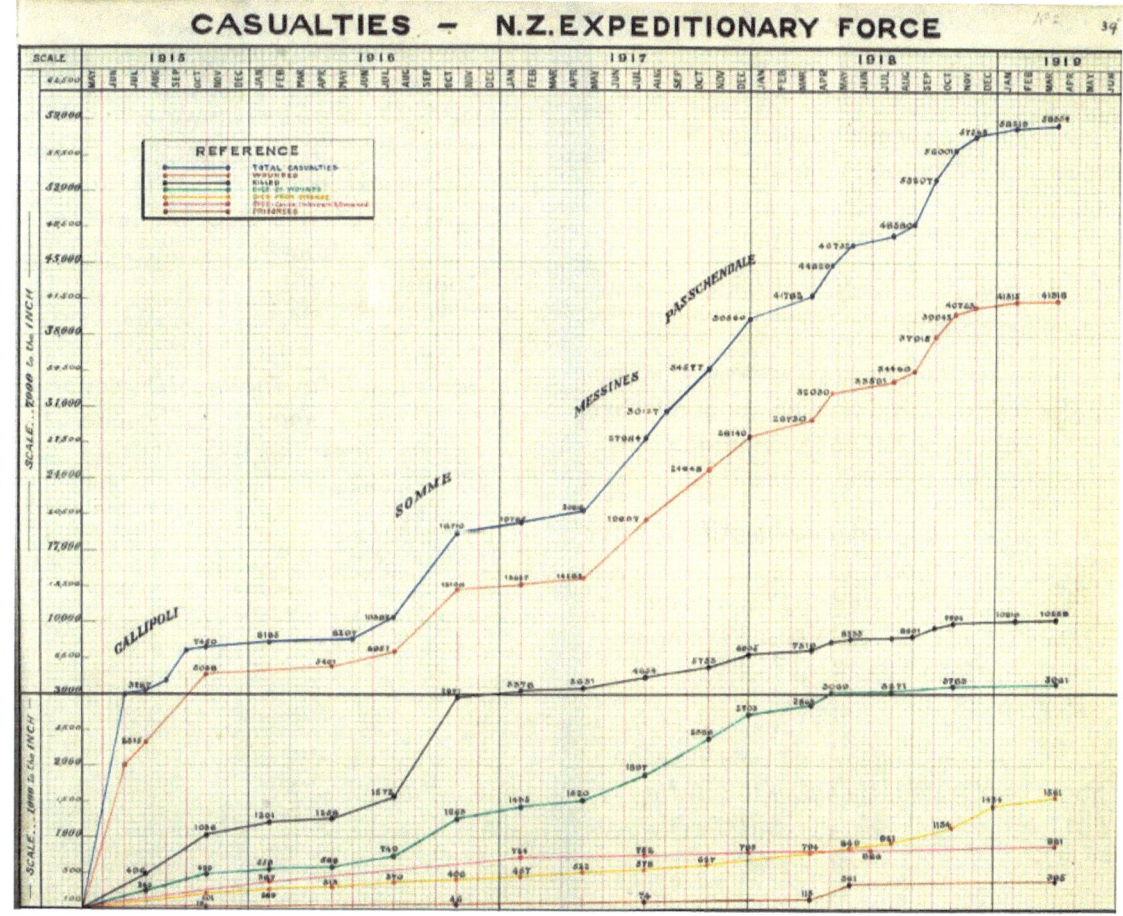

FIGURE 1
Graph of casualties of the World War I New Zealand Expeditionary Force (https://nzhistory.govt.nz/media/photo/graph-nzef-casualties-1915-1919)

- as resources for a given news topic.
- Assign a topic, and ask students to research credible information sources that support both sides. Ask them to document and justify each of the CRAP criteria as they apply to data.

ALLERGY WARNING

It is important to discuss balance when using a lesson of this type and be sensitive to balance when leading discussions about controversial topics or materials.

ADDITIONAL RESOURCES

New Zealand Expeditionary Force Data

New Zealand History. "Graph of NZEF Casualties, 1915–1919." Ministry for Culture and Heritage. Last updated March 20, 2015. https://nzhistory.govt.nz/media/photo/graph-nzef-casualties-1915-1919.

New Zealand History. "NZEF Fatalities by Unit." Ministry for Culture and Heritage. Last updated March 20, 2015. https://nzhistory.govt.nz/media/photo/nzef-fatalities-unit.

Guidelines for Data Visualizations

Pantoliano, Mike. "Data Visualization Principles: Lessons from Tufte." *Moz* (blog), February 12, 2012. https://moz.com/blog/data-visualization-principles-lessons-from-tufte.

A blog post with a good discussion of Edward Tufte's rules for data visualization.

Tufte, Edward R. *The Visual Display of Quantitative Information*. Cheshire, CT: Graphic Press, 1983.

Applying the CRAP Test to Data

Robinson, Marcene. "Can't Separate Crucial COVID-19 Facts from Fake News? Turn to the CRAP Test, Says UB Digital Literacy Expert." State University of New York at Buffalo News Center, April 17, 2020. https://www.buffalo.edu/news/tipsheets/2020/028.html.

Blogs for Identifying Good and Imaginative Data Visualizations

FlowingData, https://flowingdata.com.

FlowingData is a data visualization blog with imaginative examples for using graphs and infographics to convey topical information, often related to popular culture.

Information Is Beautiful, https://informationisbeautiful.net/about.

Author David McCandless's website that promotes his books and training about effective use of data visualizations and infographics.

Blogs and Social Media for Identifying Poor Data Visualizations

Twitter and Reddit are good sources for updating bad data visualization examples—search and follow hashtags such as #BadDataVis or #BadInfographics.

Couron, Aaron. "The 9 Worst Data Visualizations Ever Created." *LivingQlik* (blog), May 2, 2017. http://livingqlikview.com/the-9-worst-data-visualizations-ever-created.

Just one of many blogs and blog posts about bad data visualizations. It includes explanations of why each example is poor.

A Good General Reference about Data Literacy and Visualizations

Gray, Jonathan, Lucy Chambers, and Liliana Bounegru. *The Data Journalism Handbook: How Journalists Can Use Data to Improve the News*. Sebastopol, CA: O'Reilly Media, 2012.

NOTES

1. Kenneth Orenic and Molly Beestrum, "The CRAP Test," Loex2008collaborate wiki, April 28, 2008, http://loex2008collaborate.pbworks.com/w/page/18686701/The%20CRAP%20Test.
2. Issues and Insights Editorial Board, "Five COVID-19 Charts Democrats Can't Let You Know About," Issues and Insights, August 26, 2020, https://issuesinsights.com/2020/08/26/four-covid-19-charts-democrats-cant-let-you-know-about.
3. Edward R. Tufte, *The Visual Display of Quantitative Information* (Cheshire, CT: Graphics Press, 1983).

Interrogating the Data
What Data Sets Can and Cannot Tell Us

Kristin Fontichiaro, Clinical Associate Professor, University of Michigan School of Information

NUTRITION INFORMATION
When novice researchers search for data sets, finding *any* data that match their search terms can be a relief. What students may not realize is that experts interrogate data to determine potential insights and limitations. In this recipe, we use a small government data set with emerging data as an object lesson to help students grapple with the questions "What can these data tell us?" and, "What can these data not tell us?"

TARGET AUDIENCE AND NUMBER SERVED
This lesson is well-suited to high school students or undergraduate learners. If class size is under 20, a whole-group discussion will be more time-efficient. For larger group sizes, consider assigning students to discussion groups. The small-group method may take longer but will have the added benefit of more student voices being represented, a positive step toward more diverse, equitable, and inclusive instruction.

LEARNING OBJECTIVES
Students will
- acknowledge that interrogating information is as important with data sets as it is with text and other media
- understand that metadata descriptions are critical to one's understanding of data
- recognize that data from a single point in time offer limited insight and that comparing data across time, context, geographies, and populations can unlock new understandings

COOKING TIME
30 to 45 minutes

DIETARY GUIDELINES
This lesson fills a gap by applying information literacy practices to data sets. By questioning available data, students can gain experience applying some or all of these frames from ACRL's *Framework for Information Literacy for Higher Education*:
- *Information Creation as a Process:* Those aggregating and sharing data in real time may make flawed decisions that create unintended gaps in understanding.
- *Information Has Value:* Researchers need to consider whether their questions can be answered with the available data.
- *Research as Inquiry:* Understanding data sets is an active process and "talking back to the data" helps reveal meaning.

INGREDIENTS
- Current State of Michigan's COVID-19 School-Related Outbreak Reporting dashboard (https://www.michigan.gov/coronavirus/stats/school-outbreak)
- Google Folder containing March 22, 2021, report of Michigan COVID-19 School-Related Outbreak Reporting data and sample collaborative document (https://drive.google.com/drive/folders/1k9DAsYngS3k_I16nJHKQtR-SYTzmlBIRv).

PREPARATION
Make a copy of the collaborative document and determine whether you will use the current data set or the archived one (see Ingredients above).

INSTRUCTIONS
1. *Give introduction and overview.* Greet students, introduce yourself, distribute your contact information, and offer to meet privately with students moving forward. Give a quick overview of the lesson activities: students appreciate knowing what lies ahead, especially when meeting a new instructor.
2. *Set context.* Say, "I know that for your assignment, you will be searching for and using data sets to _____. If you're like most students, working with data is somewhat new, so before you start your assignment, let's spend some time together working through some things to keep

Section 2. Finding and Evaluating Data

in mind as you find and work with data sets." Present copies or direct students to the URLs for the data set and collaborative document.

3. *Build on prior knowledge by capturing first impressions.* Say, "Take a few minutes to look over the data set. What do you notice? Make notes in the first table on the Google Doc." (The Doc has two columns: "I/we noticed…" and "This makes us/me wonder…") This formative assessment will reveal what details students observe as well as meanings made of those details. Additionally, beginning from their observations encourages constructivist learning in which the students, with your help, build on their own questions and curiosities.

4. *Discuss their observations.* Nudge them to consider aspects of the data they did not notice. Their observations might include
   a. There are two tables: one for new outbreaks and one for ongoing outbreaks. This makes it difficult to get a sense of the overall health of each county.
   b. This is a table but not a spreadsheet, so you cannot re-sort entries easily.
   c. This shows one week's data, so we cannot tell if the number of COVID cases are decreasing, remaining the same, or increasing over time.

5. *Get familiar with the metadata.* Now ask who started by reviewing by looking at the tables, and how many began with the descriptive paragraphs at the top. Novices often begin by jumping straight into the tables or spreadsheet rows. Often, there is a second document that explains the definitions for the data or how they were collected: the *metadata*. This data set's metadata is found in the opening paragraphs. What does the metadata tell us? Ask students to add to the collaborative document. Likely insights include
   a. School data are reported weekly, even though statewide case counts are reported daily.
   b. Case counts are narrowly defined. Schools will be added to the list only if (1) there are two or more cases AND (2) cases were transmitted on school grounds AND (3) transmission was to a different household. So if a family of seven has COVID and the children attend the same school, only one case will be reported. If an entire travel soccer team unaffiliated with the school gets COVID, those cases won't be counted at all.
   c. Extracurricular activities such as sports are considered school sites.
   d. Due to various reasons, these data are likely underreporting the total number of cases. A school could be having an outbreak and not be counted on the list.

6. *Consider how metadata impacts data.* In a large group or in small discussion groups, ask, "How did the metadata change how you saw the data?"

7. *Consider what the data cannot tell us.* While every data set can tell us something, it cannot answer everything. Ask students to list—aloud or on the collaborative document—what the data *cannot* tell them. Responses might include
   a. We don't know which schools are remote only, face-to-face, or a hybrid. A school could be off the list simply because it is not holding any on-grounds activities. It's hard to extrapolate whether certain kinds of schooling are safer than others.
   b. We don't know if most of the cases are in students or teachers because they are counted together. This makes it hard to know whether it would be safer for kids or teachers to go back to "normal" schooling.
   c. We don't know how seriously the affected communities took the state mask mandate and how that might be contributing to case numbers.
   d. We don't know whether COVID-19 is on the increase, on the decrease, or staying the same because we see only one week's data. (Follow-up: Why might that be?)
   e. We don't know how bad the situation is without knowing the school populations. Fifty cases in a 2,000-student high school would be 1 in 40 people ill; 50 cases in a rural elementary school of 300 translates to 1 in 6. This is why data experts so often convert raw numbers to ratios or percentages: it makes comparisons easier.
   f. We don't actually know how sick school-age populations are because adults and children are included in these data.

g. For all these reasons, it's hard to answer "Is it safe to reopen schools?" or "Will I be safe doing face-to-face teaching?"

REVIEWS/ASSESSMENT STRATEGY

Ask small student groups to synthesize their understanding by creating a list of tips on the collaborative document that they can refer to when doing their own research.

ADAPTING THE RECIPE

As librarians, we often feel anxiety that we have to cover so much material in a limited amount of time. However, this kind of deep dive helps build our students' ability to really use what they have found and shows faculty that our work can be deeper than database demos. You could adapt these questions to analysis of visualizations or texts as well.

ADDITIONAL RESOURCES

For more on implementing data literacy into existing information literacy instructional practices, download these books from https://quod.lib.umich.edu/m/maize/ at no charge:

Fontichiaro, Kristin, Amy Lennex, Tyler Hoff, Kelly Hovinga, and Jo Angela Oehrli, eds. *Data Literacy in the Real World: Conversations and Case Studies*. Ann Arbor: Michigan Publishing, University of Michigan Library, 2017. https://doi.org/10.3998/mpub.9970368.

Fontichiaro, Kristin, Jo Angela Oehrli, and Amy Lennex, eds. *Creating Data Literate Students*. Ann Arbor: Michigan Publishing, University of Michigan Library, 2017. https://doi.org/10.3998/mpub.9873254.

Data Zines
A Hands-On Approach to Community Curiosities

Tess Wilson, Program Manager, All of Us Training and Education Center, University of Pittsburgh, tesskwilson@gmail.com

NUTRITION INFORMATION

We create civic data every day, whether we realize it or not. When we report potholes, register our pets, and go about many of our daily activities, we help build different data sets, many of which are freely available to the public via data repositories. By taking the time to locate, analyze, and present these data to an audience in an easily distributed format like a zine, we can become more intimately familiar with our own community ecosystem—and our role within it! This workshop was first delivered at a public library in Pittsburgh, Pennsylvania, as part of summer programming for teens.

TARGET AUDIENCE AND NUMBER SERVED

Anywhere from 1 to 20 students, depending on staff capacity and availability of ingredients. This recipe is best for somewhat experienced chefs, generally over the age of 10.

LEARNING OUTCOMES

By creating these zines, students will be able to
- locate and understand civic data via local and national data repositories
- translate data into visual graphics such as charts and maps
- explore the value of civic data and the role individual citizens play in data creation and collection
- adapt the zine format for their own uses

COOKING TIME

This recipe could be adapted as a full-day activity, a week-long series, or anywhere in between! When planning, consider the time it might take to research a topic, organize findings, and create the zine itself.

DIETARY GUIDELINES

This recipe is a great introduction to larger conversations about community advocacy and citizen science. In the same way we contribute to civic data sets every day, citizen science is an exciting way for regular citizens to play an important role in scientific discovery!

INGREDIENTS

Basic Ingredients
- Paper
- Pencils
- Pens
- Scissors
- Long-arm stapler
- Computers or tablets
- Internet access
- Printer

To Taste
- Colored pencils
- Old magazines
- Stick glue
- Graph paper
- Stickers
- Stamps and ink
- Paper punches
- Rulers
- Construction paper
- Zine examples
- Reference books and websites (see Additional Resources)

PREPARATION

Guest chefs: If local reporters or researchers can speak to the value of civic data, invite them to help with this recipe! The original version of this workshop was supported by a local nonpartisan, data-driven news organization, and the students learned a lot from these experts' hands-on experience and practical advice. Plus the presence of real journalists validated the work our students were doing.

Prep work: Ingredients such as paper, pencils, and scissors can all be set out before the students arrive. Ideally, each student will have access to a computer or tablet of their own with an internet connection, but it is absolutely possible for students to share devices or use their personal phones to research.

INSTRUCTIONS

Exploring Data Resources
- Share the Guide to creating a civic data

zine (figure 1) with students to help outline the exploration process. Each student will determine a civic issue they would like to focus on. It might help to frame this as a curiosity. Examples from previous iterations include What is the demographic makeup of my community? How many parks are there in my city and who manages them? Why are there so many potholes and what's being done about that?

- Students will then explore national and local data resources to find information about their chosen issues. Look for local data repositories (often supported by state or city government) or civic data organizations. National data resources might include the National Institutes of Health, the Food and Drug Administration, the US Census Bureau, and others. For more information, reliable news sources can be consulted to find stories or articles related to a student's chosen topic.
- When relevant data sets are located, students should either bookmark or print these findings for later use.

Creating Data Stories

- Students can refer to books such as *Dear Data* (Posavec and Lupi 2016), *Make a Zine* (Biel 2008), and *Information is Beautiful* (McCandless 2009), or sites such as Google Earth Timelapse (https://earthengine.google.com/timelapse/) to understand how data can be used to tell a story.
- Using the data collected, students will determine the best way to deliver their findings. Maps, bar graphs, line charts, and other data visualizations can be used, but encourage creativity! Mona Chalabi is a wonderful role model when it comes to outside-the-box data visualization. (Check out this article by Naomi Rea about Chalabi's work: https://news.artnet.com/art-world/meet-mona-chalabi-1893221).
- Since data visualizations are most effective when accompanied by explanatory text, students will write the story of their data—their initial civic curiosity, research process, findings, and any other relevant information. In some cases, other information might include a short directory of important contacts, a conclusion based on data findings, or instructions for further research.

Presenting Our Data

- Introduce your students to zines. A zine is a simple and straightforward DIY publication, often comprised of just a few pieces of paper stapled together. Because of their low-cost, highly flexible nature, they are often used to disseminate how-to guides, manifestos, or news relevant to a certain cultural subgroup. There are different formats, all of which have their benefits. Using standard 8½-by-11-inch sheets of paper stapled together allows more space to fill with visualizations, but the one-page zine is easily replicable and highly portable. Discuss various methods with your students and see what fits their needs best. (It helps to have some examples on hand, or reference books!)
- Students will transfer their data visualizations and narrative text to their zine pages. Encourage students to use colored pencils, collage materials, stamps, and other craft tools to make their zines uniquely theirs.
- Each zine should include
  - a visualization of some kind
  - an explanation of the topic

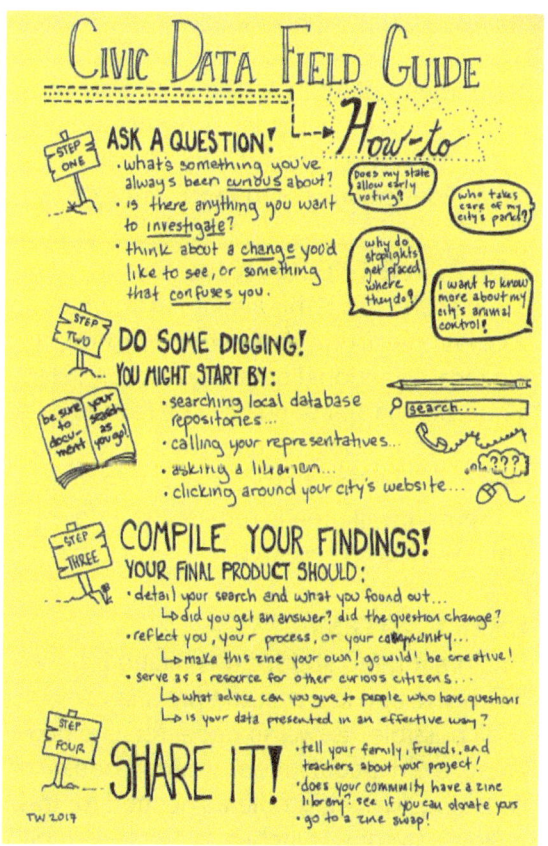

FIGURE 1
Guide to creating a civic data zine by Tess Wilson.

- When the zines are complete, do a gallery walk to share student work. Check to see if your library or community center has a zine library and submit these to be included!

REVIEWS/ASSESSMENT STRATEGY
The first iteration of this workshop was followed up with a survey distributed to participants, which revealed that teens were excited by the hands-on approach and delighted by the results of the workshop.

Assessment might include a survey that addresses basic data literacy concepts like the ones detailed in *Creating Data Literate Students* (Fontichiaro, Oehrli, and Lennex 2017).

ADAPTING THE RECIPE
If you prefer to do this activity in one day, consider preparing data sets in advance. Using local or national data repositories, select a number of topics and provide related data sets for your students' use. While this might limit the selection of topics available to your students, it will greatly reduce the amount of time needed for research.

CHEF'S NOTES
One great thing about zine making is how low waste and low cost the process can be! If at all possible, use scrap paper for zine pages and use old magazines or weeded books for collaging. Many ingredients needed for this recipe—stamps, punches, stickers—can be found at local craft supply donation centers. Plus your library likely already has the pens, colored pencils, glue, staplers, rulers, and tape required for zine construction!

ALLERGY WARNING
There are a few practical things to consider when creating a zine.
- Markers have a tendency to bleed through paper, so avoid using these.
- Remember that zines are recreated by photocopying, so carefully contemplate color contrast, materials, and size in preparation for this process.
- Keep a margin around the edges of each page for the sake of readability when stapled and copied.
- Make sure glue dries before closing any pages.

ADDITIONAL RESOURCES
Biel, Joe. *Make a Zine*. Portland, OR: Microcosm Publishing, 2008.

Chalabi, Mona. "Visualisations." https://mona-chalabi.com/data-visualisation/.

Fontichiaro, Kristin, Jo Angela Oehrli, and Amy Lennex, eds. *Creating Data Literate Students*. Ann Arbor: University of Michigan Press, 2017.

McCandless, David. *Information Is Beautiful*. New York: Collins, 2009.

Posavec, Stefanie, and Giorgia Lupi. *Dear Data*. London: Penguin Press, 2016.

Rea, Naomi. "Meet Illustrator Mona Chalabi, the Data Sleuth Making Poignant Visualizations about the Full Impact of Racism." Artnet News, July 9, 2020. https://news.artnet.com/art-world/meet-mona-chalabi-1893221.

On the Hunt
Understanding and Analyzing GSS Data Extraction for Incorporation within Sociological Research Projects

Amy Dye-Reeves, Associate Education and History Librarian, Texas Tech University, *amy.dye-reeves@ttu.edu*

NUTRITION INFORMATION

The General Social Survey (GSS) provides statistical information on American societal change. The data include employment, national spending priorities, and more. The GSS Data Explorer website, used within this recipe, allows librarians and sociology instructors to interact with GSS data for future incorporation in research projects.

The recipe introduces undergraduate sociology students to using the GSS Data Explorer repository. The following instructional plan provides an easy way to explore, analyze, extract, and share GSS data. The section Adapting the Recipe includes information on a digital alternative.

TARGET AUDIENCE AND NUMBER SERVED

Focus: Undergraduate sociology students.
Number served: 9 students (3 in each group). The number can be increased or decreased based on course enrollment.

LEARNING OUTCOMES

Each participant will
- create and build a content analysis to support specific research projects (ACRL's *Framework for Information Literacy for Higher Education*: Searching as Strategic Exploration)
- extract GSS data using tables, correlations, and regressions held within the GSS Data Explorer website (*Framework:* Information Creation as a Process)
- recognize the value of incorporating diverse perspectives from the extracted data sets (*Framework:* Scholarship as Conversation)

COOKING TIME

90 minutes:
- Introduction/GSS Data Explorer registration/pre-assessment questions (20 minutes)
- Part 1: Learning and Interacting with GSS Data (30 minutes)
- Part 2: GSS Scavenger Hunt and Using GSS Data Explorer in Group Projects (30 minutes)
- Wrap-up: post-assessment questions (10 minutes)

INGREDIENTS

For in-person instruction:
- Two large sheets of Easel Pad Size Post-it Notes
- One to four packs of small Post-it Notes

Optional:
- SMART Board
- Computers with Internet Access

PREPARATION

For in-person instruction: First acquire two large sheets from a Post-it Note Easel Pad. The sheets need to be large enough for students to attach individual Post-it Notes to answer the pre- and post-assessment questions. Second, acquire one to four small packs of Post-it Notes, depending on the class size. The rest of the assignment requires both computer and internet access for the remaining components of the lesson.

Prepare the pre-assessment questions:
1. Have you ever used the GSS Data Explorer website in previous courses? Yes/No
2. If yes, how did you incorporate the data in previous research projects?

Create an account within the GSS Data Explorer page (https://gssdataexplorer.norc.org).

Next, select a topic; for example, equity in employment. Then, select the variables (right-hand side of the page) and add them to the cart located on the right hand side of each variable. It is essential to familiarize yourself with the data analysis section. The list of

Section 2. Finding and Evaluating Data

choices includes cross-tabulation, multilevel tabulation, correlation, and regression. Prepare the post-assessment questions to be distributed using a SMART Board or printed-out copies. Post-assessment questions: (1)Was the GSS Data Explorer website easy to use? Yes/No (2)How do you plan to incorporate these data within future research projects? (3) Did you find the hunting for date to be helpful? If not, why?

INSTRUCTIONS
Part 1: Learning and Interacting with GSS Data

1. Distribute the pre-assessment questions to all students. Please use the Preparation section for the assessment questions and the Adapting the Recipe section for suggestions for offering the class digitally. The instructor should gather all the post it notes and review with the students before proceeding to get a clear picture of understanding of how to work with GSS Data.
2. The students will each need to create an account and log into the GSS Data Explorer website (https://gssdataexplorer.norc.org). Demonstrate how to create a sample project within the MyGSS account. Each student needs to create an account. This step will prove necessary during part 2.
3. Click on Project Setting (within the right-hand side of the screen). The sample project must be shared/visible and not set to a private feature as the whole group cannot view the material. Send students an invitation through the Invite Others section to view the material.

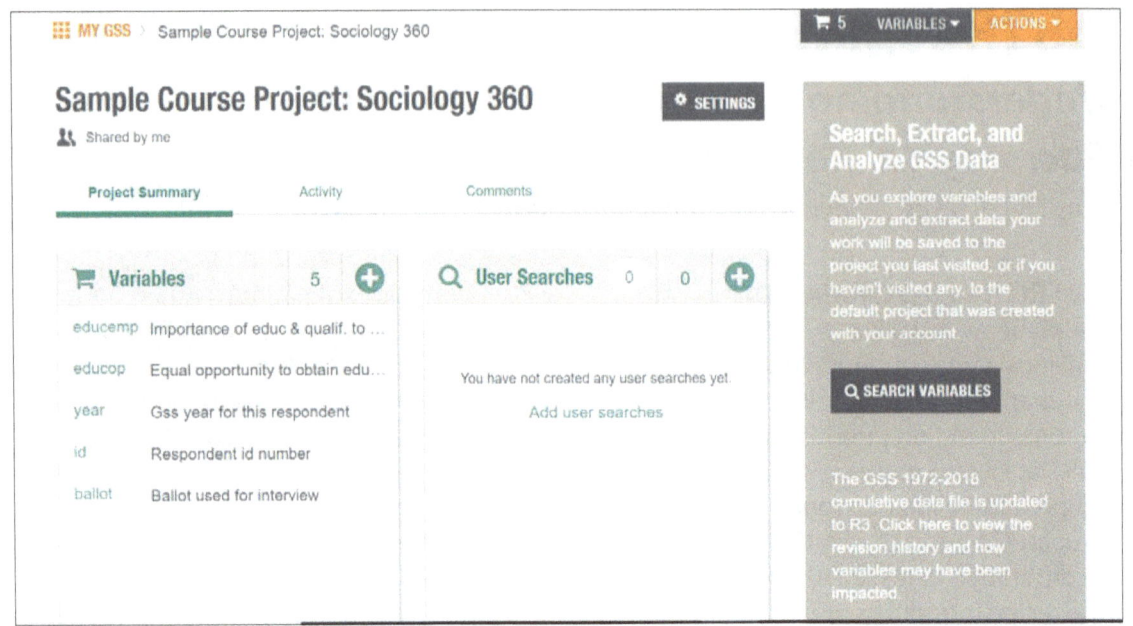

FIGURE 1
Sample: Using the topic page of "education and equality in the workplace" within MyGSS

4. Each student must log in again and view the individual project page.
5. The sample topic for this project might be "education and equality in the workplace." This is an example topic and instructors are free to choose any topics listed within "search GSS data" page. If you are using the "education and equality in the workplace" example, the students must click on the second variable (educop) within the MyGSS project (https://gssdataexplorer.norc.org/variables/241/vshow ; see figure 1).

   The second variable is entitled: "Equal opportunity to obtain education?" (see figure 2).

6. The students will continue by reading the survey question: "Does everyone in this

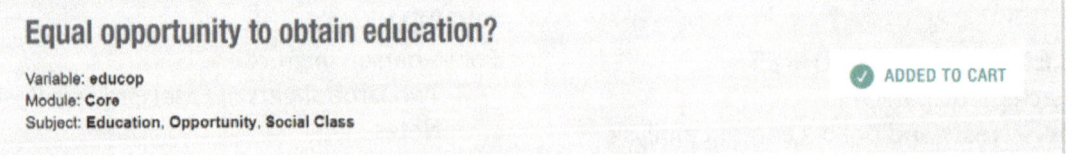

FIGURE 2
Examining the Survey and Variables within the Survey Questions

country have an opportunity to obtain an education corresponding to their abilities and talents?"
7. Have each student self-select five variables that work best with this research topic. Students must hunt through a list of possible variables on the right-hand side of the page.
8. Each student must open a blank Word document. Each student should answer the following three questions on the blank document: What five variables worked best with this research question? Why did you select these variables? How would these five selected variables be helpful within a research project?
9. The students should go back to the main GSS Explorer Data sample project home page. The students will need to select the Actions tab and click Analyze Data on the right-hand side.
10. The students should decide which type of analysis works best for their research topic given in step 6. The types of data analysis include cross-tabulation, multilevel tabulation, correlation, and regression. For the sample project, the students might select correlation when thinking about employment trends within a decade. Guide students through this process.
11. The students will need help practicing extracting the data to save for future projects. Students should find the data under Actions on the right-hand side and select Extract Data.

Part 2: GSS Scavenger Hunt and Using GSS Data Explorer in Group Projects
1. Place the students in small groups (3 to 4) depending on course size. The group will need to choose a leader for which to lead the discussion and be the manager of the myGSS project page as demonstrated in part one.
2. The students must go to the GSS Data Explorer website (https://gssdataexplorer.norc.org/).
3. Each group will need to create a new project in their MyGSS account.
4. Each group of students will need to go back to the main GSS website. The students will need to scroll down to the section entitled "available GSS" data. Students can choose anything from the "available GSS data" section.
5. Each group must begin to hunt for a list of variables that will work with their own self-selected research topic. The variables need to both inform and correlate within their current line of research. Each group will need to select five variables and add them to the cart. For verification purposes, the designated group leader must screenshot the variables and links to show project completion to the instructors.
6. The leader for each group needs to create a Word document and record answers to the following two questions: Why did you select these five variables? How do these data advance your research on your topic?

REVIEWS/ASSESSMENT STRATEGY
Provide a post-assessment to the student groups using either large Post-it Note Easel Pads or Padlet for online classes. The students should answer the post-assessment questions found in the Preparation section. Each group will need to submit this information with the required screenshots pasted into a Word Doc. Have the project e-mailed or have students upload the document through a specific content management system.
- *Supplementary student assessment strategies:* reflective learning, peer instruction, open-ended questions, qualitative and quantitative analysis
- *Supplementary instructor assessment strategies:* pre- and post-assessment, peer instructional assignment, qualitative and quantitative analysis

ADAPTING THE RECIPE
For virtual class alternatives, use Padlet to create digital Post-it Notes. The participants can leave their assessment feedback within the digital charts. The instructor can also think about using SurveyMonkey, Qualtrics, or other repositories for which to receive answers from the pre- and post-assessments.

ADDITIONAL RESOURCES
GSS Data Explorer, https://gssdataexplorer.norc.org

Padlet, https://padlet.com

Using Statistics to Define the Problem
Data and Service Learning

Amy Harris Houk, Education Librarian, Assistant Dean for Teaching and Learning, and Professor, University Libraries, UNC Greensboro, amy_harris@uncg.edu; Jenny Dale, Information Literacy Coordinator and Associate Professor, University Libraries, UNC Greensboro, jedale2@uncg.edu

NUTRITION INFORMATION
According to the National Youth Leadership Council, "Service-learning is an approach to teaching and learning in which students use academic and civic knowledge and skills to address genuine community needs."[1] Service learning is considered a high-impact educational practice.

Communication and Community is an undergraduate service-learning course in which students are required to complete a field placement at a nonprofit organization in the community. These placements have included an equine therapy program, an after-school tutoring program for children who live near the university, a food bank, and the YMCA. The main assignments for this class are reports and speeches related to these sites and the communities they serve. Librarians became involved in this class when instructors reached out for assistance with helping students find credible, relevant statistics for their reports. Some students had difficulties finding statistics beyond those on their organization's website and often struggled to figure out how to frame their organization's mission in terms of larger societal issues. By working with a librarian, students can find statistics that show why their organization is needed and how it can impact the community. In turn, this shows students how volunteering can be beneficial not only to the student but also to society as a whole.

TARGET AUDIENCE AND NUMBER SERVED
- This recipe is best served to undergraduate students.
- It can be scaled up or down as needed, but the original recipe is 20 to 30 servings.

LEARNING OUTCOMES
Students will
- identify a credible statistic for their assignment using an accepted evaluation framework
- select relevant statistics that explain community needs

COOKING TIME
- 1–2 hours of preparation before the first session, plus
- 50–75 minutes for the class

DIETARY GUIDELINES
Like most course-integrated library instruction, this recipe serves the larger purpose of helping students develop data literacy skills as well as broader information literacy skills.

This recipe connects to ACRL's *Framework for Information Literacy for Higher Education* in the frame Authority Is Constructed and Contextual. It emphasizes knowledge practices such as "use research tools and indicators of authority to determine the credibility of sources, understanding the elements that might temper this credibility" and "recognize that authoritative content may be packaged formally or informally and may include sources of all media types." The recipe ends with a searching discussion and activity, which reflects Searching as Strategic Exploration, with a particular focus on knowledge practices such as "determine the initial scope of the task required to meet their information needs," "identify interested parties, such as scholars, organizations, governments, and industries, who might produce information about a topic and then determine how to access that information," and "utilize divergent (e.g., brainstorming) and convergent (e.g., selecting the best source) thinking when searching." Students are also encouraged to develop the disposition to "seek guidance from experts, such as librarians, researchers, and professionals."[2]

INGREDIENTS

- 1 computer for each student (They can either use a lab computer or bring their own.)
- Internet access
- 1 instructor station with projector and screen
- Printed copies of the worksheet if students are not completing it electronically

PREPARATION

- Request access to the course in learning management system (LMS) if possible. This allows the librarian to have access to all relevant course materials and to see assignments in the larger context of the course. If this access is not possible, seeing the full syllabus for the course is a helpful alternative.
- Communicate with the course instructor early to confirm where students are completing their service-learning field placements.
- Update course activities to include appropriate examples for field placements.

INSTRUCTIONS

Introduction (5 minutes)

1. Welcome the students to class and ask a few of them to share the name of their field placement and briefly what the organization does.
2. Ask students to take a moment to think about why their organization is important and what need in the community the organization is trying to meet. Ask a few students to share their answers. For example, a few students each semester do their placement at a local food bank. They will likely answer that their organization is trying to solve hunger in the community.
3. With the help of the instructor, talk through the assignment and why students need to find statistics to be able to talk about their organization. Differentiate between data (individual pieces of information) and statistics (which represent observations based on data analysis).

The CRAAP Test, Part 1: Discussion (5–7 minutes)

1. Introduce the CRAAP test (see Additional Resources) to the class by explaining what each letter stands for and leading a discussion about each of the criteria (suggested questions for each criterion are below).
    a. *Currency:* Data are collected and statistics are created at various intervals of time; for example, US Census data are collected every ten years, meaning that our usual rules of thumb for currency of information may not apply to statistical information based on data. What do you think currency means? How current do you think would be current enough to be useful when trying to establish the need for your organization in the community?
    b. *Relevance:* Statistics are readily available about almost any topic imaginable. In this context, determining relevance requires an understanding of the work of your organization and the ability to connect statistics you find to that work. What kind of information would be relevant to your organization? What is the scope of the statistics that would be helpful (local, state, national)?
    c. *Authority:* Understanding authority requires you to research the person, people, or organization behind the data. You should look for information about who collected the data and who analyzed them. Who collected the data on which the statistics are based? Are they experts on the topic?
    d. *Accuracy:* Ideally, you can find statistics based on data collected in an ethical and transparent way. Look for information about what methods were used to collect and analyze the data. Where is the information from? Would you consider the source to be unbiased? Are there other sources that can back it up?
    e. *Purpose:* Data are collected and statistics are reported for many reasons. Understanding the purpose behind the data collection and how the data were analyzed can help you answer questions about the motivation for collecting and sharing this information. Why were these statistics presented? Is the purpose to inform or persuade?
2. After this discussion, the librarian should emphasize that the CRAAP test is not a checklist but a way of gathering information in order to make an informed decision about whether or not a source should be used.

Section 2. Finding and Evaluating Data

The CRAAP Test, Part 2: Activity (20 minutes)
1. The librarian shares the two activities discussed below. Students work in small groups to complete Activity 1 by answering the questions for each of the sources. After most groups have completed the chart for Activity 1 (appendix), the librarian should briefly talk the groups through the questions and ask which statistic would be the better source for their project.
2. Students then look at the chart provided for Activity 2 (figure 1). They take two minutes to brainstorm what story this chart might tell and how they could turn it into a statistic they could use in their paper. For example, students may point out that there were more Hispanic students in schools in 2017 than in 2000, which may mean that different community supports are needed.

Searching (15 minutes)
1. Using one of the issues students discussed at the beginning of the session when they considered what issue their nonprofit was trying to solve, the librarian and class brainstorm some ideas for keywords that could be used to find statistics to help situate the field placement in the community. For the food bank example, possible keywords could include *hunger, food insecurity, food bank, food pantry,* or *poverty*. They should also discuss geography, namely how adding the name of the city, county, or state may help to return locally relevant statistics. The librarian demonstrates a brief Google search and talks through some of the results using the elements of the CRAAP test, such as the website https://map.feedingamerica.org/, which would prove useful for the food bank topic.
2. The students search individually to find two relevant statistics for their own service-learning site while the librarian circulates to help as needed.

Closing (3–5 minutes)
1. The librarian brings the class back together and asks students if they were successful in finding the statistics they need and if there are any final questions.
2. The librarian briefly shows the students citation resources such as the APA Style Guide's "Webpage on a Website References" (https://apastyle.apa.org/style-grammar-guidelines/references/examples/webpage-website-references) and reminds them that their statistics must be cited in their service learning report.
3. The librarian asks each student to e-mail them the two statistical sources from the searching activity, the name of the sites,

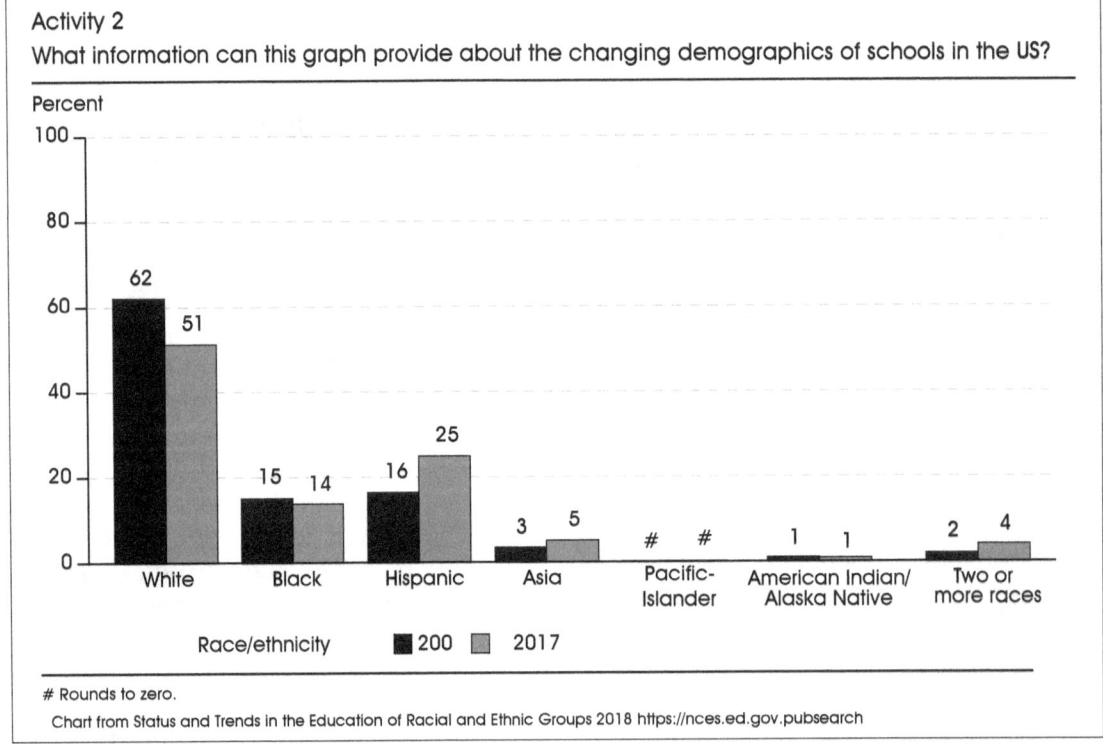

FIGURE 1
Chart for Activity 2: Percentage distribution of the US resident population 5 to 17 years old by race/ethnicity: 2000 and 2017

and the links. Students can also send their citations to the librarian for feedback if they choose.

REVIEW/ASSESSMENT STRATEGY

The librarian assesses the students' ability to find statistics that are relevant to their service-learning site and provides individual feedback to each student. If this assessment counts as an assignment or a participation grade, students are more likely to participate.

ADAPTING THE RECIPE FOR ONLINE COURSES

The original recipe was adapted for asynchronous online formats as multiple sections of the course moved online during the COVID-19 pandemic. Librarians created video content and short online assignments that allowed students to see demonstrations of skills, concepts, and tools and then to practice on their own. Students watched an instructional video and then e-mailed a relevant statistic and a brief evaluation of that statistic to the course librarian.

ADDITIONAL RESOURCES

California State University, Chico. "Evaluating Information—Applying the CRAAP Test." September 17, 2010. https://library.csuchico.edu/sites/default/files/craap-test.pdf.

Kuh, George. *High-Impact Educational Practices: What They Are, Who Has Access to Them, and Why They Matter*. Washington, DC: Association of American Colleges and Universities, 2008. https://www.aacu.org/node/4084.

NOTES

1. National Youth Leadership Council, "Service-Learning," accessed March 3, 2021, https://www.nylc.org/page/WhatisService-Learning.

2. Association of College and Research Libraries, *Framework for Information Literacy for Higher Education* (Chicago: Association of College and Research Libraries 2016), https://www.ala.org/acrl/standards/ilframework.

APPENDIX: WORKSHEET

Evaluating Statistics Using CRAAP

ACTIVITY 1

Imagine your placement is an after school tutoring center, where you are tutoring students in math.

	Currency	Relevance	Authority	Accuracy	Purpose
Questions to ask yourself.	Are the data I have the most recent data available? Are they recent enough?	Are they relevant to the topic I'm researching?	What group or organization collected the data? Do you consider it to be authoritative?	Do these data seem to be accurate? Can you tell how they were collected?	Why were these data collected?
Texas Dropout Rates—2000 All Students: 7.2% African American: 9.9% Latina/o: 11.2 % White: 4% http://go.uncg.edu/cst200data1					
2018–2019 Fifth Grade Math Proficiency: Peak Elementary: 27.3% Guilford County: 55.2% http://go.uncg.edu/cst200data2 (select Peck Elementary)					

Data and Statistics in the News and Media

Kaetlyn Phillips, Data Services Librarian, Dr. John Archer Library, University of Regina, kaetlyn.phillips@uregina.ca

NUTRITION INFORMATION
This recipe introduces students to data, in particular data found in everyday media and how they are collected and reported. Students will explore the behind-the-scenes process of data creation and analysis. Through in-class discussions, students will critically assess how they encounter data in everyday media.

TARGET AUDIENCE AND NUMBER SERVED
No more than 30 students and no fewer than 5 students, as students will work in small groups. Best for undergraduate students. Students are looking at clean data summaries and visualizations in academic publications and everyday media, not actively analyzing data or looking at complex forms of data (e.g., microdata).

LEARNING OUTCOMES
Students will
- examine the process for creating data and statistics in everyday media
- compare and discuss the changes to perception when information is packaged in different formats
- discuss how well-known authoritative sources of data should be evaluated with a critical eye and challenged if necessary

COOKING TIME
This recipe can be served in 60 minutes, but it can be extended.

DIETARY GUIDELINES
While specifically focused on data, learning objectives for this recipe center around two of the frames from ACRL's *Framework for Information Literacy for Higher Education*: Authority Is Constructed and Contextual and Information Creation as a Process. The intention is to introduce students to critical thinking about the data they encounter on a daily basis and to build confidence in their ability to assess data. Not only will this help them as they progress through their education and gain more experience with original research and research data, but it will also help them navigate data and misinformation in their everyday lives. Students will learn that data are not always neutral because they are constructed by authority figures (academics, government, think tanks, etc.) and interpreted by other authorities (media).

INGREDIENTS
- Networked computer access for all students
- Instructor station with projector (or ability to share screen)
- Space for breakout groups (or breakout room option for distance/online/remote learning)

PREPARATION
The suggested *New York Times* article (see Discussion Article under Additional Resources) may be behind a paywall, so check if you have access via a library resource or use a similar article. The 3, 2, 1 assessment can be done in Google Forms (or something similar) or using paper recipe cards.

INSTRUCTIONS
Begin by going over the learning objectives and activities for the lesson. Present that the way we usually see data is the final product, which looks flawless and perfect, but there is a behind-the-scenes process that determines if the data are of good quality. Consider using an analogy to illustrate the behind-the-scenes process, such as creating a flawless cut diamond, how a great photo is edited, or even how a film or television scene is created. The main point is that what we see when we look at data is the final product and there is a complicated behind-the-scenes process to prepare those data. This is also a great opportunity to highlight that while data seem unbiased, they are ultimately created and manipulated by humans who are biased. You can also address confirmation bias.

Next, review the CRAAP (currency, relevance, authority, accuracy, and purpose) evaluation methodology.[1] Depending on the year of the students, some may be familiar with this

method if they've already had information literacy instruction. If they haven't, having more time for the session would be beneficial. If the students are more familiar with data and information literacy, you can skip this section as the discussion questions address these concepts. If students have had information literacy instruction, highlight that the process is the same, but include data specific elements, such as

- For Currency, data typically take up to two years to be cleaned and processed.
- For Relevance, make sure that the data presented relate to the article or book.
- For Authority, emphasize that each agency (government, academic research, private firm/industry, NGO, think tank) has different biases that will shape how data collection occurs and what conclusions are drawn. Bias can also occur when independent agencies interpret the same data set.
- For Accuracy, emphasize how data collection determines the accuracy of the data. While knowing how the data were analyzed is important, if the data collection is of poor quality, then the results will be of poor quality. A simple way to describe it is "garbage in, garbage out."
- For Purpose, highlight that, like authority, the biases of the people sponsoring or collecting can impact the data's purpose.

Depending on the students' level of experience, use real data examples with the behind-the-scenes analogy and CRAAP evaluation methodology to start the discussion and do some formative assessment. If students are more familiar with data concepts, go straight to the *New York Times* article. Possible discussion points could include

- Ask students to list all the possible definitions of the word *income* (or another common variable with different definitions, such as *adult*). Provide a media headline or article on average incomes. Based on the article, discuss with the students how we know or don't know the type of income is being measured.
- Find statistics or infographics from different sources measuring the same variable (stick with income if that works). Discuss with students how we know or don't know if the sources are using the same definition for the variable. This discussion could be expanded to include questioning how the variable was measured and analyzed. For example, was mean, median, or mode used? Were outliers removed?

Take time to pause for student questions, then move into the discussion section using the *New York Times* article "Trump Suggests Virus Death Count Is Inflated. Most Experts Doubt It." If the session is longer, more time can be given to this discussion. Ask students to read the article before the session to save time. Once students have read the article, place them in small groups of three to five people, and ask them to discuss the following questions:

1. In what ways has the authority of the CDC been challenged?
2. How do the biases of the CDC and the White House affect the accuracy of the data?
3. Does the White House have a good case for changing how death statistics are collected? Consider how other countries collect death statistics and how a country's response to the virus could be judged by its reported statistics.
4. Who are you more inclined to believe in terms of accurate data, the White House or the CDC? What possible personal biases are affecting your decision?
5. What are the strengths and weaknesses of the data collection from both groups? Which do you trust more? Why?

Ask each group to present a summary of its discussion. Take time to answer questions. Conclude the lesson with a 3, 2, 1 reflection, which asks students to list 3 things they learned, 2 things they want to learn more about, and 1 question they still have. If they have no questions, they can leave feedback on the presentation.

REVIEWS/ASSESSMENT STRATEGY
A large amount of the assessment is formative, especially when discussing how data collection and analysis occur. The librarian monitors the discussion and ensures through student responses that understanding is occurring. The 3, 2, 1 reflection is also formative and provides feedback for the instructor.

CHEF'S NOTES
Depending on the class and how often you'll be seeing the students, the assessment could be anonymous or used as an opportunity to

follow up with the individual students.

ALLERGY NOTE

COVID-19 has had a global impact, and discussing death statistics may be a sensitive matter for students. A brief content warning is advised to not shock or upset students.

ADDITIONAL RESOURCES

For Explanations of Data Literacy and Common Forms of Data Misinformation

Bergstrom, Carl, and Jevin West. "Calling Bullshit: Data Reasoning in a Digital World." INFO 270/BIOL 270 course syllabus. Accessed March 29, 2020. https://www.callingbullshit.org/syllabus.html.

Best, Joel. "Damned Lies and Coronavirus Statistics." Recorded September 22, 2020, at Data in Real Life: ICPSR Data Fair, Ann Arbor, MI. Posted October 5, 2020, by ICPSR. YouTube video, 1:04:49. https://www.youtube.com/watch?v=HyL3CDKAZDw.

Glusker, Ann. "Data Engagement for the Data-Hesitant Librarian." Recorded September 24, 2020, at Data in Real Life: ICPSR Data Fair, Ann Arbor, MI. Posted October 2, 2020, by ICPSR. YouTube video, 47:19. https://www.youtube.com/watch?v=EVIEE9jw9fM.

Discussion Article

Weiland, Noah, Maggie Haberman, and Abby Goodnough, "Trump Suggests Virus Death Count Is Inflated. Most Experts Doubt It." *New York Times,* May 22, 2020, https://www.nytimes.com/2020/05/22/us/politics/coronavirus-trump-death-toll.html.

NOTE

1. Sarah Blakeslee, "The CRAAP Test," *LOEX Quarterly* 31, no. 3 (2004): article 4, https://commons.emich.edu/loexquarterly/vol31/iss3/4.

Section 3.
Data Manipulation and Transformation

61   [[Ch15]]**A Kinesthetic Approach to Data: Moving to Understand Nominal, Ordinal, Interval, and Ratio Relationship in Data**
Wendy Stephens

64   [[Ch16]]**Text Mining Charcuterie Board**
Yun Dai and Fan Luo

67   [[Ch17]]**Anyone Can Cook (R)! Open Data with R, a Five-Week Mini-mester**
Jay Forrest and Ameet Doshi

70   [[Ch18]]**Software Carpentry Al Dente: Rendering Tech Training for Online Artisans**
Peace Ossom-Williamson, Shiloh Williams, and Hammad Rauf Khan

73   [[Ch19]]**A Recipe for Improving Online Instruction for the Carpentries**
Kay K. Bjornen and Clarke Iakovakis

A Kinesthetic Approach to Data
Moving to Understand Nominal, Ordinal, Interval, and Ratio Relationship in Data
Wendy Stephens, Associate Professor and School Library Program Chair, Jacksonville State University, wstephens@jsu.edu

NUTRITION INFORMATION
This lesson on determining data characteristics will be a little different with every group because it is essentially a version of "Stone Soup." Participants won't bring ingredients for our communal pot; instead, they will share their own personal data for the group to parse and sort while the instructor elicits observations and extends learning throughout the workshop.

TARGET AUDIENCE AND NUMBER SERVED
This is a very flexible activity that can work with big and small groups and a diversity of learners and age groups. This chef has served it successfully with seventh graders, graduate students, and school librarians.

LEARNING OUTCOMES
Learners will
- interact with a variety of data types through physical movement
- observe and develop a tangible understanding of a range of characteristics of numerical data

COOKING TIME
20 to 45 minutes

DIETARY GUIDELINES
Understanding data characteristics is particularly useful to students who are learning to read research studies and need an acquaintance with statistical methods to appropriately interpret and evaluate research findings and to novice researchers who are preparing to collect data for their own scholarship or those who are making use of existing data sets, who need to understand the relationships between data, how they are characterized, and how those values can be contextualized in consideration of statistical significance.

INGREDIENTS
An eager crowd of participants, some space to spread out, such as a large room, a handheld microphone if the space is not big enough for everyone to hear the conversation without it.

PREPARATION
- Determine your physical start and end points for the group sorting based on the available space. If the whole space is available, consider the three-dimensional plotting exercise (reflected in the Adapting the Recipe section).
- Let the learners know they will be moving around the space and to leave their bags and devices at their desks or seats.
- Plan for any accessibility issues.
- Be ready to facilitate and interact with all participants.
- The goal of the lesson is to get learners talking about the different organizational methods for data as they order themselves. The process of moving around the space and situating oneself will help make the concepts resonant. Once the group is satisfied that they are sorted appropriately in each of the iterations, the instructor will walk down the line and ask each participant, in turn, to articulate and explain the value that their place in line represents. In cases where learners did not sort appropriately, the instructor should ask the group for help in re-sorting and urge them to state the "value" of the data type they represent.

INSTRUCTIONS
Step 1: Nominal Data
Part 1: Have participants organize alphabetically by first name. In most studies, a respondent's identity would be translated into a participant ID, which might be numeric. But in most cases, the identification numbers are still considered to be nominal data, with no inherent statistical value. Instead, these are considered attributes that help with organization of the data set.

Section 3. Data Manipulation and Transformation

Part 2: Have participants put themselves in order by street address. Depending on time, you can ask them to do this in various iterations—beginning with the house or apartment number, then alphabetized by the proper name. The variation in street names emphasizes the use of nominal characteristics as an organizational piece rather than as a value. This has inherent relevance for most studies.

Step 2: Ordinal Data

Part 1: Ask learners to sort themselves based on birth date. This will work with both Month-Day and Month-Day-Year; the goal is to demonstrate that birthdays—whether they fall within a calendar year or more broadly over time—are sequential. But within that sequence, there is no set relationship between the value of numbers in the group. To stress the concept, remind them that some things in life do key on ordinal data—think of the date of birth plus 16 years for driver's license eligibility, or plus 18 years for voting. Have them note that the range existing between participants remains the same in those scenarios.

Part 2: Have learners sort themselves based on the time they arrived in class today. (Alternately, you could do first semester enrolled if you are operating virtually.)

Emphasize that, when data are ordinal, it is important to preserve the order in which things occurred, something possible when you employ statistical methods using square roots, as in correlational methods.

Step 3: Interval Data

Interval data rely upon equal distances or fixed units. This is an important statistical concept across disciplines for those who are interested in measuring differences. One fundamental example is the 24-hour day, where all hours are equal in duration. Have participants order themselves by the year (or decade, if the group is less homogenous) they graduated from high school. Anticipate questions about the academic year and the calendar year and be prepared to discuss these as fixed units. The difference between any two interval values can be calculated using subtraction. This makes interval data valuable for determining the relative relationship between two data points. One example of this mathematical subtractive trait is the conversion of Celsius temperature to Fahrenheit, since a difference of 20 °C is equivalent to a difference of 36 °F, regardless of the measurement on either scale.

If you have time, and students are engaged, try the intervals exercise in Adapting the Recipe.

Step 4: Ratio Data

Ratio data are interested in proportions. Proportions ensure that you can multiply a ratio by a constant to get statistically meaningful data. Know that with ratios, there is a true zero, like the zero on the Kelvin scale. And, since that true zero exists at midnight, time can be treated as a ratio. Have learners sort themselves in the order they woke up this morning.

ASSESSMENT STRATEGY

The instructor can close this activity by asking participants, either orally or via an electronic or paper-based exit slip, to identify each of these four basic data types at the end of the lesson, in a way that reinforces the limitations of each number type. One possibility is live polling around examples of each type of data, using a true-false reveal for the review.

ADAPTING THE RECIPE

Intervals: In cases where you have more time, the group can spend some more time with intervals by using pages 9–10 of the "Fractile vs. Equal" lesson for grades 7–10 from the Federal Reserve Bank of St. Louis. This activity works best with 4 to 6 participants per group. The exercise involves two sets of 24 cards with distribution of worker ages. In an effort to organize the data to make them easier to understand, the groups explore both intervals of equal amounts and those are fractile. There are the same number of data points in each resulting group. This activity demonstrates how the underlying construction of an interval determines the distribution of the data collected; multiple sorting options are equally valid, but arranging the data differently could generate different takeaways. The cards associated with this activity can be printed on durable colored card stock that can be reused with multiple groups.

In Three Dimensions: Negotiate about the mathematical operations it is possible to carry out on data types. Emphasize that you can multiply ratios with a constant. Another instructional extension, and one that works

particularly well as an icebreaker, especially in a group with geographic distribution, can involve asking students to sort themselves based on their home location, reflecting on north to south, west to east, or plotting both relative points with all four walls of a place representing axes. This self-organization activity demands conversation involving numeracy, the common negotiation of meaning, and integration of a part into a whole. It sets students up for understanding how such disparate data points can be treated equivalently.

ADDITIONAL RESOURCES

Federal Reserve Bank of St. Louis, "Fractile vs. Equal," lesson for grades 7–10, 2014. https://www.stlouisfed.org/~/media/education/lessons/pdf/fractile-vs-equal.pdf.

Text Mining Charcuterie Board

Yun Dai, Data Services Librarian, Library, New York University Shanghai, yun.dai@nyu.edu, http://shanghai.hosting.nyu.edu/data/; Fan Luo, Digital Scholarship Technologist, Library, New York University Shanghai, fan.luo@nyu.edu

NUTRITION INFORMATION

Manipulating textual data is like curing meat; once cured, the meat can be used in a variety of ways. The same is true for text. After a series of preprocessing and manipulation steps, textual data can evolve into forms that can be ingested in more advanced modeling and analyses.

This recipe summarizes how to prepare for a workshop that is a text mining charcuterie board, a starter that equips the attendees with the essentials of text mining using R. Attendees of the workshop will see the workflow of turning texts into data for modeling and more advanced analyses and learn several text mining techniques and how to implement them using R.

This is one of three recipes that weave data literacy competencies into project-based technical workshops. See also "Web-Interfacing Data Visualization in a Rainbow Layer Cake" in the Data Visualization section and "Stuffed Shiny App with Business Intelligence" in the Data in the Disciplines section.

TARGET AUDIENCE AND NUMBER SERVED

Serves a small group of 5–10 attendees. Best for students who are familiar with R and who plan to use R for a text mining research project.

LEARNING OUTCOMES

Attendees of the workshop will be able to

- formulate a research question and its need for textual data
- identify key terms that describe information needs
- retrieve textual data from various sources with corresponding methods
- evaluate the sources of data and the quality of collected data
- recognize that information may need to be extracted from raw textual data to meet information needs
- recognize that information from various sources can be combined to produce new information
- apply preprocessing methods to clean textual data and use introductory text mining techniques to extract information from textual data

COOKING TIME

2 hours

DIETARY GUIDELINES

This exercise touches on the frame Research as Inquiry in ACRL's *Framework for Information Literacy for Higher Education*. It uses the following performance indicators in ACRL's *Information Literacy Competency Standards for Higher Education*: Standard 1, Outcome 1.a, on identifying a research topic in class discussions; Standard 1, Outcome 1.e, on identifying key concepts and terms describing information needs; Standard 2, Performance Indicator 2, on constructing and implementing effective search strategies; Standard 4, Outcome 1.d, on manipulating texts and data to transfer them from their original locations and formats to a new context; and Standard 5, Outcome 2.e, on legally obtaining, storing, and disseminating text and data.

INGREDIENTS

- TED Talks data sets from Kaggle[1]
- Sample data collected from Twitter
- R script with step-by-step instructions
- Slides that outline the workshop with (1) introduction to text mining and (2) techniques and workflow of text manipulation for text mining
- 1 instructor and 1 helper (recommended for an online session and optional in person)

PREPARATION

1. *Download TED Talks data sets from Kaggle.* Merge the data set of transcripts with the data set of each talk's metadata. Keep a small number of cases from the merged data set for use in the workshop, considering that running some algorithms on a large data set can be time-consuming.

2. *Collect data from Twitter that can talk to the subset of the TED talks data.* The Twitter data set contains the tweets from the speakers and the metadata associated with each tweet. The resulting data set, including tweets and their metadata, should contain information that can link to the TED talks subset. This sample Twitter data set is prepared in case attendees are not able to access Twitter APIs during the workshop.
3. Prepare an R script with commands to use in the demonstration.
4. Check the internet connection in the room where the workshop will take place.
5. Send e-mails to the registered attendees before the workshop with instructions to create a Twitter account and to set up the developer credentials.

INSTRUCTIONS

Steps 1–6 are a facilitated discussion followed by a lecture/presentation.
1. Invite the students to share their experience with textual data or research that uses textual mining methods.
2. Introduce what text mining is and address its challenges.
3. *Discuss two published research papers that use social media data.* Analyze how researchers apply text mining techniques in their studies (e.g., identifying the frequency of emotional language within each post using linguistic inquiry and word count; identifying latent themes within posts using topic modeling).
4. *Outline the workshop schedule.* Get ready to cure the meat to lay out the charcuterie board.
5. *Find and evaluate sources of textual data.* The first thing to do is go to a reputable butcher shop and select cuts for curing.
   a. Introduce various sources of textual data and methods to collect textual data. For this workshop, attendees learn to search data sets on online platforms such as Kaggle and to compile queries to collect data on Twitter via its APIs through an R package.
   b. Evaluate the collected data. Make sure what we get is good meat. For instance, are there copyright and licensing concerns? Is there a codebook or rich metadata that describes the data? Is the data set actively maintained?
6. *Demonstration.* Learn to build our text mining charcuterie board, from scratch.

See figure 1. The instructor uses the TED talks data set to demonstrate the workflow of preparing textual data for more sophisticated analyses. Note that real-life research is an iterative process and does not necessarily follow a linear workflow. Note also that a research project should be guided by a research question.
   a. Import text data into R. Show how to read the TED Talks data, stored in a CSV file consisting of a character vector for documents and additional vectors for document-level variables.
   b. Identify the information needs, operationalize the concepts, and locate the relevant variables. Then decide which text mining techniques to use to extract information from the variables containing text. Explain how to preprocess texts and restructure texts to formats

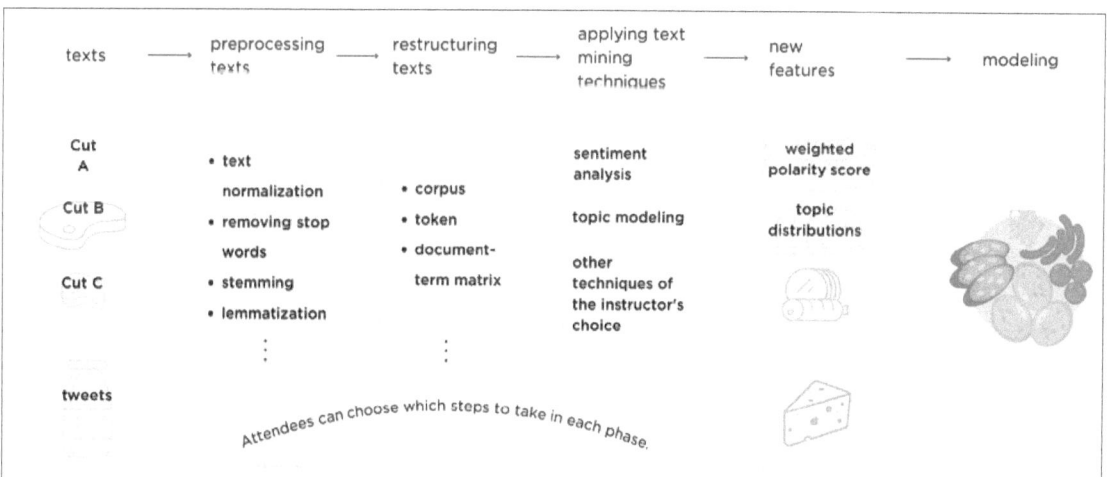

FIGURE 1
Workflow of text manipulation, created by Fan Luo and Yun Dai under CC BY-NC 4.0.

compatible with text analysis before applying a text mining technique.

The following two examples illustrate the logic and its implementation with R.

Example 1. Use one attribute *ratings* and one specific measurement in sentiment analysis, polarity score, to create a new metric that quantifies overall sentiments toward a talk. Attribute *ratings* is a dictionary that stores 14 adjectives (e.g., "inspiring," "jaw-dropping") and their frequencies associated with each talk. To calculate a new sentiment metric, for each talk, attendees follow the instructions to transform the dictionary to a data structure more friendly to R, extract the adjectives and their counts, weight the polarity score of each adjective by its count, and add up the scores of the 14 adjectives.

Example 2. Use the latent Dirichlet allocation (LDA) algorithm for topic modeling to obtain a group of features that show the per-document-per-topic probabilities from the transcripts. First, preprocess transcripts by converting texts to lowercase and removing stop words. Second, perform data structure transformations. Convert sentences to words (tokenization), stored in a data frame. Split the data frame into one-token-per-row data frame. Cast this data frame to a document-term-matrix. Next, estimate how many topics are in each transcript by calculating the average number of tags attached to each talk, stored in the attribute *tags*. Finally, feed that number into the LDA algorithm to extract the topic distributions of each transcript.

7. *More on one's own.* In this section, students work on their own project. This is the time to add some cheese to the cured meats on our charcuterie board. At this step, attendees apply what they have learned to mine the textual data on their own. They need to come up with a research question, collect data on Twitter, and then figure out a way to join the collected Twitter data with the TED Talks data set to answer the research question. The attendees are not expected to actually test a hypothesis or fit a model; this activity is introduced as a next logical step following previous procedures. If attendees have problems accessing Twitter APIs, they can use the sample Twitter data set instead.

REVIEWS/ASSESSMENT STRATEGY
Gauge attendees' tech savviness at the start of this workshop. One question could be "What are the criteria to select a data source for text mining?"

Use attendees' reactions to assess learning effectiveness. Evaluate teaching effectiveness with a feedback form at the end of the session. Questions could be "How useful did you find the workshop? Were you able to follow the workshop? What were your favorite and least favorite parts?"

ALLERGY WARNING
A project-based workshop covers many facets of data literacy, so attendees have lots to digest. It may also pose more technical barriers than an introductory or topical session. Instructors should share the class materials beforehand, both for attendees to preview and to manage their expectations.

A project-based workshop may also attract an audience of various tastes, and hence it is important to balance the flavors. For instance, in this case, a data science student may be attracted by the text analysis, and a business student may want to hear more about the sentiment analysis.

CHEF'S NOTES
It is the chef's choice regarding which data sources to use (meat to cure) and with what text mining techniques (curing methods). This recipe mentions two techniques, but there are many others that can be added to the charcuterie board.

The workshop is intended to be practical without delving into the statistics behind each text mining method we choose. The instructor may wish to explain what happens underneath a line of code. Or the instructor may discuss the rationale and trade-off of choosing one method versus an alternative.

NOTE
1. Rounak Banik, "TED Talks," data about TED Talks on the TED.com website until September 21, 2017, Kaggle, last modified September 2017, https://www.kaggle.com/rounakbanik/ted-talks.

Anyone Can Cook...R!
Open Data with R, a Five-Week Mini-mester

Jay Forrest, Data Scientist Librarian and Subject Librarian for Analytics, City and Regional Planning, Earth and Atmospheric Sciences, History and Sociology, International Affairs and Music, Georgia Tech Library, jay.forrest@library.gatech.edu; Ameet Doshi, Head, Donald E. Stokes Library, Princeton University, ameet.doshi@princeton.edu

NUTRITION INFORMATION

Open Data with R is an accessible introduction to data analysis with the R statistical programming language using the principles of open science as a framework. The National Academies of Sciences, Engineering and Medicine defines open science as encompassing "the free availability and usability of scholarly publications, the data that result from scholarly research, and the methodologies, including code or algorithms, that were used to generate those data."[1] This class covers importing, transforming, visualizing, modeling, and communicating data with R software in an open manner. Importantly, we embrace an inclusive ethos that "anyone can cook" when it comes to learning the R programming language.

This recipe is inspired by making and decorating a cake from scratch, reflecting our goal of bringing students from no or beginner knowledge of R to an intermediate skill level through a project: that is, "baking a cake." Students can choose to make a cake of their choosing (a paper, poster presentation, or oral presentation) in any flavor (open data set). A key opportunity from the mini-mester is for students to show off their delicious creation to other burgeoning chefs, or potentially get hired at a four- or five-star restaurant. We encourage our chefs in training to add this R project to their portfolio as they enter the job market.

TARGET AUDIENCE AND NUMBER SERVED

The average is 20 student cooks per mini-mester. The kitchen has limited space and requires both a sous chef and a primary chef while teaching, so we find that 20 is an optimal number. Interestingly, the course attracts a wide variety of major disciplines as well as undergraduate, masters-level, and doctoral students. In the last two mini-mesters, all six Georgia Tech colleges have been represented (business, computing, design, engineering, sciences, and liberal arts). This creates opportunities for interdisciplinary exposure, getting students out of their disciplinary silos and furthering a strategic goal that the library convene students across majors while improving data skills.

MINI-MESTER LEARNING OBJECTIVES

1. Understand and apply the principles of open science including data sharing, open data tools, and open reproducible research.
2. Input data from a variety of formats into R (readr).
3. Apply data transformation techniques in R (dplyr, tidyr, stringr).
4. Apply data visualization techniques using the grammar of graphics and R (ggplot2).
5. Analyze data through exploratory data analysis in R.
6. Analyze and evaluate basic statistical models in R (lm, rpart, glm).
7. Communicate and share data (RMarkdown language, GitHub).

COOKING TIME

This is a one-credit-hour course that meets for ten 75 minute cooking sessions over 5 weekly modules (online, synchronous, or in person). However, our culinary students have noted that their investment of time in the kitchen is highly dependent upon their prior cooking experience.

DIETARY GUIDELINES

The recipe for our course aligns with the goals of the Open Science Framework, as well as positioning the library as a center for tools-based instruction across disciplines. Furthermore, since we focus on open-source software and open data, the students learn

Section 3. Data Manipulation and Transformation

tools that can help them beyond their time at university and into the workforce.

INGREDIENTS
Basic Cake
- 3⅔ cups of R
- 1½ cups of R Studio or RStudio.cloud
- 2 tsps of the summary() function
- ¾ tsp skimr() function
- 1 tbsp plot() and hist() functions
- 1½ cups of tidyverse: tidyr and dplyr libraries
- 2 cups of lm() linear regression function
- 2 other models: glm for logistic regression, rpart for decision tree

Frosting and Decorations:
- 5½ cups of ggplot2, RColorBrewer, ggcorrplot, maps, mapproj libraries
- 1½ cups of RMarkdown and Bibliometrix
- 1½ teaspoons of MS Word or PowerPoint
- ⅛ teaspoon of shiny
- Open Textbook: *R for Data Science*: https://r4ds.had.co.nz/ [2]

PREPARATION
Mise en place: For online instruction, students use the free tier of RStudio.cloud to ensure that everyone is using the same version of R and RStudio. For in-person classes, we have a commercial kitchen (library instruction computer lab) with the software installed at each cooking station (workstation).

INSTRUCTIONS
Module 1 (mixing the cake): Every tasty cake needs to include good ingredients as its foundation. R is no different. In module 1, we lay a foundation of open science and R/Studio. Introduce students to open data and open science and sources of open data, and encourage them to select a data set that matches their interest from open data repositories such as Data.gov or Kaggle. Introduce RStudio—the four panes (source, environment, console, files), basic R syntax, and data types—by creating and importing a data/object and walk through deductive and inductive approaches to analysis with a correlation, regression, multiple regression, and decision tree analysis. Students deliver project checkpoint 1: identify a data set, develop a data dictionary, and conceive of preliminary research questions. Students also participate in a discussion on the reproducibility crisis and how open data can help.[3]

Module 2 (baking the cake): Students experiment with selection of ingredients by baking the cake. In class, demonstrate data transformation/cleaning using dplyr and tidr and exploratory data analysis (plot, ggplot, summary, skim). In project checkpoint 2, students apply their transformation and exploratory data analysis (EDA) skills to their own data sets and create a bibliography for their data and topic within their respective fields. For the second discussion, students review and discuss an article on data transformation and Climategate.[4]

Module 3 (preparing the icing and decorations): Students prepare the icing and decorations by applying appropriate data models to their data set. At a minimum, cover regression analysis (linear and logistic), regression discontinuity, and classification with decision trees. Beyond these examples, integrate other models based on student interests. See Optional Module 6: Culinary Curveball in Adapting the Recipe. In project checkpoint 3, students identify their data model and relevant R packages, and annotate references about that model and/or R packages. For the third discussion, students read Dr. Hannah Fry's article in the *New Yorker* on statistical significance.[5]

Module 4 (decorating the cake): Decorate the cake by exploring different visualization options. Focus the lessons on ggplot and the grammar of graphics, but also lay a foundation for communicating data by introducing GitHub and RMarkdown. Project checkpoint 4 includes all of the R code needed for the student's final project, and the students share this with the instructors via GitHub. For the fourth discussion, students review a paper on AI and open data.[6]

Module 5 (tasting the cake): Communicating data: The final step in cake making is tasting. Students have implemented their data model and visualization in R and now tell the story of their data. Students present their analysis to their peers and receive peer and instructor feedback. In the final lesson, cover additional topics of interest such as bibliometric analysis, text analysis, and ordinal regression. For the final discussion, students review a paper on data ethics.[7]

REVIEWS/ASSESSMENT STRATEGY

Attendance (25%): Sessions 2–10 are hands-on instructor-led coding sessions on test data sets. If the students master these workshops, they can apply similar code to their own data set.

Discussions (25%): Allow room to expand knowledge beyond the classroom and project, and tie in aspects of open science, data literacy, and data ethics.

Project checkpoints (20%): The checkpoints allow the instructors to check in on the students' progress. They are specifically designed to knit together into the final project.

Final project (30%): Students in groups of 1–3 create a communication on a data set of their choosing: written report, presentation/talk, or poster presentation. The project builds from the classroom sessions and 4 weekly deliverables: identifying a data set, preparing the data for analysis, determining a model, and applying the model using R code. The final project is instructor- and peer-reviewed. Of the project grade, 25% is the student's work, and the remaining 5% is their constructive feedback to their peer students.

Key assessment questions: Did you make a cake? How well was the cake decorated? And how did it taste? Feedback is provided after weekly tastings.

ADAPTING THE RECIPE

A five-day six-hour workshop with breaks in a not-for-credit workshop is not enough time to grade deliverables and create the final project. It is enough time to review the code and host class discussions.

Class Module 6: "Culinary Curveball": In this session we allow students to recommend which models they would like to see. Generally, we provide a discussion board for students to suggest data models relative to their field and/or data set that is due about one week before the class to allow the instructors time to prepare the additional lesson. In past courses, we covered time series, genomics models, proteomics, neural networks, multiple category regression, and ordinal regression.

ADDITIONAL RESOURCES

Looking for new or exotic ingredients? Check out these open data sources:
- Data.gov, https://www.data.gov.
- "Data Repositories," Open Access Directory (wiki), Simmons University, http://oad.simmons.edu/oadwiki/Data_repositories
- Kaggle, https://www.kaggle.com.
- *Open Science Training Handbook*, 2019, https://open-science-training-handbook.gitbook.io/book/open-science-basics/open-concepts-and-principles.
- UCI Machine Learning Repository, University of California Irvine, https://archive.ics.uci.edu/ml/index.php.

Several thousands of others!

NOTES

1. National Academies of Sciences, Engineering, and Medicine, *Open Science by Design: Realizing a Vision for 21st Century Research* (Washington, DC: National Academies Press, 2018), 1, https://doi.org/10.17226/25116.
2. Hadley Wickham and Garret Grolemund, *R for Data Science: Visualize, Model, Tidy, and Import Data* (Sebastopol, CA: O'Reilly Media, 2017), https://r4ds.had.co.nz/.
3. "Challenges in Irreproducible Research," *Nature*, October 18, 2018, https://www.nature.com/collections/prbfkwmwvz.
4. Myanna Lahsen, "Climategate: The Role of the Social Sciences," *Climatic Change* 119 (2013): 547–58, https://link.springer.com/article/10.1007/s10584-013-0711-x.
5. Hannah Fry, "What Statistics Can and Can't Tell Us about Ourselves," *New Yorker*, September 2, 2019, https://www.newyorker.com/magazine/2019/09/09/what-statistics-can-and-cant-tell-us-about-ourselves.
6. National Academies of Sciences, Engineering, and Medicine, *Open Science by Design*; "AI and Open Data: A Crucial Combination," European Data Portal, April 7, 2018, https://www.europeandataportal.eu/en/highlights/ai-and-open-data-crucial-combination.
7. David J. Hand, "Aspects of Data Ethics in a Changing World: Where Are We Now?" *Big Data* 6, no.3 (2018): 176–90, https://www.liebertpub.com/doi/full/10.1089/big.2018.008 (page discontinued).

Software Carpentry Al Dente
Rendering Tech Training for Online Artisans

Peace Ossom-Williamson, MLS, MS, AHIP, Associate Director of NNLM National Center for Data Services and Associate Curator, NYU Langone Health, peace.williamson@nyulangone.org; Shiloh Williams, GIS and Data Services Librarian, Brock University, shiloh.williams@brocku.ca; Hammad Rauf Khan, Data Management Librarian, University of Texas at Arlington, hammad.khan@uta.edu

NUTRITION INFORMATION

Librarians have become increasingly involved in supporting the research shift to the use of high-performance computing and big data technologies and the drive toward open science. This support involves teaching data literacy, offering research data services, and creating workshops for learning and growth at their institutions. The development of library data services has led to librarians' increasing involvement in The Carpentries, which began with Software Carpentry, a reproducible open science software training program founded in 1995. Since the growth of Software Carpentry, Data Carpentry, a training program that teaches data literacy, was founded in 2014, and Library Carpentry, an adaptation of these lessons for librarians to learn these skills, was founded in 2015.[1] The Carpentries became a joint organization of Data and Software Carpentries in 2018, and Library Carpentry was added as a third official lesson program later that year.[2]

This recipe was created at an institution that has taught Carpentries workshops since 2016. However, this recipe describes an alternate model for virtual learning developed in 2020 during the COVID-19 pandemic, when the Carpentries training moved online for the first time. Combining expertise in online instruction with the tenets of the Carpentries Instructor Training, which melds educational psychology and instructional design, this recipe can be recreated or adapted at your institution for teaching data and software skills to students, faculty, and staff in a virtual environment.

TARGET AUDIENCE AND NUMBER SERVED

Serves 5–40. Participants should be limited based upon the number of available helpers, with an approximate ratio of approximately 6 participants to every helper.

LEARNING OUTCOMES

Upon completion, participants will be able to
- describe the structure and components of the language being used
- navigate between directories
- structure and restructure variables
- use and write functions
- subset (i.e., a method for accessing elements from an object based upon a certain condition) tables to answer research questions
- download and keep libraries and software packages up-to-date within a programming environment
- seek and gain assistance with programming questions online

COOKING TIME

Varies: 4–6 hours a day for 2–4 days or over one or two weekends

DIETARY GUIDELINES

The Carpentries trainings allow users to (1) incorporate open-source big data analytics languages in their research, (2) incorporate methods of automating processes to make their research more efficient, and (3) identify the library as a place for seeking support and further assistance when working with their software or data. The Carpentries tie into the frames Information Creation as a Process and Research as Inquiry from ACRL's *Framework for Information Literacy for Higher Education*.

INGREDIENTS

- A virtual meetings platform, preferably a multifunctional tool such as Microsoft Teams, which allows for a meeting space with a chat function in addition to the ability to share files, provide notes, and engage in separate conversations and meetings outside of the main meeting.
- Workshop web page according to The Carpentries' template, which provides

details about the date and time, instructors and helpers, instructions for installation, and the training schedule.
- GitHub repository with the files and the lessons provided as a readable markdown file for following along with the instructor.
- Registration form for collecting important demographic information that will assist assessment.
- E-mail templates for (1) confirmation of registration, (2) reminder for installing software, and (3) day-before communication of schedule and important information.
- Welcome message to participants at the start of each day that provides links to the schedule, lessons, and files.

PREPARATION

Early preparation involves selecting the instructors and which lessons they will teach and recruiting or updating volunteer helpers. Confirm everyone's roles and availability for the dates you have selected. Next, create and update the workshop website with correct dates and registration information, and send out marketing and communications e-mails, messages, and social media posts letting your intended participant population know about the upcoming event. One to two weeks before the workshop, test the conferencing platform with instructors and volunteer helpers. Make sure to also e-mail participants reminders about what they need to do before arriving at the workshop so accessibility issues may be addressed beforehand. The instructor must prepare to teach at a reasonable (not too slow and not too fast) pace and should practice during this time. Last, the team should finalize the curriculum and make sure to insert breaks into the workshop schedule so learners can rest or ask questions.

INSTRUCTIONS

Participants should download software prior to the workshop. A best practice is to attend an installation troubleshooting session ahead of time (one to two days) to deal with any problems. During the workshop sessions, the instructor teaches using a modified version of the openly licensed Carpentries lessons that teach about a system, its purpose, why it is useful, and its basic layout and operations. The instructor should follow the set curriculum while the participants follow along, creating kinesthetic memory by typing and running the code themselves. Participants are given challenges along the way to situate their learning by testing their application of taught activities in new scenarios. These lessons are scaffolded, with the curriculum and challenges building on what was previously taught.

The virtual workshops should be taught by experienced instructors with a high helper-to-participant ratio in order to assist people in a virtual environment. In addition, because conversations can be taxing and distracting in a virtual environment, general questions should be asked during in-meeting chat, while tech support questions should be placed in the Teams channel (or a breakout room). As lessons proceed through programming instructions and challenges, participants who need assistance can attend a breakout room with a helper, sharing screens and talking through problems so that they can remain engaged and on track. It is important learners feel comfortable asking for help as soon as they have a problem, because the lessons move along quickly, and the faster one receives assistance, the faster one can return to the workshop. Throughout training, the facilitators (teachers and helpers) are able to build trust and rapport, teach and support learning effectively, and as a result, participants are able to apply newly learned skills and build a relationship with their library's research data services team (who may be helpers or instructors in this workshop) for future help with their research projects.

REVIEWS/ASSESSMENT STRATEGY

The Carpentries require use of their survey instrument for assessment of activities, which allows for ease of assessment across institutions, timelines, and other variables. This instrument is helpful for assessment of preparation, the instructor's ability to teach the concepts well, and other potential opportunities for improvement.

ALLERGY WARNING

Regardless of modality, participants often arrive without having required software installed prior to the workshop. To prevent time and error issues due to delayed installation, it is recommended that facilitators do the following:

1. Add in the sign-up form a place for participants to make the commitment to install software at least 24 hours prior to the workshop.
2. Provide a 30-minute online meeting just before the workshop begins where participants can join with installation questions, needs, or problems.
3. (optional) Set up virtual computers with required software and make these available as a last resort for participants who are running into major software or hardware problems during the workshop.

CHEF'S NOTES

Depending on the interest at your institution, registration may fill up very quickly. Typically, our registration is full within hours of announcing its availability to students, particularly in the colleges of engineering, business, and science. Therefore, adding a small cost to registration or making registration available just days before the workshop can cut down on no-shows who fill up registration but do not attend, taking the places of those who would have been interested.

ADDITIONAL RESOURCES

Carpentries. "About Us." https://carpentries.org/about/.

Software Carpentry. "Our Workshops." https://software-carpentry.org/workshops/.

NOTES

1. Rayna Harris and Tracy Teal, "Joint Future for Software Carpentry and Data Carpentry," *Software Carpentry* (blog), September 2, 2017, https://software-carpentry.org/blog/2017/09/merger.html.
2. Tracy Teal and Chris Erdmann, "Library Carpentry Is Now Officially a Lesson Program!" *Carpentries* (blog), November 2, 2018, https://carpentries.org/blog/2018/11/welcoming-library-carpentry/.

A Recipe for Improving Online Instruction for the Carpentries

Kay K. Bjornen, PhD, Research Data Initiatives Librarian and Assistant Professor, Oklahoma State University Library; Clarke Iakovakis, Scholarly Services Librarian, Oklahoma State University Library

NUTRITION INFORMATION

Over the last few years, Oklahoma State University (OSU) Libraries have offered regular, hands-on, in-person instructional workshops teaching data and computational skills based on the Carpentries model and materials. The Carpentries is a global, volunteer-led organization whose mission is to teach coding and computing skills to beginners. The COVID-19 pandemic in spring 2020 prompted a transition to online delivery. Teaching coding workshops online presents some unique challenges. Modifying the delivery method and increasing communication channels can help overcome some of these issues.

TARGET AUDIENCE AND NUMBER SERVED

Target audience: Researchers and others interested in learning coding and computing skills. Workshops are targeted at those with no previous experience.

Number served: Online workshops allow higher registration numbers than in-person workshops due to the lack of space constraints. Average workshop participant numbers are 20–25.

LEARNING OUTCOMES

Learners will
- identify the advantages of using an interactive computing environment (notebook) for online coding instruction
- modify lessons created for in-person instruction for online teaching
- create notebooks to be used for Carpentries and Carpentries style workshops

COOKING TIME

Workshops are usually taught in half-day increments, held across one to four days. Preparation time varies depending on the number of lesson modules and the instructor's experience using the technologies described in this recipe. Setting up Jupyter Notebooks in preparation for workshops requires a greater time commitment than preparing for traditional face-to-face workshops. However, once created, the notebooks are easily shared and implemented again in future workshops. A time commitment is also required by the computing center for creating user IDs and passwords for the cloud computing environment, software installation, and duplication of notebooks on the server.

DIETARY GUIDELINES

The Carpentries is a community-led organization that teaches foundational coding skills to researchers worldwide. As data skills become more and more important to research success, the Carpentries fills a niche for those who haven't received formal coding and data science education.

The Carpentries has provided an effective recipe for teaching coding and computing skills to beginners through live coding with a healthy splash of encouragement and the occasional pinch of one-on-one troubleshooting. When the COVID-19 pandemic caused instruction to move online, Carpentries workshops had to be modified so that lessons designed for in-person instruction could be taught effectively through a tiny screen, and instructors could both give and receive feedback in a virtual environment.

INGREDIENTS

- Carpentries lessons for the planned workshop—available for download from the website https://carpentries.org/.
- Slack channels to facilitate communications—Our OSU Carpentries community created a Slack workspace to communicate with instructors, volunteers, and learners before, during, and after workshops.
- Software installed in a cloud computing environment—We partnered with our university High Performance Computing Center (HPCC) to create JupyterHub access to R and Python in Jupyter Notebooks for the workshops. This allows workshop attendees to run programs and write code in their browser, without having to download anything to their

personal computer. This greatly reduces technical problems with software variations and different operating systems.
- Lessons in a notebook environment—Interactive computing environments such as Jupyter Notebooks (IPYNB files) allow lessons to be "packaged" so that they are distributed with explanatory text and live code. Paragraphs of instruction in text form are created in Markdown, interspersed with executable code chunks to demonstrate the concepts.

PREPARATION

A few days before the workshop, send invitation links to the Slack workspace and encourage participants to introduce themselves to other participants. Slack is a good place for any questions that they may have about the setup of devices, the agenda, or participation. If you will be using the university high performance computing services to host the workshop notebooks, contact them and make arrangements for providing unique workshop IDs and passwords to participants. These should be activated the day before or the first day of the workshop.

Workshop lessons (modules) are openly available for download and can be accessed from the Carpentries website (https://carpentries.org). Lessons incorporate sample code snippets as well as explanatory text. For teaching online, it can be helpful to have the explanatory text, the sample code snippets, and the environment for writing and executing code all in one place. We find Jupyter Notebooks useful for this purpose. In order to convert the lesson pages to Jupyter Notebooks (IPYNB), download the Markdown files for each lesson (which are stored on GitHub repositories) and use a Python package called Notedown to transform them into the IPYNB files. Notedown automatically renders text as Markdown and sample code chunks as executable Python or R code blocks. Notebooks should be edited to clean up any formatting errors from the conversion and also to create cells with partially entered and empty code. These partial code entries are selected so that learners can complete them or write functions in order to reinforce the instruction. The result will be a separate notebook for each lesson module (called an "episode" in the Carpentries curriculum). You can also modify the lessons by omitting or adding sections in order to fit it into the allowed time or address concepts that you think are important.

Finally, upload the completed Jupyter Notebooks and data sets to the JupyterHub cloud environment and test them.

Notebooks we have created at OSU based on Carpentries lessons are available for download from the URL in the Additional Resources section.

INSTRUCTIONS

At the beginning of the workshop, plan an icebreaker activity in Slack to get the participants comfortable with communicating with the instructor, helpers, and each other using the platform. Establish a separate channel for volunteers and instructors so that they can share information privately. The Slack workspace also creates an ongoing communication channel that encourages past participants to submit questions and volunteer as helpers for later workshops.

At the start of the workshop, ask attendees to log into their assigned JupyterHub environment and access the IPYNB files for the workshop. Because each workspace is personal, they can modify, save, and download the files during the workshop. Discuss with your university computing center how long to leave the learner accounts active. We found that a week or two post-workshop is adequate for participants to download the notebooks and set up software on their own devices.

Show learners how to navigate the Jupyter Notebook and how to use the Run button for prewritten sample code chunks to see the output. As the lesson progresses, learners should be encouraged to add their own empty code blocks to do live coding and complete exercises. Notebooks, as well as the data and any other materials, are accessible during the workshop on the left directory sidebar, allowing for easy navigation between lesson episodes.

REVIEWS/ASSESSMENT STRATEGY

The Carpentries encourages the use of pre- and post-workshop surveys for overall workshop assessment. Brief, "one up, one down" feedback surveys are distributed at the end of each day's instruction that allow more granular feedback on instruction. In an online workshop, these daily surveys replace the Carpentries' traditional "sticky feedback." The

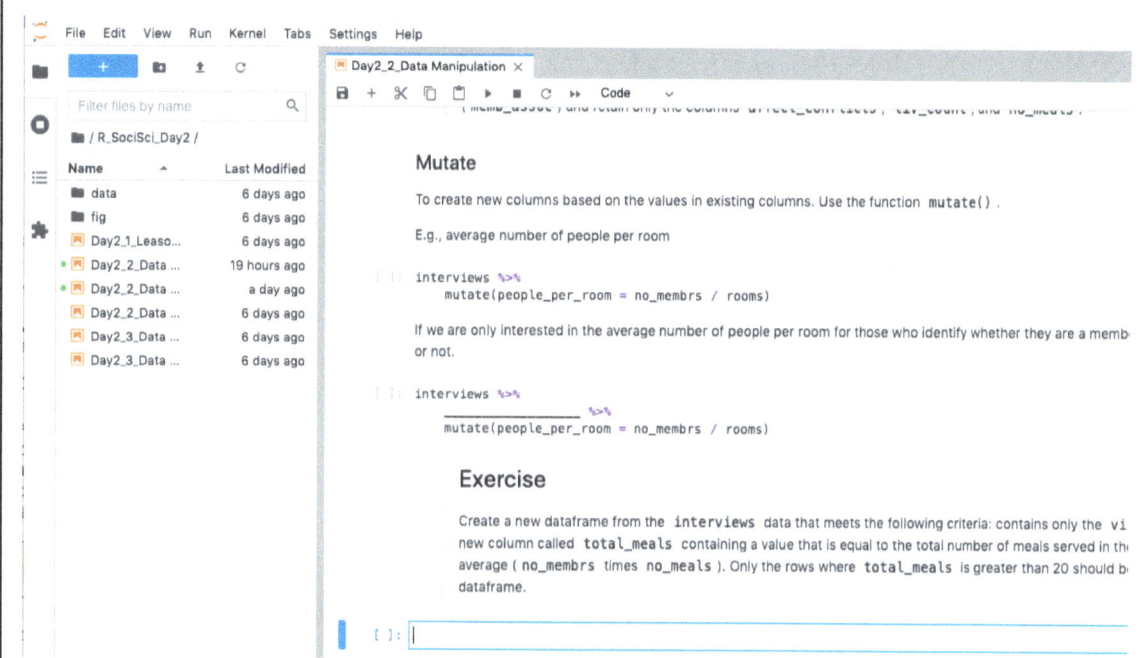

FIGURE 1
Screenshot of a Data Carpentry R lesson Jupyter Notebook

Carpentries guidelines for online workshops (https://carpentries.org) also encourage the use of polls, such as those provided in Zoom and in Slack, during the workshop for immediate feedback.

ALLERGY WARNING
The Carpentries organization had not encouraged the use of remote instruction prior to the pandemic because its instructional model was difficult to execute online. Carpentries in-person instruction is based heavily on formative assessment, where instructors monitor and solicit feedback on learners' attitude, motivation, and comprehension, and adjust their approach accordingly. This method relies on the use of visual cues, such as colored sticky notes. The instructor is able to observe when participants are not following along, frustrated, or in need of a break. However, visual cues may be unavailable or limited in online environments, as learners may keep their cameras off or be reluctant to send messages to instructors indicating their status.

The other significant difficulty with online instruction is that beginning coders are often unprepared to resolve difficulties when setting up their software. It is common at the beginning of in-person workshops for helpers to spend time with individual participants troubleshooting software downloads, administrative permissions, and connectivity issues and setting paths to working directories. These troubleshooting operations are greatly complicated by an online meeting interface.

At OSU, we realized that some of these complications can be minimized or eliminated entirely by changing how instructors communicate with learners. Utilizing a JupyterHub cloud environment and packaging lessons in Jupyter Notebooks helps minimize technical difficulties and provides self-contained modules that learners can later download and use to reinforce the lessons.

ADAPTING THE RECIPE
For those who do not have the ability to create their own local JupyterHub environments, Binder can be a good option for turning a GitHub repo into an interactive coding environment. The Carpentries has created "Scaffolds" in Binder, with code and data for some lessons (see Additional Resources).

CLEANUP
We recommend following up with learners after the workshops to assist with downloading the IPYNB files and installing the software on their own computers.

We created and used Jupyter Notebooks based on both R and Python lessons, but we recommend that instructors emphasize that the notebooks for the R lessons are being used primarily as a teaching tool (i.e., to wrap together explanatory text, prewritten code, and empty code cells). Using notebooks precludes the introduction of RStudio, a popular environment for R, during the workshop. A brief introduction to RStudio can be included when learners prepare to download software for use on their own devices.

ADDITIONAL RESOURCES

Download sample lesson Notebook at: Bjornen, Kay Kuhlemeier, Phillip Doehle, Damith Mahapatabendige, Clarke Iakovakis, Beth Jones, Lin Hua, and Nathalia Graf Grachet. "Notebook Files and Scripts for Download." OSF, Center for Open Science. March 29, 2021. https://osf.io/ptswr/.

Binder Team. "Binder Documentation." Binder User Guide, 2017. https://mybinder.readthedocs.io/en/latest/.

Carpentries. "About Us." Accessed March 29, 2021. https://carpentries.org/about/.

Carpentries. "Recommendations for Teaching Carpentries Workshops Online." Accessed March 29, 2021. https://carpentries.org/online-workshop-recommendations/.

Carpentries. "Scaffolds." Accessed March 29, 2021. https://github.com/carpentries/scaffolds.

Jupyter Project. "JupyterHub." Accessed March 29, 2021. https://jupyterhub.readthedocs.io/en/stable/.

O'Leary, Aaron. "Notedown." Last updated November 16, 2017. https://github.com/aaren/notedown.

Project Jupyter, Matthias Bussonnier, Jessica Forde, Jeremy Freeman, Brian Granger, Tim Head, Chris Holdgraf, et al. "Binder 2.0—Reproducible, Interactive, Sharable Environments for Science at Scale." In *Proceedings of the 17th Python in Science Conference*, ed. Fatih Akici, David Lippa, Dillon Niederhut and M. Pacer (SciPy.org, 2018), 113–20. https://doi.org/10.25080/Majora-4af1f417-011. See also The Binder Team. "Binder Documentation." Accessed March 29, 2021.

Section 4.
Data Visualization

79 [[Ch20]]Correlation Does Not Equal Causality: Introducing Data Literacy through Infographics and Statistics in the Media
Nick Ruhs

83 [[Ch21]]Pies, Bars, Charts, and Graphs, Oh My! A Data Visualization Appetizer
Haley L. Lott

86 [[Ch22]]Data Visualizations: The Good, the Bad, and the Ugly
Kaetlyn Phillips

89 [[Ch23]]Seasonal Visual Literacy: Using Current Events to Teach Data and Spatial Literacy Skills with Adaptable LibGuides
Jacqueline Fleming and Theresa Quill

93 [[Ch24]]To Visualize Is to Experience Data
Chelsea H. Barrett and Gerard Shea

97 [[Ch25]]Upping the Baseline for Data Literacy Instruction
Jessica Vanderhoff

101 [[Ch26]]A Literacy-Based Approach to Learning Visualization with R's ggplot2 Package
Angela M. Zoss

104 [[Ch27]]Build Your Own Data Viz Pizza: A Modular Approach to Data Visualization Instruction
Rachel Starry

108 [[Ch28]]Veggie Pizza: Choosing a Data Visualization Tool
Rachel Starry

111 [[Ch29]]Four-Cheese Pizza: Color and Accessible Design
Rachel Starry

114 [[Ch30]]Data Visualization using Web Apps in a Rainbow Layer Cake
Yun Dai and Fan Luo

117 [[Ch31]]Graphical Abstracts: Creating Appetizing Infographics for Your Research Article
Aleshia Huber

Correlation Does Not Equal Causality
Introducing Data Literacy through Infographics and Statistics in the Media

Dr. Nick Ruhs, STEM Data and Research Librarian, Florida State University

NUTRITION INFORMATION

As we move into the third decade of the twenty-first century, the amount of data being generated and accessible to the public continues to grow at a rapid pace. Data are present and utilized in different contexts by many people, including academic researchers, private companies, and government agencies. However, with greater access to data comes a heightened risk that incorrect or incomplete information will be presented as fact. Data are highly amenable to manipulation and can then be passed off in the media as fact. For these reasons, it is often difficult to determine the accuracy and reliability of certain data. To combat these issues, we can introduce critical thinking and data literacy education into our workshop curriculum. The purpose of this chapter is to introduce a workshop session and activities designed to present general data literacy principles and practices in a way that is understandable and accessible to a wide student audience.

TARGET AUDIENCE AND NUMBER SERVED

- Undergraduate students (all disciplines)
- First-year graduate students
- Anyone with little to no experience with data-related research

LEARNING OUTCOMES

By the end of this session, students should be able to

- identify different types of data and their sources
- acknowledge the importance of data literacy
- discern the relationship between data and information
- differentiate between reliable and misleading data representations or visualizations
- apply critical thinking skills to the evaluation of data representations or visualizations in academic and nonacademic sources

COOKING TIME

- Presession research: 3–5 hours
- Workshop instructional design: 4–5 hours
- Session: 1 hour
- Follow-up and assessment: 1 hour

DIETARY GUIDELINES

These objectives can be applied to two of the frames described in ACRL's *Framework for Information Literacy for Higher Education*:

- *Authority Is Constructed and Contextual*: The evaluation of data often hinges on the expertise and credibility of the source and a critical evaluation of the methodologies used to produce the data. The critical thinking skills introduced in this session equip attendees to assess data content with healthy skepticism grounded in fact.
- *Information Has Value*: A key emphasis in this workshop is that data are analyzed and presented for a specific purpose and specific audience, influenced in some cases by socioeconomic interests. Discerning the reasons behind data dissemination is inherent to determining their value.

INGREDIENTS

- 1–2 librarians or library staff members to design the session and conduct the workshop
- Computer
- Presentation and word processing software (Microsoft PowerPoint/Word, Google Docs/Slides, etc.)
- Marketing and engagement prior to workshop
- Examples of good and misleading ("bad") infographics
- Index cards

PREPARATION

1. Conduct presession research for the workshop. Utilize a variety of academic

Section 4. Data Visualization

and reputable nonacademic sources for this session, as data literacy is broadly applicable across academic and nonacademic settings. Some general resources are provided in the Additional Resources section of this chapter.

2. Design the workshop curriculum. A sample workshop outline is shown in table 1.

TABLE 1
Workshop outline

Item	Topic
1	Definitions: data and data literacy
2	Where do you encounter data?
3	Types of data
4	An introduction to data visualization
5	"Good" vs. "bad" infographics
6	Putting it all together: questions to ask about data in the media

3. Develop two activities.
   a. *Data in everyday life:* To prepare for this activity, ensure that there is an adequate number of index cards and writing utensils for those in attendance, and write up a brief explanation of the activity to place on a presentation slide. (See Instructions for a description of the activity).
   b. *Good vs. bad infographics:* To prepare for this activity, it will be necessary to source example infographics. Identify at least one or two characteristics of each infographic that can be highlighted during the discussion.
4. Create a slide show presentation to deliver the workshop content.
5. Schedule the workshop session and book a space for the session. Alternatively, you can create a virtual room on Zoom or another available platform for an online session.
6. Advertise the session via social media, e-mail, and your organization's website.

INSTRUCTIONS

1. At the start of the session, introduce the instructors and the topics for the workshop.
2. Define terminology.
   a. Provide definitions for terminology that you will be using during the session, such as *data* and *data literacy*.
   b. Establish the relevance of data literacy to the audience (see the Glusker article in Additional Resources).
3. Activity: Where do you encounter data?
   a. Briefly explain the first activity, "Data in everyday life."
   b. In this activity, workshop attendees are asked to identify two places or situations in which they have encountered or utilized data within the last week. They write these down on an index card and then share their results in a group discussion.
   c. Spend a few minutes sharing and discussing the activity, then provide a few representative examples of where and in what context one might interact with data in both academic and nonacademic settings.
4. Highlight data types.
   a. Describe the similarities and differences between different data types. Try to keep this as broad as possible (e.g., quantitative vs. qualitative, etc.)
   b. Provide examples within each category, such as those detailed in the article by Valcheva (see Additional Resources).
5. Introduce data visualization.
   a. Provide an introduction to data visualization and examples of how data can be represented. It is a good idea to provide examples of visualizations containing multiple pieces of data.
   b. This spot in the session is also a good time to explain the statement *"Correlation does not equal causation."* Emphasize that just because two pieces of data correlate, one doesn't necessarily cause the other. There is often another factor (also known as a *confounder*) that affects both sets of data. Discuss how this concept affects the evaluation of a visualization's accuracy and relevance.
6. Introduce infographics.
   a. Provide tips on how to discern "good" and "bad" graphs in preparation for the activity. Some general characteristics for analyzing graphs are listed below from

the International Association for Statistical Education:
- Characteristics of good graphs:
  » Title
  » Labeled axis
  » Key (definitions) for variables
  » Source linking back to the original data
  » Date range the data were collected
- Characteristics of bad graphs:
  » Leaving gaps or changing the scales
  » Uneven shading or colors
  » Unfair emphasis on some sections
  » Distorting areas of the graph
  » Including too much information[1]
7. Activity: Good vs. bad infographics
   a. Briefly explain the second activity, good vs. bad infographics.
   b. Present an example infographic to the audience. Ask the attendees to discern what is bad or misleading about the infographic. Give them two to three minutes to evaluate on their own.
   c. Ask the attendees to share their observations. Provide context and clarification as necessary.
   d. Present a second infographic, and repeat the above process.
   e. Note: It is a good idea to present infographics with different bad characteristics so that attendees can apply the concepts discussed earlier in the workshop.
8. Summarize the content and put all of the concepts together.
   a. Conclude the session by summarizing the key points covered and emphasizing the importance of utilizing data literacy skills to critically evaluate data.
   b. Present this summary as a series of questions that attendees should ask about a particular piece of data or data set that they encounter in the media, such as
      i. Who is collecting the data?
      ii. For what purpose are they acquiring the data?
      iii. What is the end goal of the data?
      iv. How are the data being disseminated?
9. After the session, send a follow-up e-mail thanking the attendees and including a link to a postworkshop survey and the slides used during the session.

REVIEWS/ASSESSMENT STRATEGY

Assessment for this session and the associated activities is primarily formative, with attendee understanding measured qualitatively throughout the session via their understanding of the concepts and activities. The logistics and format of the session are assessed via a short Qualtrics survey sent to the attendees immediately following the workshop.

ADAPTING THE RECIPE

This workshop session is general in nature and thus can be easily adapted for different modalities or audiences. Primarily, adaptations can be considered for the hands-on activities and the examples presented to illustrate key concepts. Two examples of adaptations, one each for modality and audience, are briefly described.
- *Virtual workshop:* Due to the COVID-19 pandemic, it became necessary to offer an online version of this session via the Zoom web conferencing platform. For virtual sessions, changes were made to the two hands-on activities: participants were invited to enter their responses in the Zoom chat and then use the Raise Hand feature in Zoom to follow up verbally. All other content was unchanged.
- *Audience:* An online version of this session was developed for academic and public librarians in the state of Florida. In addition to the changes made to the hands-on activities, new examples were added to connect with the audience's everyday work.

ADDITIONAL RESOURCES

Bartley, Anthony, and Michael Bowen. *The Basics of Data Literacy: Helping Your Students (and You!) Make Sense of Data*. Arlington, VA: NSTA Press, 2013.

Duarte, Nancy. "The Quick and Dirty on Data Visualization." *Harvard Business Review*, April 16, 2014. https://hbr.org/2014/04/the-quick-and-dirty-on-data-visualization.

Glusker, Ann. "Data Literacy: What It Is and Why You Should Care." *Region 5 Blog*, Network of the National Library of Medicine, Septem-

ber 26, 2017. https://news.nnlm.gov/pnr/data-literacy-what-it-is-and-why-you-should-care/.

International Association of Statistical Education. "Good and Bad Graphs." 2014–2015. https://iase-web.org/islp/apps/gov_stats_graphing/GoodBad/GoodBadGraphs.pdf

National Library of Medicine. "Data Services." Last modified May 24, 2021. https://nnlm.gov/guides/data-services.

Valcheva, Silvia. "6 Types of Data in Statistics and Research: Key in Data Science." Intellspot. Accessed August 12, 2017. http://intellspot.com/data-types/.

NOTE
1. International Association for Statistical Education, "Good and Bad Graphs," 2014–2015, https://iase-web.org/islp/apps/gov_stats_graphing/GoodBad/GoodBadGraphs.pdf.

Pies, Bars, Charts, and Graphs, Oh My!
A Data Visualization Appetizer

Haley L. Lott, Student Success and Engagement Librarian, Beloit College, *lotth@beloit.edu*

NUTRITION INFORMATION
Visual representations of data are powerful tools to help your audience understand a topic, but beginners too often feel that creating them will be an overwhelming process. This lesson is designed to help students understand and create different visualization formats and use them to present data in an honest and compelling manner.

This recipe includes three parts. In Part 1, students review various types of visualizations, match sample data scenarios to the most appropriate visualization type, and then depict that scenario using that visualization. This activity tests students' prior knowledge about visualization types and provides them with an example to use throughout the session. In Part 2, students participate in a "spot the inaccuracies" game that helps them better understand some of the inherent risks in organizing and manipulating data. Students learn how inadvertent or intentional data distortion can contribute to the spread of misinformation about a specific topic. In Part 3, the students apply what they have learned as they evaluate and correct their visualizations from Part 1.

TARGET AUDIENCE AND NUMBER SERVED
While this recipe is best served to those with little knowledge of visualizing data, the activities can be adapted for use with more advanced audiences by including more complex data scenarios and visualizations. This recipe can easily stretch to serve the entire crowd, whether in person or at a distance.

LEARNING OUTCOMES
By the end of this lesson, students will be able to:
- recognize various types of visualizations
- identify the appropriate visualization to use in various situations
- discuss how and why visualizations can be used to manipulate data
- critically evaluate visualizations to determine whether or not they are credible

COOKING TIME
Cooking time is between 50 and 75 minutes.

DIETARY GUIDELINES
This recipe ties into the frames Information Creation as a Process and Authority Is Constructed and Contextual from ACRL's *Framework for Information Literacy for Higher Education*. The "drawing the data" exercise in Part 1 focuses on the iterative process of creating and sharing information in a visual format. This helps students develop their own creation process and appreciate how their choices impact the overall message the visualization conveys. The discussion of data distortion in Part 2 demonstrates the different types of authority in various contexts, stresses the importance of using trustworthy sources, and helps students value their own identity as scholars, which is a critical part of becoming an authority in any subject.

INGREDIENTS AND EQUIPMENT
- Handouts (Data Visualization Definitions, Data Visualization Scenarios, Spot the Inaccuracies), available at https://www.dropbox.com/sh/8q9h83i2nyfxu3t/AAAdEdOUW1CUoo8SYeDNl91Ha?dl=0
- Paper and markers, colored pencils, crayons, or other drawing utensils, or
- Computer with internet access or access to visualization programs (such as Google Sheets, Excel, or Canva) for each student

PREPARATION
Before the instruction session, develop one or more data scenarios for students to draw in the first activity. Ensure that the art supplies or computer programs are readily available. Identify graphs, charts, and figures to use in the "spot the inaccuracies" game. I recommend image searching "misleading charts" or downloading the charts from the book *How Charts Lie* or from the blog *Junk Charts*.[1]

Section 4. Data Visualization

If you're more interested in a kit-based approach, you can access all of the preprepped materials, including the data definitions, data scenarios, and spot the inaccuracies game at the URL listed under Ingredients and Equipment.

INSTRUCTIONS

Introduction (5–10 minutes)
1. Briefly introduce the data visualization topic. Ask students what *data visualization* means, which visualizations they know of or use, and what questions they have about visualizations.

Part 1: Understanding Data Visualizations (10–15 minutes)
1. Draw the data.
   a. Provide students with a sample data scenario, such as this: "Draw a visual diagram of Kevin's top songs of the year. They were 'Hand Me Downs' by Mac Miller (372 listens), 'Forever' by Charli XCX (423 listens), 'My Future' by Billie Eilish (255 listens), 'Triptych' by Samia (198 listens), 'Kyoto' by Phoebe Bridgers (311 listens), and 'Bad Decisions' by The Strokes (160 listens)." Have them use the provided art supplies to draw the data, using whatever visualization they feel best fits the scenario. More sample data scenarios are available on the data scenarios handout.
   b. Using the think-pair-share method, assign groups and ask students to explain how their drawing represents the data and answer why they picked that type of visualization, how they made specific design choices, and how their drawing helps the audience understand the overall topic.
   c. If multiple students worked from the same prompt, how similar are their drawings? As a whole, did the class prefer a specific type of visualization?
2. Data visualizations.
   a. Share the Data Visualization Definitions handouts with students—both handouts contain the same information but are organized either A-Z or based on visualization types. Were students aware of all of the options? Does the type of visualization matter in every scenario? If so, how might one visualization be more effective than another in a particular scenario?

Part 2: Data Distortion (15 minutes)
1. How and why are data distorted?
   a. Explain the variety of reasons why data may be distorted or unintentionally misrepresented. These reasons include poor design, using untrustworthy data, using insufficient data, misrepresenting estimates or forecasts as certain, and suggesting false conclusions.
2. Risks of data distortion
   a. Discuss some of the risks in organizing or distorting data and ask the students what could happen when data are either inadvertently or intentionally distorted. Guide the students to understanding that data distortion can result in the spread of misinformation about a topic.
   b. Explain how date distortion can impact a scholar's reputation.
3. Spot the inaccuracies game
   a. Show students a few examples of bad charts or graphs (see figure 1) retrieved from the handout, *How Charts Lie*, *Junk Charts*, or other sources (see Additional Resources). You can add in some well-made visualizations for students to assess as well.

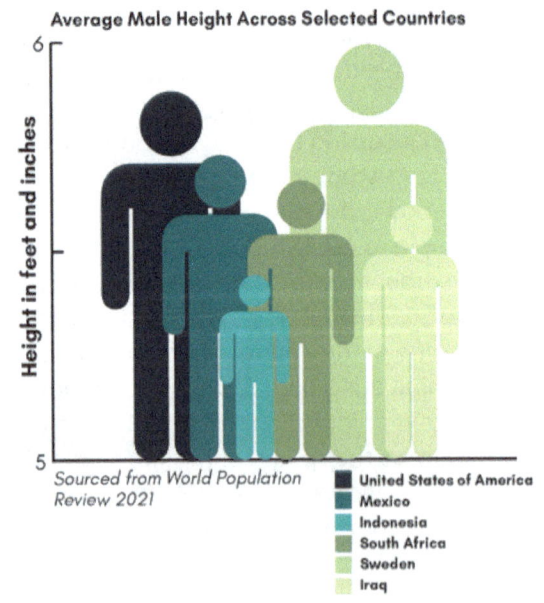

FIGURE 1
Average male height across selected countries

b. As a group, discuss each chart. What is the creator trying to convey? Were they successful? Is there evidence suggesting that these visualizations are inaccurate or misleading? How can these charts be improved?

Part 3: Evaluation and Assessment (15–20 minutes)
1. Redraw the data.
   a. Ask students to revisit their drawings from earlier in the session. Are they still confident that they picked the right type of visualization? Did they inadvertently misrepresent the data? Have students make corrections to their original drawings.
   b. Using the think-pair-share method, have students partner to discuss the changes that they made to their drawings and evaluate the effectiveness of their data visualization.

Wrap-Up and Review: (5–10 minutes)
2. End the session by answering student questions and asking students identify what they have learned about visualizing data. Share resources with students about how to find more information on data visualizations.

REVIEWS/ASSESSMENT STRATEGY
Collect the data visualization drawings from students and compare the first iteration with the final iteration to assess student learning.

As an alternative, students could do a 3-2-1 assessment survey, asking them 3 things they learned, 2 questions they still have, and 1 thing they plan to implement.

CHEF'S NOTES
- This session can be tailored to fit a more advanced group by swapping out the hand-drawn visualization activity with computer-generated visualizations using programs such as Google Sheets, Microsoft Excel, Canva, or Tableau, among others.
- This session can be extended by having small groups of students find examples of poor data visualizations on the web and then discuss how to fix them.
- Poor charts are shared across the internet every day. Keep this activity fresh and current by rotating out older examples of data visualizations.

ADAPTING THE RECIPE
This recipe can be easily adapted for in-person and distance learning. If teaching from a distance, you can create a collaborative Excel document and share it with students to create their visualizations. It works best to have each student create a tab within the workbook. Please note that this may require more time. Group work can be discussed in breakout rooms.

ADDITIONAL RESOURCES
Cairo, Alberto. *How Charts Lie: Getting Smarter about Visual Information*. New York: W.W. Norton & Co., 2019.

Fung, Kaiser. *Junk Charts* (blog). https://junkcharts.typepad.com.

Lott, Haley. "Data Visualization Scenarios." March 29, 2021. https://www.dropbox.com/sh/8q9h83i2nyfxu3t/AABPsU4ncV0TcqU2UAwpdkwTa/Data%20Visualization%20Scenarios%20%281%29.pdf?dl=0.

Lott, Haley, and Erin Gallagher. "Spot the Inaccuracies: The Data Visualization Game." March 29, 2021. https://www.dropbox.com/sh/8q9h83i2nyfxu3t/AABZr0VQqxqhpv-vInfbl6gODa/Spot%20the%20Inaccuracies.pptx?dl=0.

Lott, Haley, and Peyton Scarpaci. "Data Visualization Definitions by Category." March 30, 2021. https://www.dropbox.com/sh/8q9h83i2nyfxu3t/AAC6n3oyXwttoEgTK-6OkeuRa/Data%20Visualization%20Definitions%20by%20Category%20%282%29.pdf?dl=0.

Lott, Haley, and Peyton Scarpaci, "Data Visualizations Definitions A–Z." March 30, 2021. https://www.dropbox.com/sh/8q9h83i2nyfxu3t/AADC9TIZc_3OyWrcsCTwg7Lxa/Data%20Visualization%20Definitions%20AtoZ.pdf?dl=0.

NOTES
1. Alberto Cairo, *How Charts Lie: Getting Smarter about Visual Information* (New York: W.W. Norton & Co., 2019); Kaiser Fung, *Junk Charts* (blog), https://junkcharts.typepad.com.

Data Visualizations
The Good, the Bad, and the Ugly

Kaetlyn Phillips, Data Services Librarian, Dr. John Archer Library, University of Regina, *Kaetlyn.Phillips@uregina.ca*

NUTRITION INFORMATION

This recipe is designed to introduce students to data visualizations. The recipe will show common forms of accurate and inaccurate data visualizations that are presented in mainstream media and social media. It will also examine types of data summaries and how to identify bad visualizations. Through in-class discussion, students will critically think and assess how they encounter data visualizations in everyday media.

TARGET AUDIENCE AND NUMBER SERVED

This session works best with at least five people. Larger groups are acceptable as long as all the students are able to access Kahoot! or the quiz software of your choice. Best for undergraduate students. Students will review clean data summaries and visualizations in everyday media and are not actively analyzing data or looking at complex forms of data (e.g., microdata).

LEARNING OUTCOMES

Students will
- compare and discuss the changes to perception when information is packaged in different formats
- identify common types of biases and misinformation in data visualizations and statistics
- discuss how well-known authoritative sources of data should be evaluated with a critical eye and challenged if necessary

COOKING TIME

This recipe can be prepared in 60 minutes.

DIETARY GUIDELINES

While specifically focusing on data, objectives center on two frames from ACRL's *Framework for information Literacy for Higher Education*: Authority Is Constructed and Contextual and Information Creation as a Process. The intention is to introduce students to critical thinking about the data they encounter on a daily basis and to build confidence in their ability to assess data visualizations and navigate misinformation. Students will learn that data visualizations are not always neutral because they are constructed by people or entities. While these groups are usually reputable, some are not. Likewise, data visualizations can be rapidly shared and manipulated without accuracy checks or challenges.

INGREDIENTS
- Networked computer access for all students
- Instructor station with projector (or ability to share screen)

PREPARATION

Use the Additional Resources section to find examples of good and bad data visualizations. Visualizations can be real and bad; real and good; fake and bad; or fake and good. Make sure the visualizations you find highlight the key examples: truncated y-axis; bad visualization practices; use of mean, median, and mode; and correlation versus causation. More examples could be added if there is time. Make an online quiz in advance using Kahoot! or another online quiz software.

INSTRUCTIONS

Begin by going over the learning objectives and activities for the lesson. Before the lesson, search on social media (e.g., Twitter, Facebook, LinkedIn) for trending data visualizations or headlines featuring statistics. Choose between 1 and 3 visualizations or headlines to show the class. For each visualization or headline ask the class:
- What is the visualization or headline trying to tell us? Is the visualization or headline successful?
- Upon first glance, would you believe this visualization or headline? Why?
- Does the visualization or headline tell

us about the data collection? Who collected the data? Who made the visualization?
- Is there any way to look at the larger data set? Links? Citation information?
- Is there anything about the visualization or headline that is misleading or confusing? Why?

Pause for student questions and then show examples of visualizations that demonstrate good visualizations, bad visualizations, and ugly visualizations. Feel free to use the Additional Resources section to find examples of bad data visualizations or find your own. Just make sure you cover these misleading visualizations:

1. *Manipulation of x- and y-axis:* Manipulating the x- and y-axis is one of the most common forms of bad data visualizations. It can be done through truncating the y-axis, flipping the y-axis, or leaving information off the x- and y-axis. In particular, explain that when the y-axis is truncated, the visualization often creates an exaggerated form of the data. Highlight that this is misleading because we read a visual before we read words, so while the visualization may be accurate, the exaggeration can elicit a strong emotional response that is misleading.
2. *Use of mean, median, or mode:* Highlight that when reading a summary of research and data, always look for mean, median, or mode if an *average* statistic is presented. Explain that, depending on how the average was calculated, there can be misleading manipulation. Feel free to use the example of average income. If the mean is used, then the average income is misleading because it includes the very poor and very rich (extreme outliers). Huff's *How to Lie with Statistics* provides great examples for this section.
3. *Bad visualization practices:* Explain how sometimes the data are presented accurately, but the visualization is poorly chosen. Examples of this include poor use of scale, poor use of color, poor use of imagery, or even using the wrong type of data visualization.
4. *Correlation versus causation:* Explain that the faulty implication that correlation equals causation is frequently seen in clickbait headlines and summaries of academic research. News articles, especially headlines that explain medical research, are a great source for examples. Highlight that correlation is often presented as causation in bad visualizations, but correlations are easy to find if enough data are analyzed. The website Spurious Correlations (https://www.tylervigen.com/spurious-correlations) is a great resource for examples.

Take students through various examples of visualizations and ask volunteers to explain why the visualization, headline, or statistics might be misleading. Again, use the examples provided in the resources, but also try new examples that may connect better with students.

Depending on the size of the group, put students in pairs or small groups and have them find a data visualization of their choosing. Allow them to search as they would normally when looking for information (Google or social media searches), or they may choose visualizations that have been shared with them through social media. Give the groups time to do a quick critique of the visualization, then ask the groups to share their visualization and critique with the whole class.

The final component is to do a quick summative review of the concepts taught using Kahoot!

REVIEWS/ASSESSMENT STRATEGY
A large amount of the assessment is formative. Monitor how the students are responding to the examples and repeat or clarify as needed. Summative assessment is also done by creating a Kahoot! quiz. This is used to assess student understanding of inaccurate or misleading forms of visualization.

CHEF'S NOTES
There is a lot of flexibility in adapting this lesson to suit your audience. There are MANY examples of bad data visualizations that will illustrate the main topics in the lesson. Find ones that will engage your audience or that are humorous, or even make your own! If you're teaching in a course, you can ask students to create their own examples and explain why their visualizations are of good or poor quality.

ALLERGY NOTE
Data visualizations are not created by neutral

bodies, and fake visualizations are often used to promote bad science (e.g., anti-vaccination) or political and social agendas. Some students, depending on their beliefs, may push back if they encounter real data that disprove their beliefs or fake data that support their beliefs.

ADDITIONAL RESOURCES

For Explanations of Data Literacy and Common Forms of Data Misinformation

Huff, Darrell. *How to Lie with Statistics*. New York: Norton, 1954.

Koehrsen, Will. "Lessons on *How to Lie with Statistics*: Timeless Data Literacy Advice." *Towards Data Science,* July 28, 2019. https://towardsdatascience.com/lessons-from-how-to-lie-with-statistics-57060c0d2f19.

Parikh, Ravi. "How to Lie with Data Visualization." *Heap* (blog), April 14, 2014. https://heap.io/blog/how-to-lie-with-data-visualization.

Sawicki, Jan. "Why Is This Chart Bad? The Ultimate Guide to Data Visualization Evaluation Using GoDVE (Grammar of Data Visualization Evaluation)." *Towards Data Science,* July 28, 2020. https://towardsdatascience.com/why-is-this-chart-bad-5f16da298afa.

For Examples of Data Visualizations

"DataIsBeautiful." Reddit. Accessed March 29, 2021. https://www.reddit.com/r/dataisbeautiful/.

"Data Is Ugly." Reddit. Accessed March 29, 2021. https://www.reddit.com/r/dataisugly/.

Virgen, Tyler. "Spurious Correlations." Accessed February 1, 2021. http://www.tylervigen.com/spurious-correlations.

Yanofsky, David. "The Chart Tim Cook Doesn't Want You to See." *Quartz*, September 10, 2013. https://qz.com/122921/the-chart-tim-cook-doesnt-want-you-to-see/.

Seasonal Visual Literacy
Using Current Events to Teach Data and Spatial Literacy Skills with Adaptable LibGuides

Jacqueline Fleming, Visual Literacy and Resources Librarian, Indiana University Bloomington, jkflemin@iu.edu; Theresa Quill, Map and Spatial Data Librarian, Indiana University Bloomington, theward@indiana.edu

NUTRITION INFORMATION

This recipe is one for your whole community! Since the beginning of March 2020, the COVID-19 pandemic has changed the way we serve our library communities. One huge change has been the new emphasis on data visualizations that are publicly available and not always based on reliable information. This constant onslaught of visual information created an infodemic (information epidemic) that communities across the world were not ready to digest.[1]

This healthy and hearty recipe is for how to create a LibGuide on visual literacy and maps. The goal in creating this research guide is to give your patrons and community the nutritional benefits of data literacy and visual literacy skills. The combination of ingredients in this recipe will create a meal that gives the vitamins and healthy fats needed to recognize the different types of information sources making visualizations and understand how to read visual information and how to assess the reliability of the sources and the data presented in each visual. By focusing on current events, this recipe offers immediately actionable literacy skills in an easy-to-digest format.

TARGET AUDIENCE AND NUMBER SERVED

This recipe is adaptable and serves a large number of people with a variety of dietary preferences. Our original recipe was developed for the palate of undergraduate students and instructors, as well as community members. We see this meal being nutritious for everyone due to the importance of data visualization and maps in our visual culture. This recipe can serve the appetites of students in any discipline related to a current event.

LEARNING OUTCOMES

People who visit the guide will be able to
- describe the importance of visual literacy for reading maps, charts, and graphs to comprehend the significance and reality of current events, specifically related to the COVID-19 pandemic
- critically analyze data visualizations to determine the appropriateness of visual representations, including color schemes, symbols, map projections, data validity, and data aggregation

COOKING TIME

45 minutes

DIETARY GUIDELINES

This recipe fits into the first three frames of ACRL's *Framework for Information Literacy for Higher Education*: Authority Is Constructed and Contextual, Information Has Value, and Information Creation as a Process. By following this recipe, others can create a guide, similar to our original, that emphasizes the importance of recognizing creator bias, as well as the value and impact of the information represented in maps and data visualizations labeled as authoritative sources in our everyday lives. We hope that this recipe gives readers the nutritional basis needed to create researched and trustworthy maps and data visualizations as well as the skills needed to question maps and data visualizations shared every day.

INGREDIENTS

- Digital research guide platform (our institution uses Springshare's LibGuides)
- Tips for analyzing and evaluating data visualizations and maps
- Resources that discuss data visualizations and maps
- Examples of the topical data visualizations and maps you are focusing on in your guide

Section 4. Data Visualization

- Data resources related to the data visualizations and maps you are discussing

PREPARATION
- Collect reliable information about data literacy, data visualization, and data visualization maps. This is general information about critically evaluating data visualizations and maps, using your current event as an example.
- Decide on a format for your research guide. Figure out the best structure for the information you want to present.
- Prepare supporting data related to your data visualizations and maps. May substitute for seasonal current event.
- Connect with relevant subject librarians for potential collaboration or assistance in promotion.

INSTRUCTIONS

Don't reinvent the wheel! If you've prepared sufficient information, you should be mostly assembling authoritative information. Your job is to emphasize the importance of visual and data literacy for everyday consumption of information and to make connections between academic writing on visual literacy and practical tips for reading maps and charts. We found that Twitter (see the North American Cartographic Information Society @NACIS for cartographers to follow) was an especially helpful resource for finding examples that connect theory with practical information and viral maps. Because this recipe focuses on current events, quick response time is important for staying relevant, which might mean that formal, peer-reviewed literature and visualizations on this topic are not yet available. This is where a librarian can be especially helpful in guiding users around dubious visualizations while not discounting reliable informal sources. In the case of COVID-19, government agencies quickly created maps that contained authoritative information, but were displayed in a way that was confusing or didn't conform to traditional cartographic conventions. Walk users through how to investigate the data source behind the visualization by asking, "When was the visualization last updated? Are the data current and authoritative? Are the data appropriately normalized, or is this a choropleth map using totals? Does the map use appropriate symbols? What geography is being mapped? Do the data match the geography?" Be critical of design choices that can impact how a map is read, such as color scheme, categorization, and projection. (See appendix examples.) Fold in some visuals that illustrate your thesis, such as maps that normalize data contrasted with maps that do not.

Make sure your research guide is organized and adaptable. For example, a main page with general information on visual and data literacy and sub-pages for specific current events, such as COVID-19. For emergent data events, be sure to indicate a "best by" date of when your guide was last updated. Our guide is organized with a general landing page containing accessible academic writing on visual literacy and map reading. Sub-pages for specific current events go into more detail about the data and display examples of visualizations about that event. At present, our guide contains pages on COVID-19, presidential election maps of 2020, and Brood X cicadas (see figure 1). While the sub-pages of our research guide cover specific topics, we encourage readers to create visual and map literacy guides that cover topics important to their library communities. Our recipe is not meant to be treated as a set of rules. You won't ruin the meal if you decide to add some extra sugar and spice! Other example topics that might work for a visual literacy and map research guide include how to read public health infographics and data visualizations in sports media.

There are several ways that you can assess the effectiveness of your guide and gain feedback from patrons. If you use Springshare's LibGuides, you can collect and see viewing data. This information can help you understand if your advertising of the guide is effective. If you do not have the ability to collect viewing data, you can gain feedback from viewers of the guide through a survey or provide an e-mail address where they can send feedback.

ADAPTING THE RECIPE

While our recipe focuses on the importance of visual and map literacy in relation to COVID-19 maps and data visualizations, you can use this recipe for any topic related to visual information. For example, ahead of the 2020 US presidential election we created a page in this guide to address the challenges of creating and reading US election maps and finding authoritative data.

4. Data Visualization

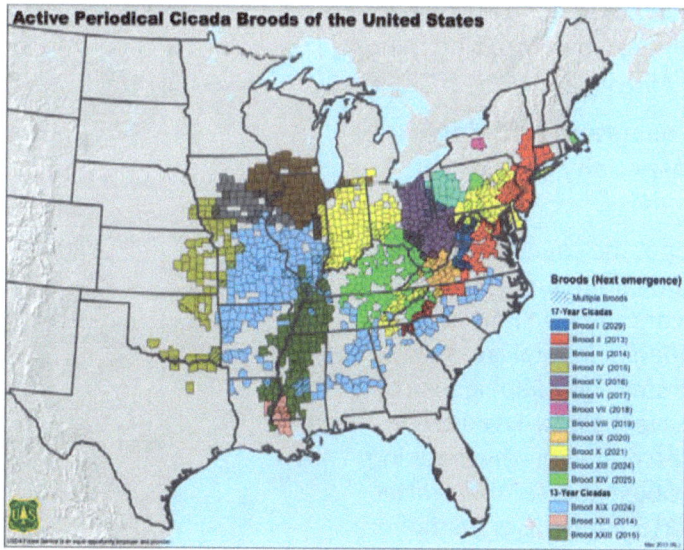

FIGURE 1
A screenshot of the landing page for the LibGuide on cicadas, including a US Forest Service map of periodical cicada broods in the United States Reviews/Assessment Strategy

ADDITIONAL RESOURCES

Our Visual Literacy and Map LibGuide
Quill, Theresa, and Jackie Fleming. "Visual Literacy and Maps." LibGuide, Indiana University Bloomington Libraries. Last updated November 1, 2021. https://guides.libraries.indiana.edu/visualliteracyandmaps/Home.

Information on Creating Information Research Guides
Puckett, Jason. *Modern Pathfinders: Creating Better Research Guides* (Chicago: Association of College and Research Libraries, 2015).

Sullivan University Library. "A How-to Guide for Research." LibGuide. Last updated October 29, 2021. https://libguides.sullivan.edu/howtoresearch.

Information about Reading and Analyzing COVID-19 Maps and Data Visualizations
Field, Kenneth. "Mapping Coronavirus, Responsibly." *ArcGIS Blog*, February 25, 2020. https://www.esri.com/arcgis-blog/products/product/mapping/mapping-coronavirus-responsibly/.

Wiseman, Andrew. "When Maps Lie: Tips from a Geographer on How to Avoid Being Fooled." *Bloomberg City Lab* (blog), June 25, 2015. https://www.bloomberg.com/news/articles/2015-06-25/how-to-avoid-being-fooled-by-bad-maps.

APPENDIX: TIPS AND TRICKS FOR READING COVID-19 MAPS

It is important to remember that all maps are created by someone with a bias and specific intentions in creating that visual. This is why it is essential to analyze and critique every map that you see. This is especially true with COVID-19 maps and data visualizations because everyone is talking about the pandemic and new information is being published every day. Below are specific aspects of COVID-19 maps that you, as a visually literate individual, will want to question.

1. **When was the visualization last updated? Are the data current and authoritative?**

    The pandemic situation is changing rapidly, so having up-to-date information is critical. A good COVID-19 map will tell you when the data were updated and where they come from.

2. **Is this a choropleth map using totals?**

    A choropleth map is a thematic map that uses colored, shaded, or patterned areas to show values. This is a common type of map, though it is not without shortcomings. One of the most common mistakes with choropleth maps is mapping total numbers of an occurrence per area. For example, a map showing the number of Coronavirus cases per county does not take into account that counties in the United States have widely varying populations. In a map like this, populous counties will appear to be disease hot spots, even though they might have a similar RATE of infection per person. A more helpful way to map this information would be to normalize the data by population (cases per 100,000 people, or percent population infected, etc.).

3. **Does the map use appropriate symbols?**

    A choropleth map can also make large areas appear more important because they take up more visual real estate on the map. For example, San Bernardino County in California is the largest county by area in the US, but it has low population density. Because of its large area, this can make it appear visually more prominent in a choropleth map than counties with smaller areas but higher populations. A proportional symbol or dot-density map solves this problem when mapping things like cases of COVID-19.

4. **What level of geography is being mapped? Be especially wary of zip code maps.**

    COVID-19 maps at the country level might not give enough information about where the disease is concentrated (the scene looks much different in rural areas than cities). Maps of small geographies, such as zip codes, have their own issues. For an excellent description of why mapping zip codes is problematic, see this Twitter thread from cartographer Lyzi Diamond: https://twitter.com/lyzidiamond/status/1238168498486022144.

NOTE

1. Department of Evidence and Intelligence for Action in Health, Office of the Assistant Director, *Understanding the Infodemic and Misinformation in the Fight against COVID-19*, fact sheet, Pan American Health Organization, CC BY-NC-SA 3.0 IGO, https://iris.paho.org/bitstream/handle/10665.2/52052/Factsheet-infodemic_eng.pdf.

To Visualize Is to Experience Data

Chelsea H. Barrett, Librarian and Assistant Professor, University Libraries, Seton Hall University, chelsea.barrett@shu.edu; Gerard Shea, Librarian and Assistant Professor, University Libraries, Seton Hall University, gerard.shea@shu.edu

NUTRITION INFORMATION

As data continue to grow and the capability to analyze data increases, there has been a growing need to be able to analyze data in a quick, efficient, and coherent manner. Though examining data in a textual fashion has been utilized for quite some time, staying with this modality may not be effective in the foreseeable future as data become a driving force in decision-making for companies, organizations, and scholars.

Visual presentations of data, known as *data visualization,* have served to combat some of the inefficiencies of textual data as they help to contextualize oral or written forms of data.[1] Data visualization is not a new technique, but there is now a smorgasbord of data visualization tools that are free and easy to use to conceptualize large amounts of data. One of the more important features of newer data visualization tools is the ability to show patterns and trends that are not as easily seen in textual or oral data.[2]

The most essential ingredient in this lesson is Tableau Public, a free data visualization software program that can transform data into interactive visualizations to be shared and published. This workshop teaches participants how to create significant visualizations that display insights in an effective and understandable way.

TARGET AUDIENCE AND NUMBER SERVED

This workshop was concocted for faculty, staff, and students in higher education who have little to no knowledge or experience with data visualization. It can be used for any person who has an interest in learning how to use a data visualization tool.

LEARNING OBJECTIVES

Participants will be able to

- strategically find, use, and aggregate secondary data to create intentional data visualizations
- express data-driven insights in a clear and concise way through the creation of unique visualizations
- apply analytical thought through comprehension of trends and patterns present in formulated visualizations

COOKING TIME

The workshop should ideally simmer for a minimum of 40 minutes and a maximum of 60 minutes, though times can vary. Below is a potential timing breakdown for the session:
- Lecture (5–10 minutes)
- Data prep/cleaning (5–10 minutes)
- Tableau demonstration (15–20 minutes)
- Participant hands-on work time (15–20 minutes)

DIETARY GUIDELINES

The essential flavor of this recipe comes from the frame Information Creation as a Process from ACRL's *Framework for Information Literacy for Higher Education*. The lesson exemplifies for participants the significance of their participation in the information creation process. Seeing that significance enables students to understand their agency in this creative process and leads to more sophisticated choices as their comprehension increases with their information needs.[3] Using Tableau is a transferable skill that can be applied to their lifelong learning endeavors in data visualization. The skills learned in this workshop will help make each participant a master chef in their own information kitchen.

INGREDIENTS

- Computer/laptop
- Tableau Public
- Tableau Desktop
- Prepared and cleaned data set for demonstration (a link to a prepped data set is available in Additional Resources!)

PREPARATION

Before you start the cooking process, the instructor should complete these steps:
1. Please view this Tableau Public Overview tutorial to get familiar with the platform: https://public.tableau.com/en-us/s/

resources?video=AZ-cy67GJck.
2. Have participants download Tableau Desktop.
3. Have participants create a Tableau Public account.
4. Make sure each attendee will have access to a laptop or other computer whether it is provided by the instructor or brought by the participant.
5. Make sure there will be internet access for the class session.
6. Make sure that the presenter has proper presentation equipment.
7. Ready a prepped data set to use for the Tableau demonstration.
8. Optional: Urge participants to bring in their own data to analyze.

INSTRUCTIONS

After the preparations are complete, it is time to start! To begin, the instructor will explain data visualization and why it is important. To get the participants immediately interested, feel free to sprinkle in an already finished Tableau dashboard! To see this dashboard, please refer to figure 1.

After the lecture, show the participants where to find data and how to clean those data. Use the provided data set to show the class how to properly clean the data so that they are ready to go into the metaphorical "pot" of Tableau. Show participants appropriate search methods and sites to search for secondary data. Work with a predetermined data set that is either cleaned or has minor errors, depending upon length of workshop or class.

Once the data have been properly prepped and cleaned, demonstrate how to import the data into Tableau and then begin to create pertinent charts and meaningful dashboards using those charts. For steps to using Tableau, please refer to the appendix.

After learning how to create data visualizations with Tableau, the participants will try to prepare their own meal using this tool. The last activity ("main course") of this workshop will allow the participants to complete the entire process of finding data, cleaning data,

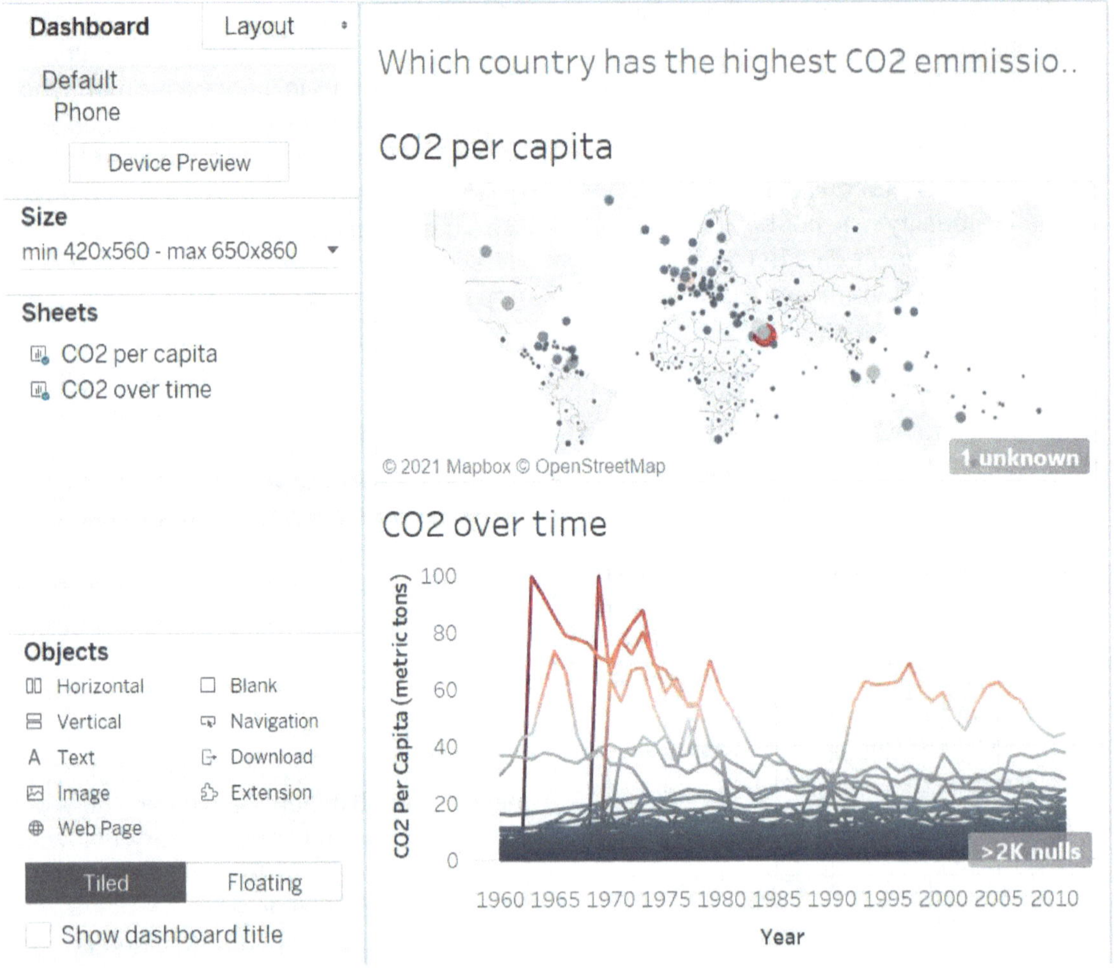

FIGURE 1
Tableau visualization dashboard

importing the data, and creating meaningful visualizations. The participants have the option to use their own data to work with during this part of the session. Help the participants during this time with any data visualization questions that may arise during the hands-on portion of the session.

REVIEWS/ASSESSMENT STRATEGY
A few indicators of success are as follows:
- Participants create unique visualizations using secondary data, which demonstrates grasped data literacy.
- Participants incorporate data visualizations into reports, scholarly presentations and other formats to present findings.

ADAPTING THE RECIPE
As we have learned, circumstances change, and then the ingredients need to be adapted. The original recipe called for an in-person session. However, after the COVID-19 pandemic hit, change was on the menu. The instructor worked with a faculty member to create a video tutorial for the material they initially planned to cover in the in-person session. The purpose of the tutorial was to show the students how to use data to create visualizations and a meaningful narrative using Tableau. In this video tutorial, the instructor showed the participants how to use data from the World Bank and demonstrated how to use that data set to create visualizations using Tableau. The video was made accessible to the students by adding it to the instructional materials in the Blackboard course shell. A URL for the video and other reference materials to help you build your own recipe is available in Additional Resources.

In addition to the World Bank data used for this activity, another option to explore could be ICPSR (Inter-university Consortium for Political and Social Research), a place to get social and behavioral science data.[4] Though it is not a completely free resource, there is a free open-access option for participants looking to work with different data sets. The instructor presented this as an alternative to participants prior to the activity and provided instructions on how to get started using ICPSR. A PowerPoint presentation with the detailed instructions is included among the reference materials listed in Additional Resources.

ADDITIONAL RESOURCES
Link to Tableau reference materials including the instructional video: https://drive.google.com/drive/folders/1naGUx-8hVlXxpWR6bXsvypMspxDmyRLX?usp=sharing.

Tableau. "Tableau Public." https://public.tableau.com/en-us/s/.

Other free data visualization tools that can be used for the activity are Qlik, Flourish Public, and Microsoft Power BI. Data visualization tools vary between Power BI and Tableau depending on what the workshop or class session requires.

Flourish. "Flourish Features." https://flourish.studio/features/.

Flourish Public is a free, easy-to use data visualization and storytelling tool that allows users to create interactive and animated visuals to share and publish. This visualization tool has a plethora of visualizations, ranging from charts and graphs to quizzes and surveys.

Qlik. "QlikView Personal Edition." https://help.qlik.com/en-US/qlikview/April2020/Subsystems/Client/Content/QV_QlikView/QlikView_Personal_Edition.htm#:~:text=Qlik%20offers%20a%20free%20version,runs%20without%20a%20license%20key.

QlikView Personal Edition is free for students, individuals, smaller businesses, and others. This business intelligence tool allows for interactive data visualizations and is primarily used for personal use.

Microsoft Power BI. "What is Power BI?" https://powerbi.microsoft.com/en-us/what-is-power-bi/

Microsoft Power BI is a data visualization tool that is like Tableau in terms of building visualizations. However, Power BI is typically used to answer business-related inquiries. This tool "allows business representatives to easily view the insights of their business in the form of interactive reports."[5] There is currently a free version of Power BI Online and Power BI Desktop that can be used for this type of a workshop.

APPENDIX: STEPS FOR USING TABLEAU PUBLIC

Once you have your data and have cleaned/edited your data in Excel or another platform, you are now ready to use Tableau Public.

IMPORT

1. Open Tableau Public where you will be prompted to 'Connect To a File'. Under Connect to a File click Microsoft Excel and open the Excel file you would like; the dataset should load right into Tableau Public.
2. Tableau will connect to the data file and you can find the worksheet you would like to use under Sheets. Select the worksheet and drag it into the "Drags table here" area. Tableau will create a preview of your data set.

CREATING CHARTS/VISUALIZATIONS

3. If you are happy with the preview of the data, you can open your first worksheet by clicking on Sheet 1 at the bottom left of the page.
4. Once the data is loaded, you will be able to create visualizations which are populated on the canvas in the middle of the page. There are different ways of creating visualization in Tableau, one of the easiest is by simply dragging and dropping the fields listed on the left of the page onto the canvas in the middle of the page.
5. Walk the participants through how to create charts. Demonstrate how to add more information and make manual adjustments. For example, you can add color to the visualization by dragging the variable to the color tile. This is a chance for the participants to think about which visualizations fit best to answer the proposed questions. The presenter and any other facilitators should be assisting participants exhibiting difficulties during this time.
6. Demonstrate how to rename the sheet by clicking on the worksheet icon at the bottom left of the page. Ask the participants to give their chart an appropriate name. To create another chart, you can click on the new worksheet icon at the bottom left of the page. Let's look at the same field, CO2 emissions per capita, for worksheet 2. You will double click to add CO2 emissions to the canvas. And you will also double click Year to see how it changes over time. You can see the rate of CO2 emissions has been increasing. You will add Country to the canvas. You can now see that the country Qatar has the most emissions. However, it does show a downward trend over the years.
7. When you are happy with the chart. You can polish it off with a little formatting. You can add CO2 per capita onto the color tile and make sure the colors match your first chart. Before you move on, you are going to give the sheet a new name so that you do not confuse it with your first chart.
8. Next, you want to combine the two charts you have created into a dashboard. You can create a new dashboard by clicking on the middle icon at the bottom left of the screen. You can see the charts you created on the left-hand side of the screen. You can drag and drop them onto the canvas to start combining them into a dashboard.

PUBLISH/CREATE DASHBOARD

1. And when you are happy with how your visualization looks, you can publish it to the web by clicking on File and selecting Save to Tableau Public.

Overview video on how to use Tableau Public: https://public.tableau.com/en-us/s/resources?video=AZ-cy67GJck

© 2021. This work is licensed under a CC BY 4.0 license

NOTES

1. Allison Hahn, "Data Visualization," *Salem Press Encyclopedia* (Amenia, NY: Salem Press, 2018), EBSCOhost.
2. Syed Mohd Ali, Noopur Gupta, Gopal Krishna Nayak, and Rakesh Kumar Lenka, "Big Data Visualization: Tools and Challenges," in *2016 2nd International Conference on Contemporary Computing and Informatics (IC3I)* (Greater Noida, India: IEEE, 2016), 656–60.
3. Association of College and Research Libraries, "Information Creation as a Process," in *Framework for Information Literacy for Higher Education* (Chicago: Association of College and Research Libaries, 2016). http://www.ala.org/acrl/standards/ilframework#frames.
4. "About ICPSR," ICPSR, accessed March 15, 2021, https://www.icpsr.umich.edu/web/pages/about/.
5. Suren Machiraju and Suraj Gaurav, *Power BI Data Analysis and Visualization* (Boston: Walter de Gruyter, 2018), 4–5.

Upping the Baseline for Data Literacy Instruction

Jessica Vanderhoff, Director of Assessment and Planning, West Virginia University Libraries, Jessica.Vanderhoff@mail.wvu.edu

NUTRITION INFORMATION

Across all academic disciplines, there is an ever-increasing expectation placed on students to interpret and create data visualizations. Yet many university faculty, constrained by limited class time, prescriptive learning outcomes, and generally dire teaching loads, find it difficult to incorporate complementary learning opportunities, particularly with regard to information and data literacy. While librarians' responsibilities are no less substantive, we are well equipped and, dare I say, accustomed to developing and delivering curricular support that helps align course expectations with course deliverables.

I originally designed this three-part data literacy and visualization series for an honors college program. While the program outcomes do not specifically identify the use or visualization of data, the majority of students find that they need to incorporate and present data as part of their research outputs. The curriculum falsely assumes that students possess a baseline ability to work with data and create high-quality visual aids,[1] a key deliverable of any oral presentation. This series of three 60-to-90-minute workshops was developed in response to that curricular gap.

Participants are guided through the processes of critically evaluating data visualizations; organizing, sorting, and aggregating data in Excel; and creating basic visualizations in Tableau. The learning objectives are widely transferable to a range of course offerings that expect basic data visualization.

TARGET AUDIENCE AND NUMBER SERVED

Limit to approximately 25 participants (but no fewer than 10) of a similar skill level or engaged in a similar course project to allow for greater customization and hands-on interaction. As this is a three-part workshop, it is critical for participants to commit to the entire series.

LEARNING OBJECTIVES

Upon completion of each session, participants will be able to

Session 1: Consuming Data: A Critical Look at Data Visualizations
- Define data, data visualizations, and infographics
- Evaluate data visualizations for fundamental design and content principles

Session 2: First Steps to Data Visualization Using Excel: Data Sources, Prep, Summary, and Charting
- Identify reliable data sources from freely accessible and university subscription databases
- Discuss the importance of cleaning and organizing data in preparation for basic analysis and charting
- Create summary tables and descriptive statistics in Excel using common functions and pivot tables
- Generate commonly used charts in Excel and apply fundamental design principles

Session 3: Visualizing Data with Tableau
- Describe how to connect to and preview data in Tableau Desktop
- Use the Tableau Desktop workspace to build views and select chart types
- Combine visualizations into a dashboard
- Export and share visualizations using common software applications, including Microsoft PowerPoint

COOKING TIME
Prep time: 2–4 hours per session
Cook time: 1–1.5 hours per session

DIETARY GUIDELINES
This series demonstrates the process of information creation through the lens of data. Before students can be expected to create new data-centered information products (i.e., data visualizations, infographics, etc.), whether based on their own primary data or from reliable sources of secondary data, they need to understand the processes behind visualizing data and, in particular, the interdependence of these processes.[2]

Section 4. Data Visualization

INGREDIENTS
Each session to include the following:
- Presentation slides including agenda, definitions and key concepts, multiple real-world examples, and links to feature resources. (For recommended readings on key concepts and featured resources, see the *Additional Resources* section.)
- Applied demonstration of key concepts using data, software, or web applications.
- Small-group, hands-on activities to introduce or reinforce application of key concepts using data, software, or web applications.
- Session assessments, including live (formative) polls and a (summative) survey.
- Technology
  - Computer lab or access to a personal device.
  - Microsoft Excel (recommended) or Google Sheets to display data in tabular format.
  - Tableau Desktop (recommended) or Tableau Public for data visualization.
  - Internet connection.
  - Assessment platforms may include Padlet, Kahoot!, Google Forms, or Qualtrics.
  - Zoom or other videoconferencing software for synchronous online sessions.

PREPARATION
Preparation for this sequence is time-consuming: it requires the careful selection of an audience-appropriate data set and existing or generated visualizations. To prepare, identify one theme or topic that will likely resonate with and be familiar to the participants. For undergraduates, examples may include social media, sports, or retail. For graduate students and faculty, themes about current events, scholarly communications, or trends in higher education are appropriate. If the series is designed around an experiential course or program where participants are engaged with a local business or community organization, locate data relevant to the company, industry, or cause.

Locate an example data set and data visualizations on the theme to help participants visualize concepts. You will use this data set and its visualizations as the basis of your conversation and demonstration in each of the three sessions. Choose wisely! One helpful hint is to locate a data set that has been visualized in a variety of ways by a variety of sources—for instance, a Pew Research or ICPSR data set that has been visualized in a research report as well as in a news source such as the *New York Times* or the *Economist*. If the data have not been visualized, take the time to create examples of high quality and poor visualizations. Reference these throughout the series.

In most cases, the raw data set will be large and too overwhelming for participants to fully grasp in the session. While it is important to maintain and show the original data set, take the time to clean, organize, and edit the original data to include only the variables you plan to analyze. Discuss why it is important to complete this step before moving forward.

Marketing materials should emphasize that while previous data skills are helpful, they are not a substitute for the critical appreciation of reliable data sources and mindful data organization and visualization.

INSTRUCTIONS
Each session contains three components:
- short lecture to introduce key concepts and software
- small-group activity to gain hands-on experience
- reporting out and assessment activity

Session 1: Consuming Data: A Critical Look at Data Visualizations
Part 1: Lecture (20–30 minutes)
Introduce the theme of the series. Ask participants to describe visualizations they have encountered on the topic. Once participants have shared their experiences, introduce your visualizations and, secondly, the corresponding data set. Discuss. Once participants are familiar with your example, define, compare, and contrast key concepts using the data set and visualizations as visual aids.

Part 2: Small-Group Activity (~30 minutes)
Break student into groups of two to four. Using specific evaluation criteria, such as the "Data Visualization Checklist," ask participants to evaluate visualizations based on best practices (see the Additional Resources section for the checklist). In addition to the evaluation criteria provided, ask participants to identify additional criteria that can be incorporated into a future checklist.

Part 3: Reporting, Review, and Assessment (15–20 minutes)

Ask groups to share their work. Present their findings on a shared screen and critique as a group. All small-group deliverables should be saved and uploaded to a shared folder for additional assessment. Review any concepts that need reinforcement. Share a link to the workshop assessment and encourage participants to respond before leaving the session.

Session 2: First Steps to Data Visualization Using Excel: Data Sources, Prep, Summary, and Charting

Part 1: Lecture (20–30 minutes)

From Session 1, review the theme of the series along with the data set and corresponding visualizations. Discuss how you located the information and provide links to other credible data sources, such as subscription databases or Data.gov. Next, introduce Excel, including the basics of the interface and functionality. Using a shared folder, access the cleaned Excel workbook containing the original data set as well as separate worksheets for each of the features and functions you plan to demonstrate. Depending on your participants' command of Excel, demonstrate one function at a time (i.e., measures of central tendency, COUNT and IF functions, and pivot tables) using your sample data set, and supplement the live demo with detailed instructions for participants to reference later.

Part 2: Small-Group Activity (~30 minutes)

Break students into groups of two to four and distribute the shared folder with the corresponding workbook and activity prompt. Participants will need to download the cleaned workbook to their local machine and execute specific tasks (i.e., measures of central tendency, COUNT and IF functions, and pivot table) using the data set in Excel.

Part 3: Reporting, Review, and Assessment (15–20 minutes)

See Session 1, Part 3, and repeat.

Session 3: Visualizing Data with Tableau

Part 1: Lecture (~30 minutes)

Introduce Tableau. Before demonstrating the software, show participants a selection of Tableau-generated visualizations to help them quickly recognize its capabilities and distinguish it from Excel. Next, introduce Tableau, including the basics of the interface and functionality. Be sure to explain common data types, including string, date, numerical, geographic, and so on and how these values determine what visualizations are possible with the data at hand. Once again, using the cleaned Excel file, demonstrate how to connect to data. Depending on your participants' command of Tableau, demonstrate one visualization at a time (i.e., text or highlight table, bar chart, maps, etc.) using your sample data set, and supplement the live demo with detailed instructions for participants to reference later. As time permits, demonstrate formatting options, filters, and dashboards.

Part 2: Small-Group Activity (~30 minutes)

Break students into groups of two to four and redistribute the shared folder with the corresponding workbook and activity prompt. For a novice group, the group activity may include the process of connecting to data, creating and formatting a simple table and bar chart, and combining the two visualizations into a dashboard. Have students export and upload their final products as a PowerPoint slide into the shared folder. Repeat step-by-step demonstrations as necessary.

Part 3: Reporting, Review, and Assessment (15 minutes)

See Session 1, Part 3, and repeat.

REVIEWS/ASSESSMENT STRATEGY

In addition to the learning objectives, assessment strategies should consider the session's time constraints, audience, and general data needs. A three-pronged approach is suggested.

- *Baseline:* Collect baseline information from participants in the form of a quick poll or survey at the start of each session or via a registration or RVSP form. Data to be collected may include participants' comfort level, confidence, or self-reported proficiency in locating reliable data, creating visualizations, or using Excel or Tableau for particular tasks.
- *Formative and summative:* Discuss and collect examples of in-class, small-group work during each session to gauge participants' understanding of key concepts

and skills. Create a shared folder for participants to easily access content, edit, and upload their work.
- *Summative:* Distribute a postworkshop survey to measure participant satisfaction, self-reported learning outcomes, and transferability of skills.

ADAPTING THE RECIPE

While the learning experience is best as a sequence, each lesson can stand alone, as well as being customized for novice through advanced learners from a variety of disciplines. It is important, however, to target participants who are actively engaged in (primary or secondary) data collection, analysis, or visualization. Always remember to tailor examples and learning objectives based on audience skill level and interests.

ADDITIONAL RESOURCES

Key Concepts

Cairo, Alberto. *How Charts Lie: Getting Smarter about Visual Information*. New York: W. W. Norton & Co., 2019.

Cairo, Alberto. *The Truthful Art: Data, Charts, and Maps for Communication*. San Francisco: New Riders, 2016.

Camões, Jorge. *Data at Work: Best Practices for Creating Effective Charts and Information Graphics in Microsoft Excel*. San Francisco: New Riders, 2016.

Herzog, David. *Data Literacy: A User's Guide*. Thousand Oaks, CA: Sage, 2016.

Krum, Randy. *Cool Infographics: Effective Communication with Data Visualization and Design*. Indianapolis, IN: John Wiley & Sons, 2014.

Murray, Daniel G. *Tableau Your Data! Fast and Easy Visual Analysis with Tableau Software*. Indianapolis, IN: John Wiley & Sons, 2013.

Data Visualization Best Practices and Guides

Dullaert, Michiel. "Design a Chart." Chart. Guide. Accessed April 1, 2021. https://chart.guide/design/.

Evergreen, Stephanie, and Ann K. Emery. "Data Visualization Checklist." Evergreen Data: Intentional Reporting and Data Visualization. Accessed April 1, 2021. https://stephanieevergreen.com/wp-content/uploads/2020/12/EvergreenDataVizChecklist.pdf.

Krum, Randy. "Data Visualization Reference Guides." Cool Infographics. Accessed April 1, 2021. https://coolinfographics.com/dataviz-guides.

Data Sources and Data Visualization Examples

ICPSR. https://www.icpsr.umich.edu/web/pages/ICPSR/index.html.

Krum, Randy. Cool Infographics. Accessed April 1, 2021. https://coolinfographics.com.

McDandless, David. Information Is Beautiful. Accessed April 1, 2021.

Pew Research Center. "Download Datasets." https://www.pewresearch.org/download-datasets/.

Tableau Public. "Viz of the Day." https://public.tableau.com/app/discover/viz-of-the-day.

https://informationisbeautiful.net.

Yau, Nathan. FlowingData. Accessed April 1, 2021. https://flowingdata.com.

NOTES

1. Association of American Colleges and Universities, "VALUE Rubrics—Oral Communication," accessed May 26, 2021, https://www.aacu.org/value/rubrics/oral-communication.
2. Vetria Byrd, "Using Bloom's Taxonomy to Support Data Visualization Capacity Skills," In *E-Learn: World Conference on E-Learning in Corporate, Government, Healthcare, and Higher Education*, (Association for the Advancement of Computing in Education, 2019) 1039–53.

A Literacy-Based Approach to Learning Visualization with R's ggplot2 Package

Angela M. Zoss, PhD, Interim Head of Assessment & User Experience Strategy, Duke University Libraries, angela.zoss@duke.edu

NUTRITION INFORMATION

This recipe describes the process for delivering a 2-hour workshop on visualization in R using ggplot2. The workshop is designed to promote literacy around ggplot2 and data visualization creation. Over the course of the workshop, participants learn about the theory behind the ggplot2 package and its approach to visualization, the typical syntax used to create a visualization with ggplot2, and a few features for visualization design offered by ggplot2. The workshop combines conceptual material with hands-on exercises to build participants' vocabulary, expose participants to the range of features available in ggplot2, and reinforce learning with repeated practice.

All teaching materials and several videos of past workshops are available publicly at https://github.com/amzoss/ggplot2-workshop.

TARGET AUDIENCE AND NUMBER SERVED

This workshop is designed for individuals who have some prior experience with R. No experience with the ggplot2 package is expected. While the workshop has largely been taught to higher education audiences (especially students), the materials may also be appropriate for advanced high school students.

The workshop works best when participants are able to follow along with the instructor's pace and ask clarifying questions, which tends to be easier with a maximum of about 30 attendees. Having helpers available for technical problems is especially useful for larger groups.

LEARNING OBJECTIVES

Students will

- recall and arrange the basic components of the code required to produce a ggplot2 visualization
- use ggplot2 documentation and cheat sheets to look up syntax information and additional helpful chart components
- critique the output of ggplot2 based on an understanding of the data set and a general knowledge of best practices for visualization design
- compare different visualization tools based on their properties and their suitability for a particular project
- identify patterns and useful keywords in warnings and errors generated while developing code for ggplot2 visualizations

COOKING TIME

This workshop runs for two hours, but it may not be possible to cover all exercises in that time. It's important to gauge the needs of the audience in real time and prioritize activities based on the topics you think will be most important or relevant to them.

DIETARY GUIDELINES

This recipe outlines a workshop that is designed to promote critical reflection on the process of creating visualizations using the ggplot2 package in R. The recipe incorporates the following frames from ACRL's *Framework for Information Literacy for Higher Education*: Information Creation as a Process (personal choices are part of visualization creation), Authority Is Constructed and Contextual (software is constructed and not neutral), Research as Inquiry (visualization assist in data-based inquiry), Scholarship as Conversation (visualization is a learned communication skill), and Searching as Strategic Exploration (there are specific search strategies for software troubleshooting).

INGREDIENTS

- Some way to share files with students (e.g., a GitHub repository, Google Drive, or Box)
- RStudio with tidyverse installed on both the instructor machine and the participant machines
- Workshop files: data sets, R Markdown exercise files and answer keys, slides, handout

Section 4. Data Visualization

- Feedback forms
- Technology to share the instructor screen with participants (e.g., a classroom with a projector, Zoom)

PREPARATION

- Finalize exercises, slides, and handout.
  - Tailor exercises to audience as needed.
  - Test all exercises to make sure they work as expected.
  - Update the answer key files with any corrections or additions.
  - Practice slides.
  - Update slides with any corrections or additions.
  - Review and update handout based on slides and exercises.
- Host final versions of files in a place that can be easily accessed by participants (e.g., GitHub, Box, Google Drive)
- Prepare feedback form for assessment.
- Set up physical or virtual meeting space (including recording, if desired).
- Send reminders to registrants, including a link to the files and instructions for installing or accessing RStudio software prior to the workshop.

INSTRUCTIONS

Begin the workshop by sharing the link to the shared workshop materials. Distribute the feedback form early in the session in case participants have to leave early.

Next, share some background on ggplot2. This gives people context for the material and also provides buffer time in case participants arrive late. After the context slides, make sure everyone can open RStudio and set up the workshop files. I prefer to walk through this slowly to make sure everyone is set up before doing any hands-on exercises, even if some participants have to wait for others to catch up. The slides also include suggestions about how to debug ggplot2 code in general and how to look for help on ggplot2 visualizations, and the subsequent activities provide many opportunities to reinforce those techniques.

When teaching ggplot2, the syntax (or the composition of the code components into a valid sequence) tends to be the most difficult material. While ggplot2 is an R package and provides functions like other R packages, the way to build a visualization in ggplot2 is different from the typical syntax in both base R and tidyverse. I split the syntax into components and give names to the different components to try to help participants remember the components more easily, a practiced inspired by techniques for explicit direct instruction.[1] I define the components as the main function (ggplot()), a geometry or shape layer (e.g., geom_point()), the data set, and the aesthetics mapping. The aesthetics mapping is the trickiest part of the syntax. I find it helpful to cover the mapping in a simple but reliable way the first time it is used and then get into more detail in later exercises.

The hands-on activities included cover different aspects of chart creation and can be delivered as work-on-your-own activities, work-together-with-the-instructor activities, or some combination. The available activities are

- *Iris:* This activity is captured in the slides, rather than an R Markdown file. It introduces the basic ggplot2 syntax with a scatterplot using the built-in iris data set.
- *Game of Thrones*: This activity uses a data set of character traits from the television show *Game of Thrones*. It reinforces the syntax for a basic scatterplot, but it goes into more detail about aesthetics mapping, adding multiple geometry layers, and searching help documentation to learn about function options.
- *Star Wars characters:* This activity explores new geometry layers by looking at numerical distribution and linear trends (optional). It also introduces the concepts of aesthetics inheritance, overriding data in a shape layer, and incorporating data manipulation inside a plot.
- *Star Wars opinions:* Using opinion survey data, this activity introduces techniques for working with categorical data. It covers geom_bar(), setting axis limits, flipping the axis of bar charts, mapping a variable to custom colors, and creating facets.
- *Gapminder:* This optional activity can be included for especially advanced groups. In this activity, a finished chart is projected and participants are challenged to build the code to reproduce it. Allot at least 10 minutes for open work time on this activity, as well as a few minutes to solve the problem with the group.

Between activities, I switch back to the slides to show a few additional concepts to prepare for the next activity. For example, before the *Star Wars* characters activity, I introduce facets

and a few useful tidyverse functions like the magrittr %>% (pipe) symbol the dplyr filter() function. I cover factors before the *Star Wars* opinions activity.

When teaching with hands-on exercises, I use best practices for live coding to scaffold the experience for participants.[2] For example, it is crucial to type slowly and verbally narrate what you are typing so that participants can type along with you without watching your screen. I also recommend always typing out the names of arguments inside the ggplot() and geometry functions—for example, ggplot(data = data_frame_name)—to help beginners understand how the elements inside the functions are interpreted by the package.

REVIEWS/ASSESSMENT STRATEGY

Use the following questions to assess the effectiveness of the workshop.
- Free-text response questions:
  - List two things that you learned in this library session.
  - List two things that you still don't understand about the information covered in this session.
  - List one thing that would improve this library session.
  - Which topics would you like to see covered in future sessions?
- Likert-type questions using a five-point scale from Strongly Agree to Strongly Disagree:
  - My understanding has increased as a result of this program/training.
  - My interest in this subject has increased as a result of this program/training.
  - I am confident I can apply what I learned in this program/training.

ADAPTING THE RECIPE

This workshop can be adapted with new activities as needed. For example, it could be tailored to an audience by selecting data sets that are targeted to a particular scholarly domain, or different features of ggplot2 could be added to existing activities. For workshops open to anyone who might wish to attend, I often select data sets related to media and entertainment, but there are many other topics of general interest (e.g., weather, sports, politics). It can be difficult to find data sets that are a good fit for visualization activities—that is, data sets that are relatively small, are easy to understand quickly, and have properties that match up with the activities you want to develop. Be sure to budget plenty of extra time for finding and preparing data sets if you adapt that part of the recipe.

ADDITIONAL RESOURCES

Posner, Miriam. "A Better Way to Teach Technical Skills to a Group." *Miriam Posner's Blog*, December 9, 2015. http://miriamposner.com/blog/a-better-way-to-teach-technical-skills-to-a-group/.

Miriam Posner's advice about teaching technical skills has always been an inspiration. She has a post from 2015 about some of her techniques, including another use for Post-it Notes, and she continues to generously share her insights.

Vaughn, Michael. *Effective Presentation Design*. Apple Books, 2016. https://books.apple.com/us/book/effective-presentation-design/id914927154.

Michael Vaughn's *Effective Presentation Design* is a great resource for presentation design. It includes an activity for designing presentations using Post-it Notes that improved the flow of this workshop immensely.

Zoss, Angela M. "Ggplot2 Workshop." Last updated March 22, 2021. https://github.com/amzoss/ggplot2-workshop.

This GitHub repository contains all of the workshop materials described in this chapter.

FUNDING ACKNOWLEDGEMENT

This project was made possible in part by the Institute of Museum and Library Services (https://www.imls.gov), RE-73-18-0059-18.

NOTES

1. Felienne Hermans and Marileen Smit, "Explicit Direct Instruction in Programming Education," in *Psychology of Programming Interest Group 2018—29th Annual Workshop* (Psychology of Programming Interest Group, 2018), 86–93. https://www.ppig.org/papers/2018-ppig-29th-hermans/.
2. Greg Wilson, "Teaching as a Performance Art," in *Teaching Tech Together: How to Make Your Lessons Work and Build a Teaching Community around Them* (Boca Raton: Chapman and Hall/CRC, 2019), 73–85, https://doi.org/10.1201/9780429330704.

Build Your Own Data Viz Pizza
A Modular Approach to Data Visualization Instruction

Rachel Starry, Digital Scholarship Librarian, UC Riverside Library, rachel.starry@ucr.edu

NUTRITION INFORMATION

This lesson enables librarians to build your own modular Introduction to Data Visualization workshop, which can be customized for different audiences. It takes a mise en place approach by preparing core and supplemental lesson ingredients ahead of time. By breaking down the components of your lesson plan into the dough, the sauce, and the toppings, you'll save prep time whenever you are ready to bake with a different group of learners. This lesson provides the core recipe for designing your own introductory data visualization (also known as data viz) workshop, while two additional lessons provide suggestions for supplementary modules you might use to tailor your workshop to different groups of learners. Those two lessons can be found in the chapters in this book titled "Veggie Pizza: Choosing a Data Visualization Tool" and "Four-Cheese Pizza: Color and Accessible Design."

AUDIENCE AND NUMBER SERVED

1 classroom of 15–20 students at beginner level (this pizza can feed a crowd!)The primary audience for this workshop is upper-division undergraduates and graduate students who need to develop basic data visualization skills and awareness of available tools and best practices.

LEARNING OUTCOMES

Students will
- develop familiarity with selecting appropriate chart types for different kinds of data
- gain experience creating an exploratory data visualization by experimenting with a particular tool

Additional learning outcomes are tied to each supplemental module.

COOKING TIME

90 minutes (3 modules at 30 minutes each)—preparation time varies

DIETARY GUIDELINES

Data visualization is an incredibly broad field of inquiry and practice, which makes deciding what to include in introductory workshops quite challenging. This modular approach empowers librarians to experiment with interactive online resources and active pedagogical strategies for teaching data visualization skills and tools, while cutting down on prep time. Each of the lesson's learning outcomes ties into ACRL's *Framework for Information Literacy for Higher Education*: iterative chart design, covered in the first core module (the pizza dough), emphasizes Information Creation as a Process, and walking students through the steps of visualizing a data set in the second core module (the pizza sauce) underscores Research as Inquiry. Ideally, the learning outcomes for the supplemental (pizza topping) modules tie in other frames, creating an engaged learning experience for students who are new to visualizing data.

INGREDIENTS

- 1 instructor
- 1 slide deck
- Activity handouts
- Pack of 3-by-5-inch index cards ("exit tickets")
- Computer access for all students

PREPARATION

This lesson consists of two core modules—the dough and the sauce—plus one supplemental module—a topping of your choice. Each module is intended to be delivered in about 30 minutes, leaving time for question-and-answer periods and short breaks. Preparation starts with becoming familiar with the elements you'll need to build your pizza.

Every pizza needs a delicious foundation: the dough is the core part of the recipe that introduces students to essential definitions and concepts, addressing the question "What is data visualization?" and providing an overview of commonly used chart types for different kinds of data sets and use cases. In addition to the many available print publications on the subject of data visualization, a number of websites and experts' blogs provide research-backed information about data visualization and data viz pedagogy that may help you build familiarity with this topic, such as *Storytelling with Data*—a blog by Cole Nussbaumer Knaflic; *Visualising Data*—a website by Andy Kirk; *FlowingData*—a blog by Nathan Yau; *EagerEyes*—a blog by Robert Kosara; or *Chartable*—a blog by Datawrapper (see Additional Resources, below).

Next comes the sauce: while you have a few options, the classic marinara is a hands-on learning activity that introduces students to using a particular data viz tool to create a simple chart. The seasoning of your sauce is up to you. Depending on your classroom environment and access to campus software licenses, you may opt for using a browser-based tool such as Datawrapper or RAW-Graphs or walking students through creating a chart in Microsoft Excel or Tableau. Whichever tool you choose, plan to spend a few minutes explaining the features of the tool's interface before you dive into chart creation with a preselected sample data set.

Finally, a pizza needs toppings: Use the last 30 minutes of the workshop to explore a supplemental topic. Examples of toppings include

- color and accessible design (outlined in this volume in "Four-Cheese Pizza: Color and Accessible Design")
- choosing a data viz tool (outlined in this volume in "Veggie Pizza: Choosing a Data Visualization Tool")
- preparing data for visualization ("tidy data")
- how not to lie with charts

Having a variety of toppings in your fridge that you can pull out for different audiences or occasions makes your Intro to Data Visualization workshop more flexible and repeatable. Springshare's LibGuides Community includes a number of high-quality library research guides that assemble data viz tools and resources. It is a recommended resource for finding examples of ways other librarians have taught these and other supplemental visualization topics.

INSTRUCTIONS

1. Mix the dough.

   Start with a brief definition of data visualization. A useful metaphor for explaining the difference between exploratory data viz—the process of making sense of our data—and explanatory data viz—the process of sharing information with an audience—is provided by Cole Nussbaumer Knaflic in her book (and website), *Storytelling with Data*. She explains that exploratory data visualization is like turning over 100 rocks to find one or two gems, while explanatory data visualization is telling a story about those one or two gems. It can be helpful to juxtapose an example such as Anscombe's quartet with a data journalism piece or infographic to reinforce the concept that data visualization can have different purposes: exploratory visualizations may bring to light patterns in data that are hidden by summary statistics, while careful visual formatting can highlight important patterns in explanatory visualizations.

   Next, discuss the importance of making an informed decision when choosing a chart type. The goal is for students to understand that not all charts are appropriate for all types of data, and many, such as pie charts, have limitations or are commonly misused. Resources such as the Data Visualisation Catalogue and From Data to Viz provide chart descriptions as well as typologies and flowcharts for selecting a chart. You might explore with students the following list of things to consider when choosing a chart type, using examples from one of the above resources:

   » What is your purpose: exploring or explaining?
   » What type of data are you working with? (text, numeric, geospatial, network, etc.)
   » How many variables do you want to visualize at once?
   » Are your variables sequential or categorical?
   » What type of pattern are you trying to explore or show? (comparison, proportion/composition, part-to-whole, hierarchy, change over time, etc.)

2. Making the sauce.

   To give students an opportunity to practice selecting an appropriate chart type and building a chart from scratch, the second core module is a step-by-step demo of chart creation, during which students can follow along on their own devices. In many cases, it's more effective to choose a simple online tool for this activity rather than desktop software, for two reasons: (1) students can more easily access the tool from their personal devices during and after the workshop, and (2) online tools tend to have simpler interfaces that can be introduced quickly. Datawrapper, for example, works well for this activity, as it is free to use and does not require participants to create an account. While not open-source, it balances robust functionality with ease of use and user privacy.

For this activity, select one or two sample data sets that participants can download or access online; using CSV or another plaintext file format is recommended for interoperability. You may also prefer to use a tool's built-in sample data sets, when available. Planning which sample data set you will use and which types of charts you will walk students through creating is key to the success of this activity. For in-person workshops, distributing a handout with written instructions can also be helpful.

3. Adding your toppings.
   Finally, this lesson can be extended and tailored for different audiences by wrapping up with a third module on a topic that complements the information covered in the two core modules. Your topping module should have its own learning objectives or outcomes.

See the recipes in these related chapters for more information: "Four-Cheese Pizza: Color and Accessible Design" and "Veggie Pizza: Choosing a Data Visualization Tool."

ASSESSMENT STRATEGY

A simple strategy for collecting immediate participant feedback is to pass out 3-by-5-inch index cards as you wrap up the workshop. Ask participants to jot down a few words in response to a small number of prompts; customize these "exit tickets" using your organization's standard workshop assessment questions, or use them to solicit feedback to help you refine future iterations of the workshop. Make sure to write the prompts on a whiteboard or display them on a slide at the end of your presentation. Example prompts:
1. What is one concept that you feel you now understand better?
2. One topic that was completely new to you?
3. One question you still have about today's material?

ADAPTING THE RECIPE

Italian tomato pie: If you have only 60 minutes, leave off the topping module and go for a classic dough-and-sauce tomato pie.

Stuffed-crust pizza: In some cases, you may wish to add your sauce at the end. Experiment with what assembly order works best for you: try going straight from your core dough module right into your topping module (stuffing that crust with cheese), leaving the hands-on activity for last.

White pizza: Reduce this recipe's time from 90 to 75 minutes by opting for an alternate hands-on exercise. Rather than walking students through creating a chart, swap out the marinara sauce for garlic oil by evaluating a chart together using the "Data Visualization Checklist," an interactive online tool and downloadable worksheet developed by Stephanie Evergreen and Ann K. Emery. Select an example chart and plan to spend 15 minutes (rather than 30) walking through the checklist; this exercise helps students understand that charts consist of many components that need to be thoughtfully selected and arranged for maximum readability.

ALLERGY WARNING

Be aware that some students may have disabilities related to vision or perception. Follow accessibility best practices by providing written and spoken descriptions of the visual elements and charts you display during the lesson.

ADDITIONAL RESOURCES

Data Visualisation Catalogue, https://dataviz-catalogue.com.

Datawrapper, https://www.datawrapper.de.

Datawrapper Blog, https://blog.datawrapper.de.

EagerEyes (blog), https://eagereyes.org.

Evergreen, Stephanie, and Ann K. Emery. "Data Visualization Checklist." Evergreen Data. Accessed March 1, 2021. https://stephanieevergreen.com/wp-content/uploads/2020/12/EvergreenDataVizChecklist.pdf

FlowingData (blog), https://flowingdata.com.

From Data to Viz, https://www.data-to-viz.com.

LibGuides Community, https://community.libguides.com.

Nussbaumer Knaflic, Cole. *Storytelling with Data*. Hoboken, NJ: Wiley and Sons, 2015.

Nussbaumer Knaflic, Cole. "Exploratory vs. Explanatory Analysis." *Storytelling with Data* (blog), April 14, 2014, https://www.storytellingwithdata.com/blog/2014/04/exploratory-vs-explanatory-analysis.

RAWGraphs, https://rawgraphs.io.

Storytelling with Data (blog), https://www.storytellingwithdata.com.

Visualising Data (blog), https://www.visualisingdata.com/blog.

Veggie Pizza
Choosing a Data Visualization Tool

Rachel Starry, Digital Scholarship Librarian, UC Riverside Library, rachel.starry@ucr.edu

NUTRITION INFORMATION

This recipe describes a module that can be used on its own or incorporated into the flexible approach to developing an Introduction to Data Visualization workshop described in this book in the chapter "Build Your Own Data Viz Pizza." The assortment of veggies that top this pizza reflects the wide variety of data visualization (also known as data viz) tools that are available, and the goal of this module is to provide learners with resources for identifying which tools may be best suited to their particular needs.

This lesson equips learners with a set of questions to ask themselves as they prepare to select a data visualization tool for a particular project or course. Encouraging learners to pause and reflect on their needs and the kinds of data they will be working with helps to prevent frustration down the line caused by misalignment between their data, goals, and selected visualization tool.

TARGET AUDIENCE AND NUMBER SERVED

1 classroom of students
The primary audience for this lesson is upper-division undergraduates and graduate students who need to develop basic data visualization skills and awareness of available tools and best practices.

LEARNING OUTCOMES

- Students build awareness of the variety of available data visualization tools and platforms.
- Students become familiar with key data visualization concepts for understanding and describing both their data sets and their purposes for visualizing them.

COOKING TIME
30 minutes

DIETARY GUIDELINES

Students and researchers are often faced with the challenge of selecting from a wide variety of data visualization tools. The learning outcomes for this module primarily tie into ACRL's *Framework for Information Literacy for Higher Education* through the frame Research as Inquiry, since this lesson emphasizes the need to identify—through a process of initial examination and perhaps data cleaning and transformation—the kind of data that they have and their purposes for visualizing them, in order to select a data viz tool that will work best for their data and use case.

INGREDIENTS
- 1 instructor
- 1 slide deck
- Computer access for all students

PREPARATION

The primary preparation step for this module is to identify the range of data visualization tools and platforms that are accessible to users at your institution, including those for which enterprise software licenses are available, as well as open-source tools supported by the library or other units on campus. Develop at least a cursory working knowledge of the features, capabilities, and learning curves for each of those tools.

The following is an example list of veggies (data viz tools) you might choose from to create your own pizza:
- D3.js
- Datawrapper
- Google Charts
- MATLAB
- Microsoft Excel
- Plotly Chart Studio
- Python + Matplotlib or Bokeh
- R + ggplot2
- RAWGraphs
- SAS + JMP Pro
- SPSS
- Tableau (Desktop or Public)

INSTRUCTIONS

This recipe consists of six steps or questions to walk participants through in order to identify which data visualization tools may be the

best option for them to invest time in learning to use.

1. What *format* is my data in?
2. Do I need to *explore* my data or *explain* patterns I've already found?
3. Are there particular *chart types* I want to create?
4. What tools do I have *access* to or will continue to have access to?
5. Do I just need to create a *few simple charts* quickly, or to create *many complex charts* that share a visual theme?
6. How will people *interact* with my chart?

Question 1 introduces the concept of *tidy data*, or data stored in "long" versus "wide" format. Different tools require data to be organized in different ways, but tidy data—a concept popularized by Hadley Wickham—is a format that works well for most visualization tools. Other researchers, including Robert Kosara, have framed the concept of data organization for visualization using various terminology, such as "spreadsheet thinking" versus "database thinking." Briefly discuss the concept of data formatting and either describe or demonstrate the process of pivoting data from wide to long format. If users frequently need to pivot their data before visualizing it, consider selecting a tool with that feature built in, such as Tableau or Datawrapper. Alternatively, you might recommend OpenRefine as a powerful tool for data transformation!

Question 2 reinforces the idea that exploratory visualizations require a different creation process than explanatory visualizations. For example, if users need to quickly reproduce a lot of simple exploratory charts, a programming solution such as R or Python is ideal. If users need to create charts that highlight particular patterns or data values and want to add annotations or other graphical elements to their charts, Plotly, Datawrapper, and Tableau Public are good options in addition to R or Python.

Question 3 emphasizes the need to select appropriate chart types for the kinds of variables you want to visualize and checking to see what chart options are available in different tools or platforms can help narrow down your choice. Resources like the Data Visualisation Catalogue and From Data to Viz (see the Additional Resources section) provide typologies and flowcharts for selecting appropriate chart types and can be excellent examples to share during this discussion.

Question 4 encourages participants to weigh the pros and cons of proprietary versus open-source software when deciding which tool to invest their time in learning. Briefly discuss the importance of considering what options will be available to them in the long term: will particular licensed software be available should they leave their current institution? Does there appear to be an active community of support for a new open-source tool?

Question 5 prompts students to consider using advanced statistical software (Stata, SAS, MATLAB) or code-based solutions (R or Python) if they need to create complex charts or a series of visually consistent graphs, as opposed to browser-based tools such as Datawrapper or RAWGraphs, which enable you to create simple charts quickly and easily.

Question 6 helps participants think through the publication and sharing process they intend to use: will readers encounter their charts in static form, either in print or on a screen, or do they want to create interactive charts to be viewed either online or offline? Creating static charts is a best use case for some tools such as Microsoft Excel and certain R or Python libraries, while tools such as Plotly, Datawrapper, Google Charts, Tableau Public, and many JavaScript libraries (such as D3.js) are designed to create interactive charts.

CLEANUP

It can be helpful to wrap up a module like this by sharing locally created library resources such as written tutorials or LibGuides for students to reference after the workshop. Also consider sharing a handout or quick reference sheet that lists all the tools and platforms discussed, with URLs to access each tool and its relevant help pages or community forums.

ADDITIONAL RESOURCES

Bokeh (Python package for interactive visualizations) Documentation, https://docs.bokeh.org/en/latest/.

Data Visualisation Catalogue, https://datavizcatalogue.com.

Datawrapper, https://www.datawrapper.de.

Section 4. Data Visualization

D3: Data-Driven Documents, https://d3js.org.

From Data to Viz, https://www.data-to-viz.com.

ggplot2 (R package for visualizations), https://ggplot2.tidyverse.org.

Google Charts, https://developers.google.com/chart/.

Kosara, Robert. "Spreadsheet Thinking vs. Database Thinking." *EagerEyes* (blog), April 24, 2016. https://eagereyes.org/basics/spreadsheet-thinking-vs-database-thinking.

Matplotlib (Python package for static, animated, and interactive visualizations), https://matplotlib.org.

OpenRefine, https://openrefine.org.

Plotly Chart Studio, https://chart-studio.plotly.com/create.

RAWGraphs, https://rawgraphs.io.

Tableau Public, https://public.tableau.com/en-us/s/.

Wickham, Hadley. "Tidy Data." *Journal of Statistical Software* 59, no. 10 (2014): 1–23, https://doi.org/10.18637/jss.v059.i10.

Four-Cheese Pizza
Color and Accessible Design

Rachel Starry, Digital Scholarship Librarian, UC Riverside Library, rachel.starry@ucr.edu

NUTRITION INFORMATION

This recipe describes a module that can be used on its own or incorporated into the flexible approach to developing an Introduction to Data Visualization workshop described in the chapter titled "Build Your Own Data Viz Pizza." The four cheeses that top this pizza reflect four simple design principles for creating accessible data visualizations, and learners are also introduced to current best practices for using color in charts.

AUDIENCE AND NUMBER SERVED

The primary audience for this lesson is upper-division undergraduates and graduate students who need to develop basic data visualization skills and awareness of available tools and best practices.

1 classroom of 15–20 students at beginner level (lesson geared toward upper-division undergraduates and graduate students)

LEARNING OUTCOMES

Students will
- evaluate the colors used in charts in terms of legibility to readers with color vision deficiencies.
- distinguish between color palettes and color scales for use with categorical and sequential data.
- identify simple design principles for creating visualizations that are accessible to people with varied vision abilities and disabilities.

COOKING TIME

30 minutes—preparation time varies

DIETARY GUIDELINES

Accessibility is a key concern related to the perception of information. The learning outcomes for this module primarily tie into ACRL's *Framework for Information Literacy for Higher Education* through the frame Scholarship as Conversation, as we encourage students to think about the ways they structure even simple charts to communicate information to readers with varied vision abilities and disabilities. Being thoughtful in our selection and arrangement of chart components can increase the readability of our data visualizations for all audiences. Just as physical curb cuts (ramps built into sidewalks) ease the passage of wheelchair users, they also increase usability for people with other mobility disabilities as well as bicyclists, people with child strollers, and others. Applying principles of universal design to the creation of data visualizations enables us to convey information more effectively to all readers.[1]

INGREDIENTS

- 1 instructor
- 1 slide deck
- Computer access for all students

PREPARATION

This recipe includes two main steps: the first step is introducing students to the process of selecting appropriate colors for charts by outlining the differences between color palettes and color scales. For this step, prepare or select a variety of example charts that use different kinds of color palettes and scales, as well as visuals to help explain a bit about the color theory. Our choice of hue, saturation, and brightness values make up a color. Websites such as Visualizing.org or the US Census Bureau's "Data Visualization Gallery" offer many examples. An excellent explanation of color palettes and scales is provided by Alan Wilson in "The Power of the Palette" (see Additional Resources).

The second main step in this recipe is sharing four principles that can guide students in making accessible design choices when creating data visualizations: simplicity, context, legibility, and contrast. For this step, select an example chart to demonstrate each design principle in action. Recommended resources for finding good example visualizations are included in Additional Resources.

Section 4. Data Visualization

INSTRUCTIONS

Part 1: Color in Data Visualization

Start by introducing the key concept that color scales are used for sequential or continuous data, while color palettes are appropriate only for categorical or discrete data. You may want to illustrate common chart types used to plot sequential versus categorical variables, using examples from the Data Visualisation Catalogue or From Data to Viz. For instance, show color scales in action in a choropleth map, heat map, or cartogram, and discuss the value of selecting palettes that maximize the distance on the color wheel between different hue and brightness values.

Lisa Charlotte Rost discusses best practices for creating accessible color palettes and scales, including selecting hues and brightness values that are visually distinguishable by people with color vision deficiencies. These have high contrast and follow established conventions (e.g., darker colors for high data values, lighter colors for low data values).[2] The National Eye Institute (branch of NIH) provides information on the known types of color blindness, which primarily informs the color gradients we should avoid. Viz Palette, created by Elijah Meeks and Susie Lu, is a very useful tool to demonstrate the process of selecting and testing color palettes and scales for accessibility. Take a few minutes to demonstrate the Viz Palette tool interface, then walk through the following list of things to consider when choosing visually accessible colors for charts:

- Do you need a color scale or a color palette?
- How many different colors are necessary? (ideally, < 7)
- Can you consistently use the same colors for the same variables across multiple charts?
- How might you use the color gray for data that are not central to the pattern you wish to highlight?
- Are you following established standards, such as dark colors for high values and light colors for low values?

Part 2: Design Principles for Accessible Visualizations

Top off this cheesy pizza by walking students through the following principles to follow when designing data visualizations to be understandable by all readers. For each principle, include a slide with your selected example chart that illustrates the principle in action. You might additionally contrast that with an example of what not to do.

1. *Simplicity:* Displaying fewer marks is generally best. Common chart types are more easily understood, but don't shy away from less common charts that convey information well!
2. *Context:* Use clear descriptive text in titles and captions, and always summarize your chart using image alt text. Whitney Quesenbery provides suggestions in the slide deck "Writing Great Alt Text": If the image contains text, repeat the words. (Use the chart caption or title.) If it contains visual information, explain it. (What is the key takeaway?) If it contains sensory information, describe it. (What patterns in the data are shown?)
3. *Legibility:* For charts in English, horizontal text is more readable than vertical or slanted text, and don't rely on legends—directly label your data whenever possible.
4. *Contrast:* Use colors (for chart elements and backgrounds) that differ from each other on the light-to-dark spectrum, and avoid using color alone to distinguish between values—also use shapes, fill patterns, size, or labels.

CHEF'S NOTES

Data visualization is a broad field of research, and the current state of knowledge changes constantly. The suggestions in this recipe represent one perspective, based on current understandings of accessibility. Please use this as a starting point to keep learning, and adapt your own recipes as best practices evolve. See the CFPB "Data Visualization Guidelines" and W3C Web Accessibility Initiative for additional guidance.

ADDITIONAL RESOURCES

CFPB Design System. "Data Visualization Guidelines." Consumer Financial Protection Bureau. Last updated May 24, 2022. https://cfpb.github.io/design-system/guidelines/data-visualization-guidelines.

Data Visualisation Catalogue, https://datavizcatalogue.com.

From Data to Viz, https://www.data-to-viz.com.

National Eye Institute. "Types of Color Blindness." June 26, 2019. https://www.nei.nih.

gov/learn-about-eye-health/eye-conditions-and-diseases/color-blindness/types-color-blindness.

Quesenbery, Whitney. "Writing Great Alt Text." Presentation slides. September 10, 2014. https://www.slideshare.net/whitneyq/writing-great-alt-text-38937551.

Rost, Lisa Charlotte. "What to Consider When Choosing Colors for Data Visualization." *Datawrapper Blog*, May 29, 2018. https://blog.datawrapper.de/colors/.

Section508.gov. "Universal Design and Accessibility." US General Services Administration. October 2020. https://section508.gov/create/universal-design.

US Census Bureau. "Data Visualization Gallery." https://www.census.gov/dataviz/.

Visualizing.org, https://www.visualizing.org.

Viz Palette, https://projects.susielu.com/viz-palette.

White, Kevin, Shadi Abou-Zahra, and Shawn Lawton Henry, eds. "Designing for Web Accessibility." W3C Web Accessibility Initiative. Last updated January 9, 2019. https://www.w3.org/WAI/tips/designing/.

Wilson, Alan. "The Power of the Palette: Why Color Is Key in Data Visualization and How to Use It." *Adobe Blog*, February 27, 2017. https://blog.adobe.com/en/2017/02/27/the-power-of-the-palette-why-color-is-key-in-data-visualization-and-how-to-use-it.html.

NOTES

1. Section508.gov, "Universal Design and Accessibility," US General Services Administration, October 2020, https://section508.gov/create/universal-design.
2. Lisa Charlotte Rost, "What to Consider When Choosing Colors for Data Visualization," *Datawrapper Blog*, May 29, 2018, https://blog.datawrapper.de/colors/.

Data Visualization using Web Apps in a Rainbow Layer Cake

Yun Dai, Data Services Librarian, Library, New York University Shanghai, yun.dai@nyu.edu, http://shanghai.hosting.nyu.edu/data/;
Fan Luo, Digital Scholarship Technologist, Library, New York University Shanghai, fan.luo@nyu.edu

NUTRITION INFORMATION

This recipe illustrates how to make a project-based technical workshop on data visualization using web apps that address the real-world complexities of reporting with data. In this workshop, attendees work on mini-projects that approach a multilayered problem from different directions. Attendees preprocess data for visualization, utilize Shiny (an R package for building interactive web applications) facilities to build up the "layers" of the web application, draw plots to be the "fillings," and assemble Shiny components and plots to form a layer cake!

The workshop integrates thinking skills and doing skills of visual data literacy. Thinking skills are the abilities to critically evaluate decisions made at each step in building these data visualizations. Doing skills are the abilities to employ a toolbox of visual vocabularies, data analysis tools, and web technologies to present the interactions of relationships in the data set.

This is one of the three recipes on weaving data literacy competencies into a project-based technical workshop. See also "Text Mining Charcuterie Board" in the Data Manipulation and Transformation section and "Stuffed Shiny App with Business Intelligence" in the Data in the Disciplines section.

TARGET AUDIENCE AND NUMBER SERVED

Serves a small group of 5–10 attendees. Recommended to those working on a data-driven project who want to create graphics using Shiny (or other tools) to present project findings by visual means.

LEARNING OBJECTIVES

Attendees of the workshop should be able to
- Recognize the complexities in presenting visual information using data. Specifically, recognize that how data are collected, structured, and stored may affect how data can be visualized; recognize that different plot or chart types characterize different data relationships and therefore may require different data types and structures.
- Identify the key concepts, techniques, and procedures in the creation of the visual components.
- Develop design strategies to produce graphical representations of data.
- Apply basic web design principles and use technologies to build an interactive web application that organizes visual information, conveys messages, and facilitates user exploration.
- Evaluate the visual representations of data for their accuracy, reliability, and effectiveness.
- Communicate arguments and narratives effectively within the end product for the intended audience in relation to the project goals.

COOKING TIME

2 hours

DIETARY GUIDELINES

Thinking skills align with the frame Research as Inquiry in ACRL's *Framework for Information Literacy for Higher Education*.

This recipe uses the following performance indicators in the *ACRL Visual Literacy Competency Standards for Higher Education*: Standard 1, Performance Indicator 1, on defining and articulating the need for an image; Standard 6, Performance Indicator 1, on producing visual materials for projects and scholarly uses; Standard 6, Performance Indicator 2, on using design strategies and creativity in image and visual media production; Standard 6, Performance Indicator 3, on using various tools and technologies to produce images and visual media; and Standard 6, Performance Indicator 4, on evaluating personally created visual products.

It also uses the following performance indicators in ACRL's *Information Literacy Competency Standards for Higher Education*: Standard 3, Performance Indicator 3, on synthesizing main ideas to construct new concepts; Standard 4, Outcome 1.a, on organizing the content to support the purposes and format of the product; Standard 4, Outcome 1.d, on manipulating data to transfer them from their original locations and formats to a new context; and Standard 4, Performance Indicator 3, on communicating the product effectively to others, considering the communication medium, information technology, principles of design and communication, and style.

INGREDIENTS

This workshop grew out of a real-life project that used a Shiny app to build graphical representations of data to report findings from a survey. Hence, most ingredients are already in storage but need processing for use in the workshop. These ingredients include

- A slice of the raw survey data with layers (e.g., website or workshop attendance) and categories (e.g., major) that could become parameters in a dashboard
- Slices of clean data sets for exercises
- R script templates for preprocessing data and graphing
- Shiny app script template
- Slides that outline the workshop with a table describing the sample data and workflow of creating a Shiny app
- One instructor and one helper (recommended to have in an online session and optional in person)

Before the workshop, send e-mails to remind registered attendees to bring cooking utensils (a laptop with RStudio installed) and the packaged ingredients (class materials in a shared folder) to the workshop.

PREPARATION

1. *Chop the raw data and preserve a lean version of it.* Keep the smallest portion of the raw data set that is able to exemplify the variety of survey question types and structures of data stored in columns. Remove personally identifiable information. Make a copy of this data set to process further in the following step.
2. *Sift through the copy of the raw data.* Filter duplicates and rows with missing data. Recode variables where needed.
3. *Carve the clean data into slices.* Each slice consists of data collected for a group of survey questions that can be molded into a type of plot.
4. *Prepare sample layers of the cake in a script.* These are higher-level structures of a Shiny app. The code blocks lay out the UI, add widgets to prompt users for input values or interactions, and render outputs.
5. *Prepare sample fillings in another script.* These are R commands to preprocess data and draw plots, which can be adapted by the audience to use in their mini-projects during the workshop (see figure 1).

INSTRUCTIONS

Steps 1–5 are a presentation or lecture. Here is where the instructor works through the key steps of transforming the data so that they are ready for processing.

1. Situate the workshop in a scenario where a project team is convened to report findings from a survey. Set the project goals. Explain how to use the class materials to accomplish the goals.
2. Lead the audience to read the raw survey data. How were the data collected? For what purposes? In what format is the data set stored? How are its rows and columns structured?
3. Inspire the audience to conceive an end product that tells the story of the data. What steps might be in between the end product and the data at hand?

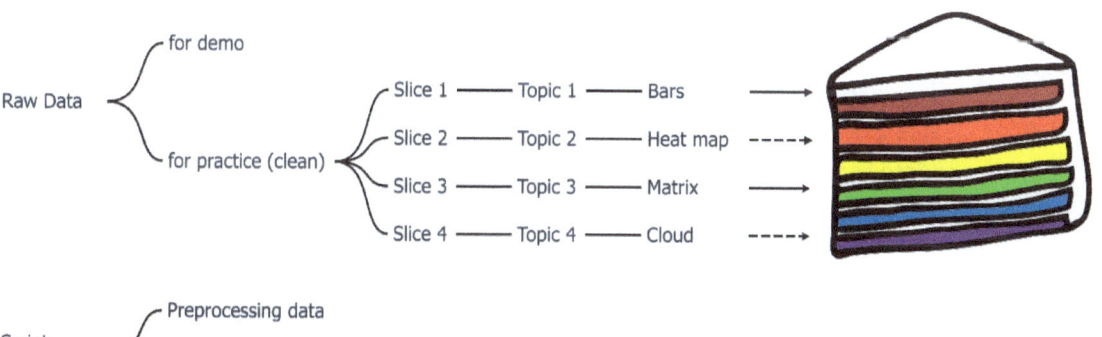

FIGURE 1
Workflow of workshop preparation, created by Fan Luo and Yun Dai under CC BY-NC 4.0.

4. Introduce the workflow of producing a visual product on web interfaces. Demonstrate the workflow with a slice of clean data set and sample scripts. Review the critical steps of cleaning data for visualization, creating plots, and establishing the UI and server sides of the web application.
5. Share and analyze design techniques that enhance user interactivity and the efficiency of organizing information in a web application.
6. Guide the audience to pick a clean data set from the shared folder to probe. What data relationships (e.g., distribution, correlation, ranking) exist in this data set? Which plot types are used to display the relationships? Do the current arrangements of columns and rows support the required data structure of the plot? For instance, before graphing, do we need to aggregate data, reshape data from wide to long, transform texts into summary tables, or combine information from several columns?
7. Then, ask members of the audience to bake one's own piece of layer cake! Find script templates in the shared folder to restructure data, draw plots, and construct graphics using the Shiny app. Accommodate those scripts to one's own mini-project. Leave time for testing and troubleshooting.
8. In groups, ask the audience to reflect on the technical and aesthetic details of the finished product.

REVIEWS/ASSESSMENT STRATEGY

At the beginning of the workshop, conduct a poll to estimate the technical proficiency of the audience and what they hope to learn from the workshop. During the workshop, assess learning effectiveness by how attendees solve problems in their mini-projects. At the end of the session, evaluate teaching effectiveness with a feedback form. Questions could be "Did you find the workshop useful? Were you able to follow the workshop? What are your favorite and least favorite parts?"

ALLERGY WARNING

A project-based workshop covers many facets of data literacy, so attendees have lots to digest. This type of workshop may also have more technical barriers than an introductory or topical session. Instructors should share the class materials beforehand, both for attendees to preview and to manage their expectations.

A project-based workshop may also attract an audience of various tastes, and hence it is important to balance the flavors. For instance, in this case, a data science student may be attracted by the visual and web design ingredients of our cake; an interactive media arts student may expect to hear more of the data parts.

CHEF'S NOTES

This project-based workshop could be a dessert served after a main dish for those who have a good appetite. The main dish can be a visual literacy session that is a combo of several of the following topics: why visualize data; what is a good or bad visualization; perception in the design of a visual media; strategies of organizing visual elements; visual variables in a plot; plot types and their functions; overview of data visualization on web interfaces; and misleading graphs.

Due to the time limit, attendees of the workshop will be able to dig into only one aspect of a project that is one piece of a large cake. Attendees may see the full picture in the web posts listed in Additional Resources ("The Dirty Work—Reshaping Data for Visualization," "The Jigsaw Puzzle Pieces—Creating Graphs with ggplot2," and "Assembling the Pieces—Creating R Shiny App") where we documented the creation of the cake as a whole. Attendees may also follow those tutorials to explore other parts of the project using the class materials.

The cooking utensils to make the layers and fillings of the cake can be replaced with other tools that specialize in data wrangling and visualization and that offer a framework for building web applications (e.g., Tableau, Power BI).

ADDITIONAL RESOURCES

Dai, Yun. "Assembling the Pieces—Creating R Shiny App." October 2018. http://shanghai.hosting.nyu.edu/data/r/case-1-3-shinyapp.html.

Dai, Yun. "The Dirty Work—Reshaping Data for Visualization." September 2018. https://shanghai.hosting.nyu.edu/data/r/case-1-1-preprocessing.html.

Dai, Yun. "The Jigsaw Puzzle Pieces—Creating Graphs with ggplot2." September 2018, https://shanghai.hosting.nyu.edu/data/r/case-1-2-ggplot2.html.

Graphical Abstracts
Creating Appetizing Infographics for Your Research Article
Aleshia Huber, Engineering Librarian, Binghamton University, hubera@binghamton.edu

NUTRITION INFORMATION
The graphical abstract is a simple infographic designed to give readers an immediate visual impression of the main findings of a research paper. It aims to make the research more accessible for readers. Thus, it is important to consider design aspects that go into creating a clear and concise graphical abstract. Some journals require a graphical abstract for article publication, while others make it an option. This session is best for attendees who study a particular discipline and come prepared with an abstract, preferably of their own research, but they may also practice creating a graphical abstract with a published paper of their choosing.

If you wish to learn more about graphical abstracts and their design, under Additional Resources is a list of references that provide guidance on creating clear graphical abstracts as well as a Tumblr blog that frequently posts poorly designed (and funny) graphical abstracts.

TARGET AUDIENCE AND NUMBER SERVED
Graduate students and faculty, 20–25 maximum

LEARNING OUTCOMES
By the end of this session, attendees will

- describe the impact of presenting research visually
- apply the best practices in presenting abstracts in a visual format
- construct a clear and concise graphical abstract

COOKING TIME
90 minutes. Prep time will vary depending on the presenter's comfort level with graphical abstracts and editing software.

DIETARY GUIDELINES
Creating a well-designed graphical abstract involves the ability to convey and share research findings in a concise image, which is a different format from what most researchers may be familiar with. This session will help attendees design graphical abstracts that are accessible to readers and successfully summarize their research findings.

INGREDIENTS
- Projector, laptop or computer, and screen for instructor.
- Laptops or computers for attendees.
- Paper and pens or pencils so attendees can sketch ideas for their graphical abstract on paper first. Markers, colored pencils, etc. provide more brainstorming flexibility.
- Example graphical abstracts from published scholarly articles.

PREPARATION
Survey attendees ahead of time in order to gauge their research areas and publication interests. This will allow you to prepare example graphical abstracts from relevant journals or publishers.

Notify attendees to bring a research paper abstract for which they wish to create a graphical abstract. Preferably this is for a research paper that they have authored, but if they do not have one prepared, they may bring a research article whose main findings they are comfortable with discussing to practice.

INSTRUCTIONS
1. Start with an introduction to what a graphical abstract is and its purpose in academic literature.
2. Discuss design elements that make an effective graphical abstract.
3. Provide some examples of graphical abstracts. Include the title of the article next to the image. These should be tailored to your audience so they may be better able to discuss the examples. Three example abstracts is the recommended serving

size, with one underbaked (not enough information), one overbaked (cluttered or hard to read), and one baked to perfection (ideal). Ask attendees what information they are able to glean from it and what their opinion is on its design and effectiveness. Figure 1 is an example of an ideal graphical abstract. Note the well-designed components: it is simple, is easy to read, and has a balance of image and description that make it easy to discern what the article is about. The arrows provide a flow to designate how the elements are all related. You may compare this image with its textual counterpart, which is listed in the figure caption.

4. Show how to check for publisher requirements for graphical abstracts. These can be found on the publisher's or journal's website.
5. Provide information on the copyright issues related to incorporating and using existing images into graphical abstracts. Discuss the ability to use public domain and Creative Commons images, demonstrate a search for finding them, and list additional resources for locating them.
6. End the presentation with an overview of tools available to create graphical abstracts. You can showcase a tool with which they are most familiar.
7. Distribute paper and writing tools. Direct attendees to start a sketch of a graphical abstract for their prepared abstract and apply what they have learned. Encourage attendees to work in groups so they can brainstorm and discuss their designs with others. Attendees who finish their paper sketch may wish to move on to creating their graphical abstract electronically if time remains.

REVIEWS/ASSESSMENT STRATEGY

Follow-up survey via paper or online, whichever is most convenient. Questions:
1. What did you find the most helpful about this session?
2. What did you find the least helpful about this session?
3. What do you wish this session covered?

CHEF'S NOTES

For tools, I showcase PowerPoint as it is one I have the most experience using to create infographics. Generally, attendees are also aware of at least the most basic graphic design capabilities (insert shapes, images, etc.) of PowerPoint, and I can help them with more advanced features. If they wish to try another tool, I encourage them to look for YouTube tutorials to learn the basics. Many YouTube

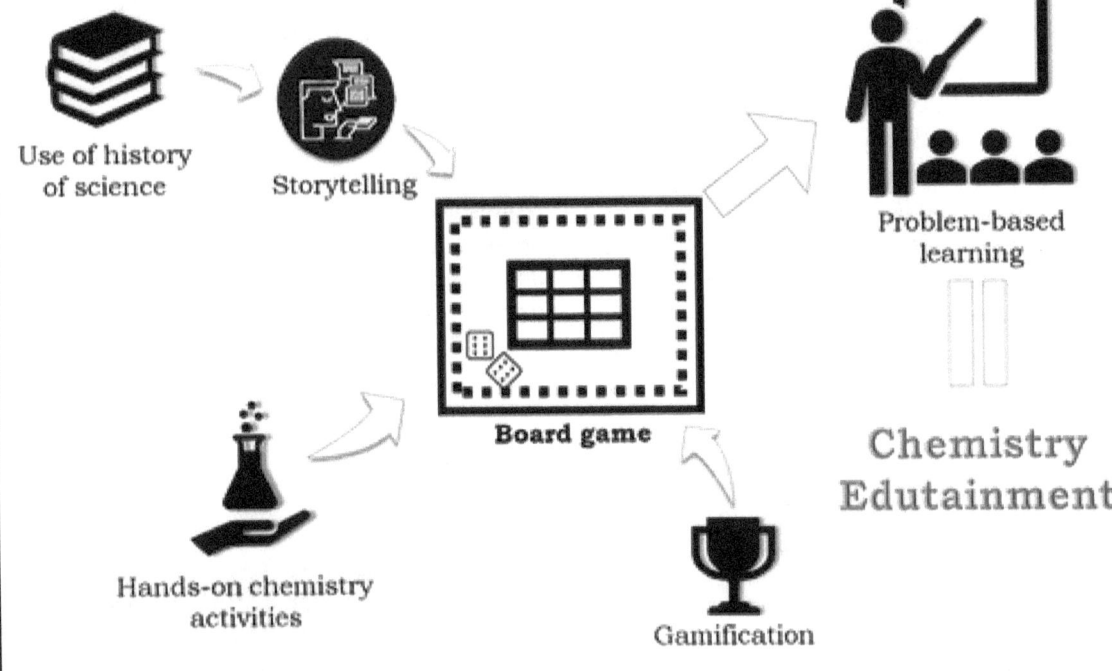

FIGURE 1
Example of an ideal graphical abstract. Reprinted with permission from Dias, Dora, José Ferraz-Caetano, and João Paiva. "'Ethics against Chemistry': Solving a Crime Using Chemistry Concepts and Storytelling in a History of Science–Based Interactive Game for Middle School Students." *Journal of Chemical Education* 98, no. 5 (2021): 1681–90. https://doi.org/10.1021/acs.jchemed.0c01469. Copyright 2021 American Chemical Society.

tutorials exist for creating graphical abstracts, infographics, and more.

ADAPTING THE RECIPE

I have written the recipe without a specific discipline in mind, but given that example graphical abstracts are discussed, it is much smoother when attendees are from the same or very similar disciplines. When attendees are creating their graphical abstracts, they can ask their peers for feedback and ideas.

The session may be shortened to an hour if you follow steps 1–4 and skip to 7. You may wish to provide a handout with a list of free image search engines and graphic design tools for attendees to reference on their own time (see the article by Tea Romih in Additional Resources for possible graphic design tools). You may also offer a follow-up session that goes into more depth on the electronic creation of the graphic abstracts, following steps 5-6 and allowing more time for attendees to create their electronic version.

ALLERGY WARNING

In the example abstracts section, if attendees cannot understand the abstract, prompt them to discuss basic design elements, such as text readability, amount of text, color scheme, amount of positive or negative space, and so on. This will help them make a better connection to the design elements when they go to create a graphical abstract on their own.

If an attendees failed to bring a prepared text abstract, ask them to partner with someone.

ADDITIONAL RESOURCES

ACS Publications. "Guidelines for Table of Contents/Abstract Graphics." Last updated July 30, 2020. http://pubsapp.acs.org/paragonplus/submission/toc_abstract_graphics_guidelines.pdf.

Cheng, Karen, Yeechi Chen, Kevin Larson, and Marco Rolandi. "Proving the Value of Visual Design in Scientific Communication." *Information Design Journal* 23, no. 1 (2017): 80–95. https://doi.org/10.1075/idj.23.1.09che.

Elsevier. "Graphical Abstracts." Accessed March 25, 2021. https://www.elsevier.com/authors/tools-and-resources/graphical-abstract.

Romih, Tea. "Which Software Can I Use to Create Graphical Abstracts?" *Seyens*, July 29, 2016. https://www.seyens.com/graphical-abstract-software/.

"TOC ROFL." Tumblr. https://tocrofl.tumblr.com.

Wong, Bang. "Points of View: The Overview Figure." *Nature Methods* 8, no. 5 (2011): 365. https://doi.org/10.1038/nmeth0511-365.

Section 5.
Data Management and Sharing

123 **[[Ch32]]Making File Names for Digital Exhibits**
 Kate Thornhill and Gabriele Hayden

126 **[[Ch33]]Data Management Failures: Teaching the Importance of DMPs through Cautionary Examples**
 Richard M. Mikulski

131 **[[Ch34]]Low-Fat Research Data Management**
 Elizabeth Blackwood

134 **[[Ch35]]Managing Qualitative Social Science Data: An Open, Self-Guided Course**
 Sebastian Karcher and Diana Kapiszewski

136 **[[Ch36]]Seven Weeks, Seven DMPs: Iterative Learning around Data Management Plan Creation**
 Emma Slayton and Hannah C. Gunderman

140 **[[Ch37]]Equitable from the Beginning: Incorporating Critical Data Perspectives into Your Research Design**
 Jodi Coalter, David Durden, and Leigh Amadi Dunewood

Making File Names for Digital Exhibits

Kate Thornhill, MLS, Digital Scholarship Librarian, University of Oregon, kmthorn@uoregon.edu; Gabriele Hayden, PhD, Research Data Management and Reproducibility Librarian, University of Oregon, ghayden@uoregon.edu

NUTRITION INFORMATION

Creating a digital exhibit, like cooking a meal and serving it to your guests, is most successful with some advance planning. There are many elements to consider, including the themes you will research and write about, the historical materials you will showcase as evidence to support your writing, the digital exhibit technology you will use to host user experiences, and the digital project plan you will use to pull everything together. Preparing ingredients is a must when cooking, and the same applies to building a digital exhibit using digitized historical resources.

Would you add whole uncut carrots or celery to a soup, or would you slice them to a predetermined size and sauté them first? Like carrots or celery, digital files representing historical materials need to be prepared with standardized care. Part of that process is developing a digital file naming standard. Messy digital file names can lead to inconsistencies in digital exhibit workflow management or potential confusion about where to place files within a digital exhibit project. This recipe focuses on teaching how to design clear, consistent file names so that participants are prepared to contribute to the formal build of a digital exhibit project.

TARGET AUDIENCE AND NUMBER SERVED

This recipe works best for up to 15 participants. It is for anyone learning how to build a digital exhibit using historical digital resources, or anyone interested in file naming best practices. It targets online learners, but it could also be taught in person.

LEARNING OUTCOMES

- Create a file naming convention that is human- and machine-readable.
- Create a file naming standard that can be interpreted and represented within a digital exhibit development workflow.
- Communicate how a file naming convention supports developing a digital exhibit.

COOKING TIME

30 minutes to review reading, and 60 minutes for live instruction

DIETARY GUIDELINES

The frame Information Creation as a Process in ACRL's *Framework for Information Literacy for Higher Education* aligns with this lesson plan. Participants experience the ways that data (digital images) description and organization can influence the process of information creation. More broadly, proper file naming conventions are essential to managing research data across disciplines. The guidelines suggested in this recipe, though specific to digital exhibits, are generalizable to any research project that generates many data files, from a small undergraduate project to a large grant-funded team effort.

INGREDIENTS

- Handout, "How to Develop a File Naming Convention for a Digital Exhibit," available as a PDF or Word document at https://doi.org/10.7264/ewzj-sh59.
- 1 lead instructor and 1 assistant to manage breakout rooms and other technical issues.
- Collaborative online worksheet structured for the learning activity, "Creating Digital File Name Conventions." Can be created using the document template available as a PDF or Word document at https://doi.org/10.7264/ewzj-sh59.
- 3 different types of digital images collections from either the Library of Congress Digital Collections, the New York Public Library, the Digital Public Library of America, Digital Commonwealth, your own library's digital collections, or another digital library. We suggest the following digital collections:
  - "Activism of the 1980s Photograph Collection, 1985–1987." Digital Com-

monwealth Search Results. Special Collections and University Archives, W. E. B. Du Bois Library, University of Massachusetts Amherst. Accessed March 29, 2021. https://www.digitalcommonwealth.org/search?f%5Bcollection_name_ssim%5D%5B%5D=Activism+of+the+1980s+Photograph+Collection%2C+1985-1987&f%5Binstitution_name_ssim%5D%5B%5D=University+of+Massachusetts+Amherst+Libraries+Special+Collections+and+University+Archives.
- "March on Washington Resources." Dig DC. Special Collections, DC Public Library. Accessed March 29, 2021. https://digdc.dclibrary.org/islandora/object/dcplislandora%3A263420.
- "National American Woman Suffrage Association Records." NYPL Digital Collections. MssCol 2097. Manuscripts and Archives Division, New York Public Library. Accessed March 26, 2021. https://digitalcollections.nypl.org/collections/national-american-woman-suffrage-association-records.
- Zoom or another web conferencing tool that allows breakout rooms

PREPARATION
- Set up web conferencing tool and make sure all learners have the link.
- Share the reading "How to Develop a File Naming Convention for a Digital Exhibit" (https://doi.org/10.7264/ewzj-sh59) and ask learners to complete it before class.
- Ask learners to be prepared to discuss the reading and engage in an activity based on it.
- Make a copy of the "Creating Digital File Name Conventions" worksheet (https://doi.org/10.7264/ewzj-sh59) and make sure everyone has access and edit permissions to it via a shared drive.

INSTRUCTIONS
Instructor and class introductions/welcome.

Activity 1—Pre-assessment (15 minutes)
3, 2, 1—Individual Activity: Ask individual learners to take 5 minutes to write down their responses to the questions about the assigned reading and share them in the chat. Ask one question at a time to give learners pace and focus. Pause to clarify concepts as they come up for students.
Questions for learners:
- 3 things you found interesting about the resource you reviewed before class?
- 2 things you learned about the resource you reviewed before class?
- 1 thing you still have a question about the resource you reviewed before class?

Activity 2—Groups Work (20 minutes)
- Share collaborative worksheet document that has been structured for each breakout room.
- Explain the activity to the learners.
- Learners have 15 minutes to complete the activity in breakout rooms.

Activity 3—Facilitated/Group Discussion (25 minutes)
- Learners report back on their file name convention decision-making. Each group reports on its decision-making.
- *Assessment:* Ask students to respond in the chat or using the worksheet to this question: How would you apply what you learned about file name standardization if you had to collect digital resources for a digital exhibit?

REVIEWS/ASSESSMENT STRATEGY
Formative assessment will be determined through class discussion. Summative assessment will be determined by the documented file name standards each group makes. The following should be considered when determining learning success:
1. Student creates with their peers a file naming convention that attempts to be human- and machine-readable:
   » File names are all lowercase and sections separated by underscores.
   » Naming convention includes a file name limit.
   » Students indicate if the standard is institutional or project-based.
2. Students articulate how file names support workflow:
   » File names will be sortable.
   » File names will appear by default in a sequential, meaningful order.
   » Student documents decision-making in collaborative document.
3. Student engages in conversation about why they have chosen their file naming convention:

- » Student articulates who will use these file names and how the file names can be meaningful to different users at different points in the file's life cycle.
- » Student identifies and justifies any trade-offs.

ALLERGY WARNING

Be prepared for learners to come without having done the reading. During the first activity, reinforce file naming best practices from the reading.

ADDITIONAL RESOURCES

"How to Develop a File Naming Convention for a Digital Exhibit" (assigned reading, activity 1)

"Creating Digital File Name Conventions" (worksheet for group work, activity 2)

Both available at Hayden, Gabriele, and Kate Thornhill. "Making File Names for Digital Exhibits [Data]." Scholars' Bank, University of Oregon Libraries. 2021. https://doi.org/10.7264/ewzj-sh59.

Data Management Failures
Teaching the Importance of DMPs through Cautionary Examples
Richard M. Mikulski, Instruction & Research Librarian, The College of William & Mary, rmmikulski@wm.edu

NUTRITION INFORMATION
Researchers frequently express frustration when confronted with data management plan (DMP) requirements, particularly when drafting or completing a grant application. This sense of annoyance is further fueled by a too-common view that the DMP is yet another hurdle that researchers need to confront during the grant writing process. Once researchers and students understand the purpose and utility of DMPs, however, many of these reservations and frustrations subside.

The purpose of this exercise is to demonstrate the importance of DMPs by giving examples of large research projects that suffer due to poor data management planning. Each case provides examples of otherwise well-organized projects in which users suffered because long-term data use and access were not considered. In each case, a DMP would have identified and remedied the issues. While these examples do not address all the ways a project may suffer from insufficient data planning, they illustrate how the lack of a DMP can damage even a well-designed and well-executed research project.

Each of the cases include a brief overview of the project, a summary of its data failings, lessons learned from the oversights, and follow-up questions. As a visual aid, a supplemental presentation slide deck has been created to accompany this exercise, which can be downloaded and amended (see Additional Resources).

TARGET AUDIENCE AND NUMBER SERVED
This exercise is for faculty, grant writers, researchers, and advanced students. There is no cap on the audience size.

LEARNING OUTCOMES
Students will learn the importance of ensuring long-term data access through data management planning. Students will consider and discuss significant data management questions, such as those relating to data collection methodology, software use, file type selection, and storage medium. Finally, students will identify how these specific projects suffered due to poor data management planning, and they will recognize why DMPs are important and frequently required for research projects.

COOKING TIME
Preparation requires 15–20 minutes for instructors to prepare notes and amend slides to meet the interests and needs of their audience.

DIETARY GUIDELINES
In addition to answering the question "Why are DMPs necessary?" this recipe reinforces ACRL's *Framework for Information Literacy for Higher Education*. The cautionary examples show that Authority Is Constructed and Contextual by demonstrating that disciplines have unique data standards and requirements that must be identified. The examples also illustrate Information Creation as a Process by showing the need to address data creation, revision, and reuse policies early in projects. The failure of these examples to anticipate the value of collected data beyond the immediate project demonstrates that Information Has Value. The need to anticipate the development of new questions and needs throughout the data life cycle, which these examples failed to consider, shows Research as Inquiry and Searching as Strategic Exploration. Finally, the need to ensure data are accessible for future researchers reinforces Scholarship as Conversation.

INGREDIENTS
Live internet connection; freely downloadable presentation slides.

PREPARATION
This demonstration, which typically takes

~20 minutes, is meant to preface a practical hands-on DMP discussion or tutorial. The goal is to swiftly demonstrate the need for data management before jumping into a more detailed and nuanced discussion of data management plans. This demonstration works best with a live internet connection and the supplementary slide deck (CC-BY) that has been created (which instructors can adopt, adapt, and amend as needed).

INSTRUCTIONS

Discuss the following examples of data collection projects, each of which demonstrates how flawed long-term data planning ultimately created significant access issues. The examples were selected because they were well-funded projects, and their data continue to be extremely important to current researchers. Each example provides overview, failures, and lessons learned. The corresponding slides, which include screenshots and URLs, are indicated for each example.

SERVING INSTRUCTIONS

The goal of this module is to give real-world examples of why data management planning is vital. However, it does not provide detailed discussions about the content of a data management plan.

Example One, US Department of Education Longitudinal Data

The US Department of Education has collected longitudinal education data for decades, making them an invaluable source of information to education researchers. Before the internet, these data were shared on disks and CD-ROMs, which were mailed to libraries as part of the Federal Depository Library Program. In the early 2000s, these data became accessible online through the National Center for Education Statistics (NCES) website, at which point participating libraries were told to discard the outdated physical material.

Failures: (1) The original format (disk) is outdated and unreadable. Opening files on these disks is extremely difficult and time-consuming. (2) The original README files are in a proprietary software (WordPerfect). (3) The updated data hosted on the NCES website are also a proprietary file type (Microsoft Access 2000) that is outdated and unreadable without appropriate software. (4) With the physical disks discarded, access is limited to the NCES website. During the recent government shutdowns, these data became inaccessible. This dependence on the NCES web portal threatens long-term access. See figure 1.

Discussion questions: What lessons can we learn from the NCES reliance on disks and website storage solutions? How could these weaknesses have been avoided?

Lessons/possible answers: Be mindful of long-term storage options like disks and websites, which can become obsolete. Online data are not truly permanent, so do not discard physical backup storage. It is also important to remember that proprietary software may not be readable in the future. Pick storage and file types that will be accessible five, ten, twenty years into the future.
Slides: 3–7.

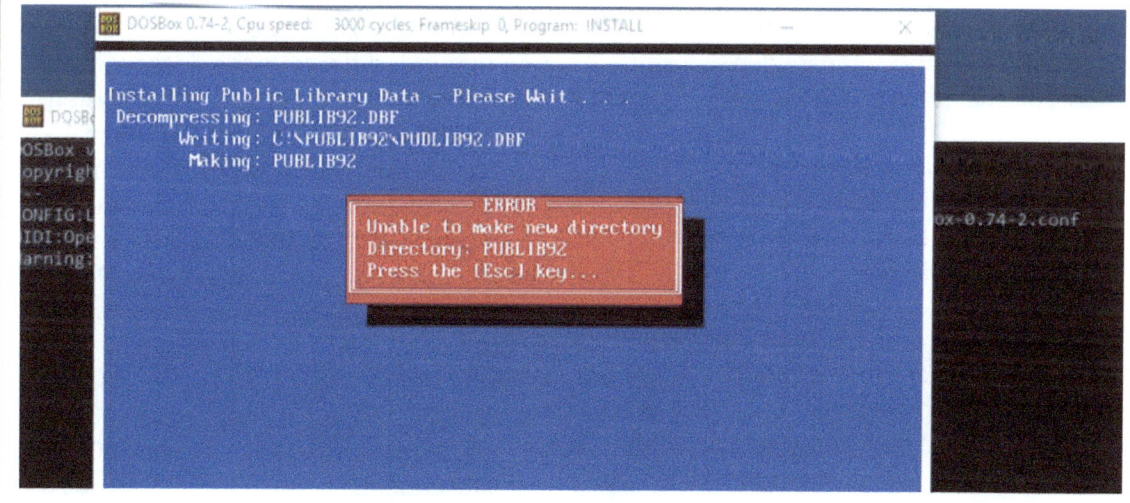

FIGURE 1
Pictured: An attempt to access US Department of Education Longitudinal Data from 1992, which is saved on a 3.5-inch disk. Even with appropriate equipment and backward-compatible software, the data are effectively useless.

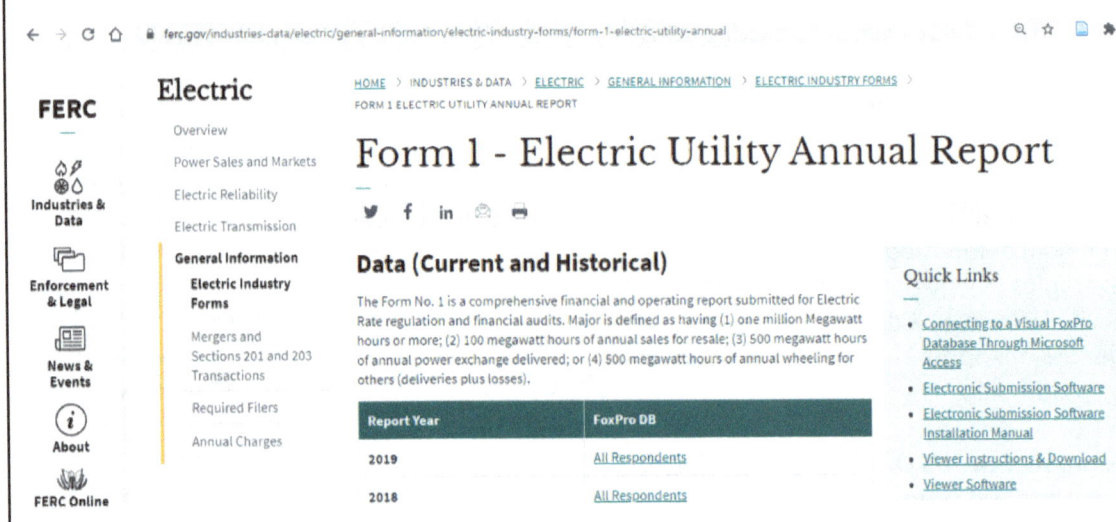

FIGURE 2
The Federal Energy Regulatory Commission "Form 1—Electric Utility Annual Review" data page. Even recent data must be viewed using FoxPro, which is proprietary software that was discontinued in 2007.

Example Two, Federal Energy Regulatory Commission (FERC)

Since 1994 the FERC has shared electric utility annual data, which is extremely valuable to energy policy researchers. Of particular note is the annual "Form 1—Electric Utility Annual Report" that provides comprehensive financial and operational overviews. Unfortunately, to open these data files researchers must use Visual FoxPro, which was popular software in the mid-1990s. If data files are opened with newer software, metadata may be lost. The issue is further compounded because the FERC continues to post current data using FoxPro, which has been discontinued. The final release was in 2004, and support was discontinued in 2015. Researchers must download this outdated and unsupported software to use this vital data.

Failures: (1) Using proprietary software that has become obsolete. (2) Continuing to use outdated tools and software due to institutional inertia. See figure 2.

Discussion questions: What lessons can we learn from the FERC's dependence on proprietary software? How could these weaknesses have been avoided?

Lessons/possible answers: Proprietary software, even top-of-the-line software, becomes obsolete. If the project will collect and produce data over a span of decades, be prepared to adopt new technologies. When possible, use open-source and nonproprietary software. Use open file formats that can be opened without proprietary software. Some examples include .txt or .rtf instead of .docx or .pages; .csv instead of .xls; .odp instead of .pptx; .tiff instead of .bmp.
Slides: 8–9.

Example Three, American Community Survey 2005 One-Year Estimates

In addition to the decennial census, the Census Bureau has also published one-year, three-year, and five-year surveys to provide timely and reasonably accurate demographic estimates about the US population. These American Community Surveys (ACS) pull data from much smaller samples and are not as detailed as a decennial census, but they are nevertheless an excellent source for up-to-date data. The ACS one-year estimate, which was first conducted in 2005, is particularly useful for finding the most recent data. See figure 3.

Failure: The 2005 one-year ACS data collection methods differ from the decennial census and other ACS reports. The inconsistencies in data collection and tabulation make the 2005 ACS incompatible with past and future census studies. As an example, unlike other census studies, the 2005 one-year ACS did not collect group quarters population samples (e.g., residents in dorms, nursing or group homes, barracks, correctional facilities, etc.). For this reason, data from the 2005 ACS cannot be compiled or aggregated with data from other census reports, and the 2005 ACS has subsequently been omitted from Social Explorer.

Discussion questions: What lessons can we learn from the 2005 ACS survey design flaws?

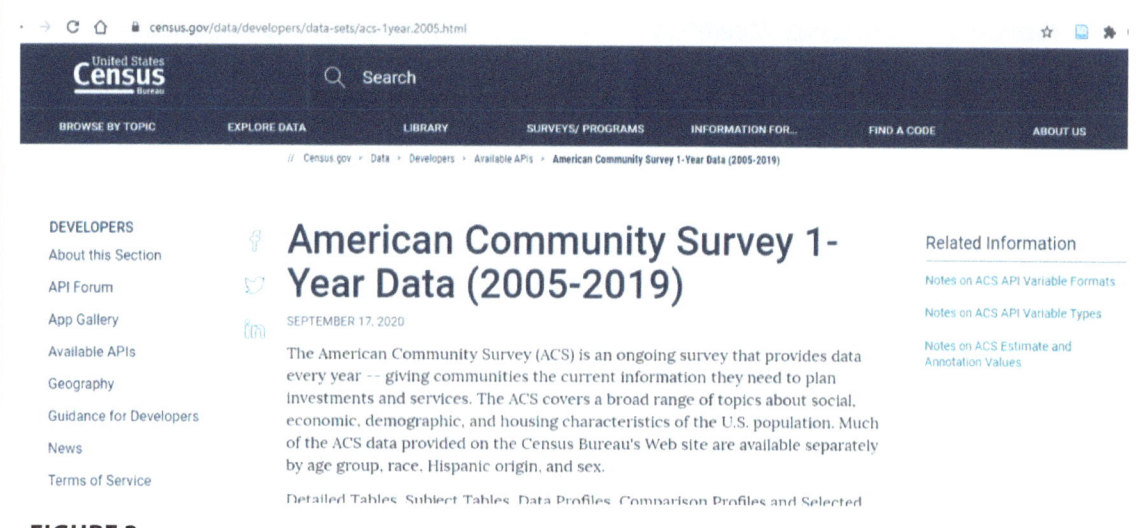

FIGURE 3
The official webpage of the US Census Bureau's American Community Survey 1-Year Data. The first survey (2005) utilized data collection methods that make the findings incompatible with all previous and subsequent Census studies.

How could these weaknesses have been avoided?

Lessons/possible answers: It is vital to adopt data collection methods that are compatible with other studies. Data have a life span beyond the life of the immediate project. Data collection methods that promote future data use once the initial project concludes should be selected. Consider how data may be reused in future projects.
Slides: 10–11.

Closing Discussion/Summary
The instructor should reiterate that these are well-organized studies with extensive planning and preparation. The data failures of each were not the product of incompetence or negligence, but rather because researchers did not consider issues of access, organization, and data compatibility beyond the immediate life span of their project. The instructor should further emphasize that a data management plan is designed to encourage researchers to consider long-term data questions that may not otherwise be discussed while designing a project.

Discussion questions: What are some broad lessons we can learn from these three real-world examples of data management errors? How can we prevent similar pitfalls in our own research projects? How do these examples demonstrate the importance of data management planning?

Possible answers: Even well-funded, well-planned, and well-executed projects can suffer from long-term data access issues. These problems can be avoided by drafting a data management plan, which forces researchers to consider questions of data storage, file type and software selection, and data collection methodology. In each of these examples, data collection was considered only within the immediate scope of the project, which subsequently resulted in data access issues. A data management plan allows researchers to plan access well into the future while also considering broader questions about the life cycle of their data.
Slide: 12.

REVIEWS/ASSESSMENT STRATEGY
The success of this module can be assessed by the content of student comments during the discussion sessions. In the closing discussion, students should be able to articulate the importance of data management planning in ensuring long-term access to data created in research projects. More specifically, they should be able to identify specific data access issues relating to each of the three examples.

ADAPTING THE RECIPE
This exercise was designed as a ~20 minute introduction to an hour-long data management planning workshop. After providing these examples of why data management is necessary, the instructor can spend the remainder of the session discussing the practical aspects of a data management plan, or they can provide a tutorial of the DMPTool. This module can also be used to demonstrate issues relating to long-term preservation and access to government data.

Section 5. Data Management and Sharing

Mikulski

ADDITIONAL RESOURCES

Companion slides have been prepared to supplement this session. Instructors are welcome and encouraged to adopt and adapt these materials as needed. The slides are hosted on PDXScholar with a CC BY-NC 2.0 license. https://doi.org/10.15760/libdata.2.

Low-Fat Research Data Management

Elizabeth Blackwood, Digital Curation and Scholarship Librarian, California State University, Channel Islands

NUTRITION INFORMATION

Research data management (RDM) is for everyone, not just folks at R1 or research-intensive institutions. This recipe details a cost-free research data management workshop for undergraduate students that prioritizes equity for those on the margins. The workshop is most successful when librarians partner with instructional faculty, but it can be executed as a one-shot session.

TARGET AUDIENCE AND NUMBER SERVED

This recipe serves any class size; it is regularly executed with a class of 30 or more. The workshop is most successful with undergraduate or beginner students who are working on a larger research project, such as a capstone project or a thesis, that involves the use or collection of data.

LEARNING OUTCOMES

- Recognize the impetus and importance of research data management
- Evaluate the management needs of a data set
- Evaluate curated files and file organization
- Identify and define the components of a data management plan

COOKING TIME

60–90 minutes

DIETARY GUIDELINES

This recipe not only introduces students to research data management within an academic context, but also fosters concrete skill building for personal data management outside of the classroom.

INGREDIENTS

- Classroom with computers for each student (workshop can also be performed remotely)
- A relevant data set, ideally from an open-access institutional repository (for example, the Schaffer-Smith and Swenson data set from the Duke Digital Repository, listed in the Additional Resources section)
- A web location to host the workshop documentation and messy data set, such as a shareable Google Drive or a website
- Undergraduate students with a significant research project
- Faculty partnership (optional)
- Data management plan integrated as a course assignment (optional)
- Assisting faculty member or workshop helper (optional)

PREPARATION

The workshop consists of two core activities, a data curation activity and a data management plan activity. Only the data curation activity requires preparation.

1. To prepare the data curation activity, select a course-relevant, well-curated data set, ideally from an open-access digital repository that includes good metadata and proper organization. For example, one could choose a data set that includes geographic and lidar data, if the students in the course are familiar with these file types.
2. Download and make a copy of the data.
3. Make the copy of the data messy. For example, dump files out of their organization, rename files so they do not follow a schema, etc.
4. Host the new, messy data in a place that students can access during the session.

INSTRUCTIONS

1. Introduce the topic.
    a. *What data should we manage?* Ask students to list the formats and software that they will be using in their projects.
    b. *Why should we manage this data?* (Use recent examples for each.)
        i. Benefits to their field of study
        ii. Grant compliance
        iii. Personal and professional success
2. Begin data curation activity (performed in groups).

Section 5. Data Management and Sharing

a. Ask students to navigate to the hosted, messy data and download them.
b. Assist the students with unzipping the files.
c. Ask students to explore the data and answer the following questions:
   i. How many files did this study produce?
   ii. What is the topic of this study?
   iii. What types of data did the researchers collect?
   iv. What software would you use to open or use these files?
   v. What information would you need to replicate this study?
d. Discuss their findings as a group.
e. Show students the institutional repository record for the original files and discuss the differences.
f. Remind students to move the files to their machine's recycling bin.

3. Introduce data management plans (DMPs), a solution to the messy data.
   a. Define *DMP*.
   b. Explain why DMPs are important.
      i. For organization of the project
      ii. For grant requirements
   c. List and define the parts of a DMP using the wizard tool from DMPTool.org.
      i. Data collection
      ii. Documentation and metadata
      iii. Ethics and legal compliance
      iv. Storage and backup
      v. Selection and preservation
      vi. Data sharing
      vii. Responsibilities and resources

4. Begin DMP activity (activity performed in groups).
   a. Ask students to navigate to DMPTool.org.
   b. Assist students with creating an account.
   c. Assist students with creating their first plan.
   d. Divide students into groups and ask them to read through one part of a DMP, answer the provided questions (from Data Curation, part 2C) for their data, and report back out to their group.

5. File name best practices
   a. Be consistent.
   b. Be unique.
   c. Avoid special characters.
   d. Demonstrate tools for bulk file renaming.

6. Introduce DMP assignment (optional).

If partnering with discipline faculty member, a DMP can be included as an assignment for the course.

REVIEWS/ASSESSMENT STRATEGY

Assessment strategies vary depending on whether the recipe is performed as a one-shot session or as a partnership with faculty that is integrated into a course. For the one-shot, try scheduling time to discuss the overall outcomes with course instructors. For the course integration, try administering a survey at the beginning of the semester and at the end.

ADAPTING THE RECIPE

Research data management instruction is often reserved for those who are well funded. In this workshop be sure to prioritize equity and inclusion throughout by asking hard questions during the planning stage. Below are several areas to consider in order to prioritize equity:

- *Storage*: Storage for backup files often incurs costs that not all students can afford. Additionally, not all students own a personal computer, and not all campuses provide dedicated server space for student projects. These factors may force some students to access and manage their data through affordable or free cloud storage. To mitigate these issues, do not skimp on best practice instruction, but rather specifically explain the importance of multiple backups and ask students to discuss how they would manage two to three backups either in different platforms, external storage, or personal computers when working in groups. Refer back to these discussions when students begin their DMPs.
- *File naming*: Students who create or collect data that are uploaded directly to cloud platforms will face issues with file naming or renaming due to the proprietary nature of cloud interfaces. There is no easy or granular way to re-

name files in bulk once in these systems without knowledge of a programming language, such as Python, or an external application or software. Students using university-owned or loaner equipment typically do not have the credentials to download software or make changes to the machines at the command line. Instruct students to think through their data collection workflows and document how digital data files will be created. Students using university-owned equipment should recognize that they will need to create file naming schemata prior to data collection and set up their tools to produce files with names to match their schemata.
- *Time*: RDM is time-consuming at its best and painstaking at its worst. It is important to make this clear so that students can plan accordingly. Remember that many students have burdens outside of their academic lives, including commuting to school from work or home and balancing child or elder care with their studies. Students may not have regular computer access at home or live on campus with direct access to library computers. Be realistic with students so that they can be set up for success.

ADDITIONAL RESOURCES

DMPTool: Build Your Data Management Plan. Accessed May 7th, 2021. https://dmptool.org.

Schaffer-Smith, Danica, and Jennifer J. Swenson. "Migratory Shorebird Response to Non-tidal Wetland Dynamics at an Internationally Important Inland Pacific Flyway Stopover," data and scripts. September 13, 2018. Duke University Libraries Digital Repository. https://doi.org/10.7924/r47w6b882.

Managing Qualitative Social Science Data
An Open, Self-Guided Course

Sebastian Karcher, Associate Director of the Qualitative Data Repository (qdr.syr.edu) and Research Assistant Professor of Political Science, Syracuse University, skarcher@syr.edu; Diana Kapiszewski, Senior Research Affiliate, Qualitative Data Repository and Associate Professor of Government, Georgetown University, dk784@georgetown.edu

Nutrition Facts	Lessons	Serving	% Daily Value
4 modules per course 3-4 lessons per module	Data Management Planning:	3 lessons	100%
6-8 exercises per module	Data Management:	4 lessons	100%
	Sharing Data:	4 lessons	100%
	Writing with Data:	3 lessons	100%

FIGURE 1
Nutrition facts: Modules and lessons

NUTRITION INFORMATION
Managing Qualitative Social Science Data (https://managing-qualitative-data.org/) is a self-guided online course about managing and sharing qualitative social science data.

TARGET AUDIENCE
- *Chefs*, i.e., data management instructors looking for exercises or teaching strategies.
- *Diners* can follow the course independently or together with others.
  - Hungry academics, from graduate students in the social sciences about to embark on their first major research project to senior scholars who are unfamiliar with data management and sharing.
  - Hungry practitioners seeking to learn about data management or improve their data management practices.

LEARNING OBJECTIVES
Managing Qualitative Social Science Data emphasizes practical advice for managing, sharing, and deploying the wide variety of qualitative data used in social science research.
Learners who take the course will
- practice correctly formatting and carefully organizing collected data and tailoring both processes to the research topic and questions
- test their knowledge of copyright law and realize the ethical imperative of participant protection and how it can constrain data sharing
- apply general principles of good data management to their own work
- acquire hands-on strategies for and experience with increasing the value of their own data through improved documentation and organization

COOKING TIME
- Serving diners who just want a quick data-management snack? Offer an individual lesson and its related exercises.
- Serving diners looking for a data-management meal? Offer all 14 lessons and 27 exercises, which follow the data life cycle, for the full menu fixe.

DIETARY GUIDELINES
Effectively managing research data—carefully organizing and kindly caring for data—allows a researcher to more effectively deploy data and thus makes their research better and healthier! Well-managed data can also be shared with others—to help them understand the research more fully, answer other questions, and teach methods and substance. Well-managed data improve a researcher's own creations, and when set

free, can be reused in many other tasty intellectual recipes.

INGREDIENTS

Module 1: Planning the Management of Qualitative Data

https://managing-qualitative-data.org/modules/1/

Ingredients: information on the research data life cycle; the key elements of planning for effective data management; how to write a useful data management plan; preparing data for sharing; and informing participants in a research project about plans to share data

Module 2: Managing Qualitative Data

https://managing-qualitative-data.org/modules/2/

Ingredients: information on the key requirements for documenting, organizing, and transforming qualitative data; the challenges of managing qualitative data; transcribing and digitizing data; and good practices for keeping data safe and secure

Module 3: Sharing Qualitative Data

https://managing-qualitative-data.org/modules/3/

Ingredients: information on the benefits and challenges of sharing qualitative research data; strategies to address data-sharing challenges; de-identifying qualitative data; evaluating copyright and getting permissions from an archive to share digitized sources; and identifying a data repository for a project

Module 4: Writing with Qualitative Data

https://managing-qualitative-data.org/modules/4/

Ingredients: information on effectively deploying qualitative data to make and support written claims and conclusions and make research more transparent; organizing qualitative data for writing; strategies for research transparency with qualitative data; and good practices in citing data

All Modules Include Side Dishes and Dessert
- Three or four thematic lessons
- Overview boxes
- Text-based discussion
- Diagrams, tables, and figures
- Exercises and solutions
- Links to further resources

PREPARATION

Chefs: Assign a specific lesson or module as part of a class on field research, research methods, or qualitative methods. Several of the exercises (e.g., on de-identification) lend themselves well to class discussion. The exercises are a core feature of the course and are designed to reinforce learning. You can use any of them as part of a data management class.

Diners: We recommend that learners take the course with a specific research project in mind. Several of the exercises will help them to better organize the data for their project.

Learners can use the exercises to assess progress and review sections as needed. Several of the exercises have free-form answers without a clear right solution. If possible, learners can work through the course with some fellow learners and exchange (and comment on) each other's answers.

REVIEWS/ASSESSMENT STRATEGY

The exercises that accompany the lessons help learners assess how well they're acquiring the skills being discussed. The best way to evaluate how well they understand data management, however, is to ask them to manage some data! Ask learners to put their skills to work immediately with the projects they are currently working on. See how well they can explain what they've learned to others!

SHARING AND ADAPTING THE RECIPE

The course and all its contents are licensed under a CC-BY-SA license so that chefs can freely adapt and reuse the material. The course runs on a static Jekyll site hosted on GitHub (https://github.com/QualitativeDataRepository/ssrc-qdr-data-management) and can be forked to create a custom version.

Seven Weeks, Seven DMPs
Iterative Learning around Data Management Plan Creation

Emma Slayton, Data Curation, Visualization, and GIS Specialist, Carnegie Mellon University Libraries, ORCID: https://orcid.org/0000-0003-2230-3101; Hannah C. Gunderman, Independent Scholar, ORCID: https://orcid.org/0000-0002-7710-7055

NUTRITION INFORMATION

Data management plans (DMPs) are an essential part of any healthy research diet. This recipe is used in the Discovering the Data Universe course, a seven-week undergraduate course taught by data librarians at Carnegie Mellon University (CMU), which includes teaching students the proper techniques for cooking up a flavorful and filling data management plan. This recipe is useful for anyone teaching foundational data literacy concepts in an academic library or others who want to leverage data librarians' expertise to provide data literacy instruction and outreach.

TARGET AUDIENCE AND NUMBER SERVED

The target audience is undergraduate students in data literacy instruction settings, with a recommendation of at least 1 chef per 25 students. (If there are more students, bring more chefs to the kitchen!) One chef will suffice if resources are limited.

LEARNING OBJECTIVES

As with the work of any good chef, success in this recipe is measured by how delicious the food is, and also the skills they teach to the sous-chefs (or students). This includes teaching what knife is needed for each kind of cutting technique a recipe requires. Specific knife skills (learning objectives) we want our sous-chefs to come away with include these:

- *Recognize data:* Identify sources of data (both research-based data sets and data from everyday life), and describe how those data fit into different possible narratives and perspectives in stories.
- *Communicate data:* Interpret meaning from data and explain the ways in which data can be manipulated. Use data to answer key questions.
- *Manage data:* Plan and perform proper data preparation and storage (including the use of the DMPTool and Open Science Framework).
- *Critique data stories and themes:* Identify common data narrative structures in a variety of examples including their classmates' (and their own) work; critically evaluate and convey feedback about a classmate's work.

COOKING TIME

Allow 7 weeks for marinating, with approximately 1–2 hours each week set aside to check DMPs for temperature and flavor and adjust spice levels.

DIETARY GUIDELINES

This recipe prepares students for future academic research projects and for cooking after graduation. Learning how to prepare a DMP helps students think through project design and connect that to what information is needed to answer research questions. To properly carry out a DMP, students must recognize what elements of data need to be explained, communicate those aspects effectively, and prepare for storing the data in a way that is accessible. They have to find the right recipe for their discipline, recognize and find the ingredients, and be able to describe those ingredients and how to use them.

Two learning objectives from this assignment that map well to ACRL's *Framework for Information Literacy for Higher Education* are "communicate data" and "manage data." These learning objectives fit most strongly to the frames Information Creation as a Process and Authority Is Constructed and Contextual. As the course progresses and students learn more about data literacy topics, their expertise and credibility in the subject increase, and this is reflected in their DMP updates throughout the course.

INGREDIENTS

- DMPTool (https://dmptool.org/). If you can't access this particular ingredient, any word-processing platform will suf-

fice. However, the DMPTool provides templates, so your recipe will definitely flourish by using it.
- Information from at least 3 funding agencies requiring DMPs. We recommend the National Science Foundation, National Institutes of Health, and National Endowment for the Humanities.
- Well-described, open data sets available for download. Our favorite data sets are from the US Census, the city of Pittsburgh, and from our institution.

PREPARATION

In preparation for this class, facilitate a scenario much like the cooking show *Chopped*. Prepare all the ingredients (or data sets) for students in advance, and ask them to create a meal (DMP) using these ingredients.

To prep the data ingredients, look for different examples of data. These include qualitative and quantitative data, and public and private data sets. The open data sets are on a national level (US Census data), and local level (Pittsburgh police incident reports from the Western Pennsylvania Regional Data Center). The private data set is from our university archives (a historical photograph data set). Students are individually assigned one of these data sets to use as their main ingredients in the course.

INSTRUCTIONS

The assignments task students with creating and iterating upon a data management plan (DMP) that is based on the assigned data set they receive in the first week of class. Below are step-by-step instructions for guiding students through their DMP.

Week 1

Topic: Sample DMPs, introducing the DMPTool, and choosing a funding agency template.

Lesson specifics: Ask students to imagine they are the head chefs (creators) of the data set they are assigned. Walk them through the process of data collection, data analysis, and data management. Show several DMPs to students, using examples from the collection of public DMPs from the Digital Curation Centre (https://dmponline.dcc.ac.uk/public_plans). When showing sample DMPs, highlight areas where the ingredients have been used effectively and areas where the recipe could have been improved.

1. Guide students through the DMPTool, and highlight appropriate templates that the students can employ to create their DMP. We recommend the National Science Foundation, National Institutes of Health, and National Endowment for the Humanities templates, as they serve as excellent recipes to draw from and cover a wide range of disciplines. Ask students to choose a template that reflects the nature of their assigned data set (quantitative data, ethnographic data, etc.).
2. Help students navigate to the DMP requirements from the organization on which their template is based and ask which elements should be included in their DMP (see *Additional Resources* for URLs for guidance from the three funding agencies).

Student deliverables: Students turn in a written explanation of which template they selected and screenshots that demonstrate that they successfully logged in to the DMPTool.

Recipe critique guidelines: Verify that students chose a DMP template that makes sense for their data type and included screenshots verifying that they logged in to the DMPTool.

Week 2

Topic: Data description and data collection basics.

Lesson specifics: Students begin to work in the DMPTool with their chosen DMP template. This week's lecture focuses on basic practices in describing data and data collection processes. What variable types are present? What are the date ranges for the data collection? How do the data appear to have been collected? This will help inspire students to apply the same level of inquiry and description to their own data sets.

Student deliverables: Students turn in their first official DMP created in the DMPTool that addresses what their data look like and how they have been collected, framing the DMP as though the students are the original collectors of the data.

DMP critique guidelines: Instructors repeat this step weekly throughout the remainder of the course. Review this round of DMPs by highlighting in what areas students successfully addressed the topics covered this week and what areas of opportunity exist for the next round of revisions.

Week 3
Topic: Data management basics and introduction to metadata.

Lesson specifics: Students learn about data management basics using the DataONE recommended practices as guidelines (see Additional Resources) including storage backups, file naming, descriptive documentation, and an introduction to metadata. Through lecture, address questions such as these: What is the importance of having a good file naming scheme for my data? How many backups should I have? Should I write down the steps I take to interact with my data throughout a project? What is a README file? How do I choose a metadata standard to apply to my data?

Student deliverables: Students turn in a DMP that addresses the revisions provided by the instructor for the Week 2 content and include new content on how the student would employ recommended practices in data management and metadata to their assigned data set.

Week 4
Topic: Data analysis foundations and storytelling with data.

Lesson specifics: Students learn about types of data analyses that are suitable for different types of data and research projects (qualitative, quantitative, etc.), such as statistical tests, ethnographic inquiry, and geospatial analysis. Discuss how to choose an analysis type for a data set and how to employ that analysis to tell an effective story about the data set.

Student deliverables: Students should identify at least one data analysis technique they would like to apply to their data set and indicate how this analysis technique could change the data management strategies listed in the previous week's DMP.

Week 5
Topic: Formatting and visualizing data.

Lesson specifics: Lead a discussion on how to choose appropriate data visualization techniques to display the results of a data analysis, showing students examples of good visualizations and visualizations that could be improved. Discuss several data visualization platforms, including Tableau and ArcGIS, and programming-based visualization packages in R and Python.

Student deliverables: Students should choose at least one data visualization technique that they would like to apply to the results of their data analysis identified in Week 4 and describe how this visualization technique may change any of their data management strategies listed in the previous week's DMP.

Week 6
Topic: Ethics of data, data licensing, and data discovery.

Lesson specifics: In this multi-topic lesson, (1) guide the students through a discussion on why it is important to be ethical when collecting and working with data, (2) describe why it is important to assign data sets a license when sharing them publicly (using the Creative Commons licenses as an example), and (3) guide students through the process of identifying an appropriate data repository for their data.

Student deliverables: Students should describe the ethical considerations that exist around their data. Do the data refer to real people? Is there any identifying information within the data set? What would happen if the data set got into the wrong hands? How might their data management strategies mitigate any ethical risks? Ask students to choose an appropriate license for their data from Creative Commons and identify at least one possible data repository in which they would share these data publicly.

Week 7
Topic: Wrapping it all together: final DMP creation.

Lesson specifics: Provide students with a collaborative working session in class where they can work with instructors or students to answer any questions about the DMP revisions they have received during the course and receive instructor help and feedback as they work on their final DMP iteration.

Student deliverables: Students turn in their final DMP, which should address all revisions and lesson topics from the course.

Critique guidelines: Review this final round of DMPs by highlighting in what areas students have incorporated course concepts and how well the students have addressed the feedback received during each week of revisions.

REVIEWS/ASSESSMENT STRATEGY NOTES

As noted above, when judging the flavor of the DMPs, pay attention to how the students make use of available ingredients (information on their assigned data set), cooking techniques (how they incorporate class materials into refining their DMP), and presentation (how well they address all the elements of the DMP).

During the critique/review process, we caution instructors on using negative language or framing a DMP as bad. Given the wide range of skills, neurodiversity, and positionality that students may bring to this work, instructors can craft a more positive learning environment by offering opportunity-based critique that highlights areas where students can improve their recipe.

CHEF'S NOTES

Because students have not collected their own data and they are working with the information provided with their data set, they must take creative liberties to fill out some sections of the DMP. Some students lean into this adventurous cooking more than others! In our course we asked students to carry out their chosen data analysis and data visualization techniques on their data. In your cooking environment you may ask the students to describe what they *would* do with the data. It depends on the cooking experience you want to create!

ADDITIONAL RESOURCES

ArcGIS. Accessed May 28, 2021. https://www.esri.com/en-us/arcgis/about-arcgis/overview.

Carnegie Mellon University Archives. Accessed May 28, 2021. https://findingaids.library.cmu.edu.

Contains historical photographs used as a data set in the course.

Creative Commons. "Share Your Work." Accessed May 28, 2021. https://creativecommons.org/share-your-work/.

DataONE. "Education Modules." Accessed May 28, 2021. https://old.dataone.org/education-modules.

National Endowment for the Humanities. "Data Management Plans for NEH Office of Digital Humanities Proposals and Awards." Accessed May 28, 2021. https://www.neh.gov/sites/default/files/2018-06/data_management_plans_2018.pdf.

National Institutes of Health. "Data Management and Sharing Policy." Accessed September 12, 2021. https://sharing.nih.gov/data-management-and-sharing-policy..

National Science Foundation. "Dissemination and Sharing of Research Results—NSF Data Management Plan Requirements." Accessed May 28, 2021. https://www.nsf.gov/bfa/dias/policy/dmp.jsp.

Tableau. Accessed May 28, 2021. https://www.tableau.com.

US Census Bureau. "Commuting Data Tables." American Community Survey. Accessed May 28, 2021. https://www.census.gov/topics/employment/commuting/data/tables.html.

Western Pennsylvania Regional Data Center. "Incidents." Accessed May 28, 2021. http://tools.wprdc.org/guides/public-safety/incidents/.

Equitable from the Beginning
Incorporating Critical Data Perspectives into Your Research Design

Jodi Coalter, Life Sciences Librarian, Michigan State University, coalterj@msu.edu; David Durden, Digital Collections Specialist, Library of Congress; Leigh Amadi Dunewood, DEI Coordinator, University Corporation for Atmospheric Research (UCAR), ldunewood@ucar.edu

NUTRITION INFORMATION

A research data management (RDM) education diet is calorie-dense with technical macronutrients, but it often lacks essential micronutrients such as equity, diversity, inclusion, and accessibility. As part of a broader effort to improve the data management diets of undergraduate Gemstone Honors research teams at the University of Maryland, College Park, the authors created a series of instructional modules on equitable data collection and research design practices. The modules developed for this program enable researchers to assess the impact of their data across the research life cycle and widen their perspectives of data collection and analysis processes to consider implicit and transparent ethical, diversity, equity, and inclusion values at all stages of RDM.

Many of the concepts surrounding equity and data presented in this workshop were drawn from the book *Data Feminism*. As the authors of this book state, "we must explicitly acknowledge that a key way that power and privilege operate in the world today has to do with the word *data* itself."[1] Examples illustrating these concepts were drawn from a wealth of resources, including *Data Feminism*; for more information please review the Additional Resources section.

TARGET AUDIENCE AND NUMBER SERVED

This workshop was designed for undergraduate students, student research assistants, graduate students, and experienced researchers (such as faculty or research associates). Smaller dinner parties are best. In a traditional classroom setting, 15 participants is ideal; in an online setting, 10 is more manageable.

LEARNING OUTCOMES

- Articulate the importance of centering research data in a broader social context.
- Interrogate one's own data collection practices, particularly when marginalized communities are involved.
- Assess one's data and analyses through a critical lens to determine what impact one's research will have once a final research product has been published.
- Reflect on data access and reuse in terms of equity and accessibility.
- Develop an awareness of ethical concerns specific to data ownership and stewardship.
- Identify and address how researcher proportionality/position affects data collection, use, and analysis.

COOKING TIME

Preparation time may vary depending on whether the workshop is prepared from scratch or selected using off-the-shelf ingredients. Workshop design may take upward of 6 hours depending on the chefs' familiarity with the material. Repeating the workshops after the lessons have been prepared will take significantly less time.

Total prep time: 2–6 hours
Baking the slide deck: 1–1½ hours
Teaching time: 1½ hours

DIETARY GUIDELINES

This recipe uses a scaffolded, modular approach informed by ACRL's *Framework for Information Literacy for Higher Education*. The modules discussed align closely with the information literacy frames (1) Authority Is Constructed and Contextual and (2) Information Has Value.

Module 1, Data Literacy for Community Engagement, is designed to encourage workshop participants to (re)consider the significance and function of reflexivity in RDM; they reflect and discuss what it means to have and define one's own positionality as a researcher. Module 2, Going beyond Informed Consent, is designed to encourage workshop participants to (re)consider how the historical, systemic marginalization of individuals and groups has perpetuated the production and dissemination of *mis*-information, particularly data, and how their research can confront, dispel, and delegitimize *mis*-information.

INGREDIENTS

- 1 prepared menu: 1 or 2 modules focusing on data equity issues
- 3 to 5 real-world examples of inequity arising from data for discussion. The following examples are used to show how centering equity before data collection and analysis begin could have prevented abusive practices from happening:
    - *"The Library of Missing Datasets"*: This project, founded by Mimi Onuoha, describes data sets that should be collected but are not. According to Onuoha, "'Missing data sets' are the blank spots that exist in spaces that are otherwise data-saturated."[2] This is an example of neglect.
    - *The failure of medical research to distribute the COVID-19 vaccine equitably*: This example focuses not on incorrect collection or analysis of the data, but rather on what happens to those data after they're collected.
- 1 collaborative platform, e.g., Google Jamboard or Padlet
- Meeting space: virtual or in person

- An assessment tool

PREPARATION

The following menu provides a flexible structure around which to plan the workshop.

Salad: Module 1: Data Literacy for Community Engagement.
This module is intended for use with research that does and does not require an institutional review board (IRB). It's designed to encourage researchers to place their projects in a broad social context, to think about how their research will be used by the public and by other researchers, and how it may have unintended consequences. Examples should be taken from real-world scenarios.

We examine the Washington, DC, data portal and the city of Baltimore data portal (URLs are provided in the Additional Resources section). We examine the Baltimore portal first, showing how frustrating it is to find information, how some information is missing, and how there are very few instructions on how to correctly interpret the data. We then explore the Washington, DC, portal. While the two portals use the same platform, DC's portal is much easier to search, has multiple tutorials on how to use the data and clear and easy-to-find definitions for the data and indicates where to go for more information.

Main course: Module 2: Going beyond Informed Consent. This module caters to research that deals with human subjects.

This module is designed for research that requires an IRB but want to take data equity into consideration as well. We describe work conducted by John Aini from Ailan Awareness and Paige West from Barnard College and Columbia University in Papua New Guinea (URLs for their work in the Additional Resources section). They worked with local communities before they conducted their own research to determine what questions area residents wanted answered. After describing Mr. Aini and Dr. West's research, we ask researchers several questions to encourage them to think about how their own research might benefit from working with communities first. We encourage them to ask, "Do research participants have access to those data? Where? How? Are they able to accurately interpret the data, or do they need assistance or training? Who provides that training?"

The main course module should include one to two data examples relevant to your audience's interests and research. Workshops should be conversational; prepare examples, discussion prompts, or questions to encourage workshop participants to think about incorporating equity into each stage of the research data life cycle. Develop new modules or adapt existing modules to accommodate for specific tastes.

Digestif: 3-2-1 Survey

Set the table by picking the technologies and collaborative tools you will use throughout the workshop. If using Google Jamboard, it's a good idea to pre-populate each board with data examples, topics, and images from the

Section 5. Data Management and Sharing

modules to aid the exercises (see figure 1). If the workshop is online, set up and distribute the virtual meeting space (such as a Zoom room) in advance. Create a 3-2-1 assessment tool on whatever survey platform you typically use.

INSTRUCTIONS

The meal is best served in discrete courses; time your modules for equal distribution. Dedicate at least 30 minutes for each module, but be sure to include enough time for teaching, conversations, and assessments. Below is a step-by-step guide to conducting a completed workshop:

The material below should be completed within 30 minutes.

1. Introduce instructors. Clearly state your name and your function within the institution.
2. Introduce the topic. Describe what you will be reviewing and clearly state that this is a participatory workshop; participants will be required to answer questions regularly, to deeply assess their own positionality within research, and to reflect on how that positionality is reflected in their data.
3. Discuss why equity is important to RDM, and review examples of bad RDM practice. See the Ingredients section above for examples.
4. Introduce your collaborative platform. Encourage participants to use the platform to take notes, make connections between concepts that are presented, and post questions—creating a collective mind map of what they are learning that they can refer to later.

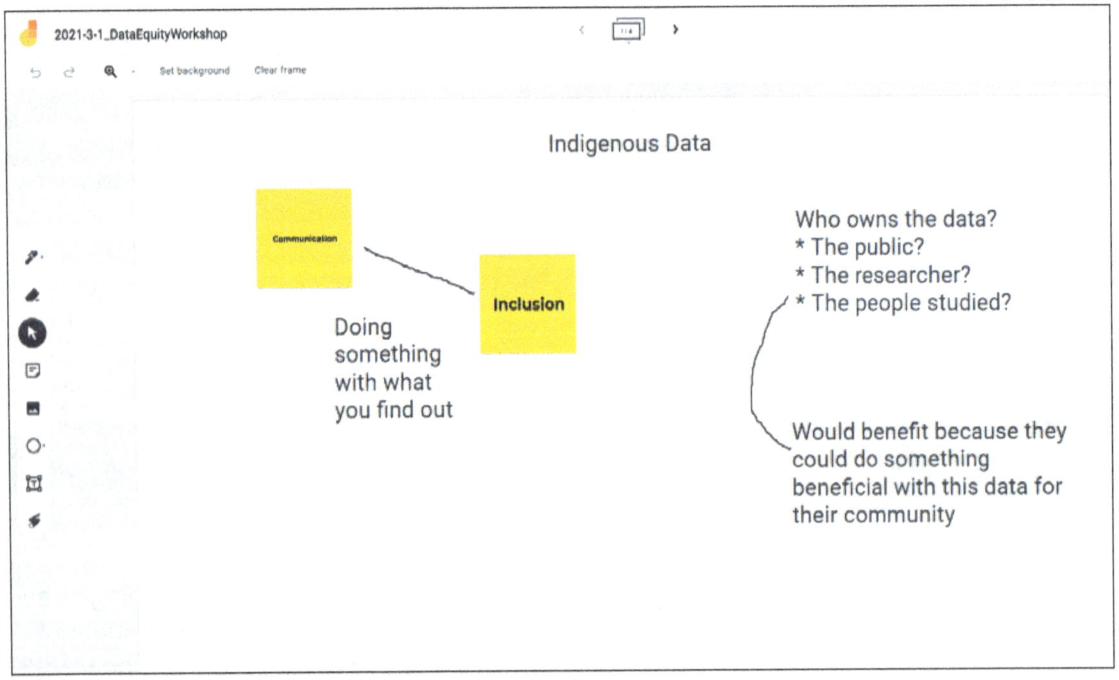

The material below should be completed within 30 minutes.

5. Begin Module 1, Data Literacy for Community Engagement.
   a. Introduce two data portals from an area close to your institution. Point out what is effective and what needs work.
   b. After introducing the portals, remind participants about the collaborative platform.
   c. Ask prepared discussion questions, encouraging participants to respond, both to the question and to the proffered answers. For this module, we ask
      i. Describe one aspect of the DC Data Portal that is beneficial to DC residents.
      ii. What aspect of either of these data portals could be improved?
      iii. After looking at how these data portals work, what changes would you make to your own data or to your data management plan?
   d. Review the mind map participants created during the discussion. Review any questions that may have been posted, and ask participants to describe what they've created.

FIGURE 1
Jamboard example created by workshop students. (CC BY-NC-SA)

6. Direct participants to a new page in the collaborative platform to take notes and ask questions for the next module.

The material below should be completed within 30 minutes.

7. Begin Module 2: Going beyond Informed Consent.
    a. Describe the importance and limitations of IRBs. An IRB is incorporated into planning—so it's a great time to think about what an IRB doesn't ask. Just because an IRB categorizes a project as low risk does not mean an absence of privacy or equity issues.
    b. Introduce your first example. We use an example of Indigenous data sovereignty. (See the Main Course section for an example of Mr. Aini and Dr. West working in Papua New Guinea.)
    c. Ask prepared discussion questions, encouraging participants to respond, both to the questions and to the proffered answers. For this module, we ask
        i. How did Mr. Aini and Dr. West incorporate Indigenous ways of knowing?
        ii. If you collect data from or about members of Indigenous groups, who owns these data?
        iii. Do research participants have access to those data? Where? How? Are they able to accurately interpret the data, or do they need assistance or training?
    d. Review the mind map participants created during the discussion. Review any questions that may have been posted, and ask participants to describe what they've created.
8. Wrap up. Encourage participants to reach out to you should they have any specific questions regarding their research.

Distribute the 3-2-1 survey and encourage participants to fill it out while they are still in the room (which will encourage them to actually fill it out).

REVIEWS/ASSESSMENT STRATEGY

Offer a postworkshop digestif to aid the digestion of the content. As a postworkshop survey, select a 3-2-1 assessment to capture participants' reflections on the workshop and comprehension of the material.

ALLERGY WARNING

The dinner party will be dining in a Brave Space. Brave Spaces are designed to be inclusive environments conducive to learning, sharing, and growing even when discussions may become uncomfortable; entering a Brave Space requires workshop participants to honor each other's experiences and opinions.[3]

CLEAN UP

One-on-one consultations may be necessary to address specific questions about integrating workshop concepts into practice.

CHEF'S NOTE

Introduction to RDM: A brief introduction to RDM is a useful module to add at the beginning of the workshop for new researchers or researchers who are unfamiliar with RDM.

ADDITIONAL RESOURCES

Aini, John, and Paige West. "Decolonizing Conservation." *Paige West* (blog), July 28, 2018. https://paige-west.com/2018/07/28/decolonizing-conservation/.

Coalter, Jodi. "Diversity, Equity, and Inclusion in Research." Research Guide, University of Maryland Libraries, 2021. https://lib.guides.umd.edu/ResearchEquity.

Data for Black Lives Blog. Accessed March 29, 2021. https://blog.d4bl.org.

Delgado, Richard, and Jean Stefancic. *Critical Race Theory: An Introduction*, 3rd ed. New York: New York University Press, 2017.

Herzog, David. *Data Literacy: A User's Guide*. Los Angeles: Sage, 2016.

I-95 Modernization Project. "Advanced Bridges." Accessed June 8, 2021. https://i94detroit.org/i94-project/advanced-bridges/.

Muhammad, Khalil Gibran. *The Condemnation of Blackness: Race, Crime, and the Making of Modern Urban America*. Cambridge, MA: Harvard University Press, 2019. https://doi.org/10.2307/j.ctvjsf4fx.

Open Baltimore. 2020. https://data.baltimorecity.gov.

Baltimore city open data hub.

Open Data DC. Accessed June 8, 2021. https://opendata.dc.gov.

Points: Data and Society (blog). Accessed March 29, 2021. https://points.datasociety.net.

Yau, Nathan. *FlowingData* (blog). Accessed March 29, 2021. https://flowingdata.com.

REFERENCES

AlHajal, Khalil. "Widening of I-94, I-75 in High-Congestion Areas Approved despite Protests." Mlive, June 21, 2013. https://www.mlive.com/news/detroit/2013/06/widening_of_i-94_i-75_in_high-.html.

3, Brian, and Kristi Clemens. "From Safe Spaces to Brave Spaces." In *The Art of Effective Facilitation: Reflections from Social Justice Educators*, edited by Lisa M. Landreman, 135–50. Sterling, VA: Stylus, 2013.

D'Ignazio, Catherine, and Lauren F. Klein. *Data Feminism*. Cambridge, MA: MIT Press, 2020.

Moodie, Susan. "Power, Rights, Respect and Data Ownership in Academic Research with Indigenous Peoples." *Environmental Research* 110, no. 8 (November 2010): 818–20. https://doi.org/10.1016/j.envres.2010.08.005.

Moutafis, Rhea. "How Bad Facial Recognition Software Gets Black People Arrested." *Towards Data Science*. June 12, 2020. https://towardsdatascience.com/how-bad-facial-recognition-software-gets-black-people-arrested-3c02738a3d54.

O'Neil, Cathy. *Weapons of Math Destruction: How Big Data Increases Inequality and Threatens Democracy*. New York: Broadway Books, 2016.

Ọnụọha, Mimi. "The Library of Missing Datasets." Mimi Ọnụọha website. Accessed May 25, 2021. https://mimionuoha.com/the-library-of-missing-datasets.

NOTES

1. Catherine D'Ignazio and Lauren F. Klein, *Data Feminism* (Cambridge, MA: MIT Press, 2020).
2. Mimi Ọnụọha, "The Library of Missing Datasets," Mimi Ọnụọha website, accessed May 25, 2021, https://mimionuoha.com/the-library-of-missing-datasets.
3. Brian Arao and Kristi Clemens, "From Safe Spaces to Brave Spaces," in *The Art of Effective Facilitation: Reflections from Social Justice Educators*, ed. Lisa M. Landreman (Sterling, VA: Stylus, 2013).

Section 6.
Geospatial Data

147   [[Ch38]]**Challenge Accepted: Introducing Geospatial Data Literacy through an Online Learning Path**
Joshua Sadvari and Katie Phillips

151   [[Ch39]]**GIS for Success Series: Learning the Basics of QGIS Workshop**
Kelly Grove

154   [[Ch40]]**GIS for Success Series: Let's Make a Map in QGIS Workshop**
Kelly Grove

157   [[Ch41]]**Statistical and Geospatial Literacy for Integrative Genetics**
Jay Forrest and Chrissy Spencer

161   [[Ch42]]**Web Map Layer Cake: Teaching Web Mapping Skills with Leaflet for R**
Sarah Zhang and Julie Jones

Challenge Accepted
Introducing Geospatial Data Literacy through an Online Learning Path

Joshua Sadvari, Geospatial Information Librarian, The Ohio State University, *sadvari.1@osu.edu*; Katie Phillips, Data Management Coordinator, The Ohio State University, *phillips.1870@osu.edu*

NUTRITION INFORMATION

The ArcGIS Online Challenge is an online, asynchronous learning path that was developed in the spring of 2020. A rapid shift to remote instruction associated with the COVID-19 pandemic caused the authors to modify our approach to offering informal learning experiences on the use of geographic information systems (GIS) in research and teaching. We focused on ArcGIS Online because of its availability and widespread use under an Esri site license at our institution, its ease of access as a web-based platform, and its lower barrier to entry compared to the steeper learning curve of desktop GIS tools.

Our motivations in designing this learning path were to (1) raise awareness of the potential of GIS beyond making maps, (2) introduce learners to the broader suite of ArcGIS Online apps, and (3) foster engagement with researchers and instructors across disciplines. By providing opportunities for knowledge transfer throughout the challenge, learners can move along a continuum from introductory education about geospatial data and tools, to focused consultations with facilitators about their particular project needs, to more self-directed applications of the knowledge and skills they have gained.

TARGET AUDIENCE AND NUMBER SERVED

The target audience for this learning path includes faculty, staff, postdocs, and graduate students from any discipline who are novice or intermediate GIS learners. Based on our experience, we recommend that two instructors co-teach the learning path, with a maximum of 40 to 50 participants per instructor.

LEARNING OUTCOMES

Through completing the ArcGIS Online Challenge, participants will be able to
- describe the potential of GIS beyond "dots on a map" and for cross-disciplinary applications
- use ArcGIS Online apps individually or in conjunction with one another for common tasks in geospatial projects
- brainstorm ways that GIS could be used for a research project or teaching assignment relevant to their own work

COOKING TIME

Approximately 2 hours per week, over the course of 5 weeks

DIETARY GUIDELINES

This activity aligns with the frame Information Creation as a Process from ACRL's *Framework for Information Literacy for Higher Education*. In successive modules, learners are introduced to aspects of the geospatial data life cycle, including finding, collecting, exploring, analyzing, visualizing, and storytelling, as enabled through the ArcGIS Online suite of apps. Participants learn about the capabilities and constraints of different apps for common geospatial data tasks and how these influence their potential uses and the resulting information products. Participants also make choices throughout the learning path about which apps to explore in greater detail, and in doing so, they are assessing the perceived fit of these technologies for meeting their own geospatial data and information needs.

Figure 1 shows the module topics and progression of the learning path, along with the skills practiced and ArcGIS Online apps used each week. Each module includes learning objectives, content, activities, brainstorming questions, and links to additional resources. Rather than developing all content and activities used in the learning path from scratch, we relied heavily on third-party resources, especially those produced by Esri. We identified, evaluated, selected, and organized these resources based on the path- and module-

Section 6. Geospatial Data

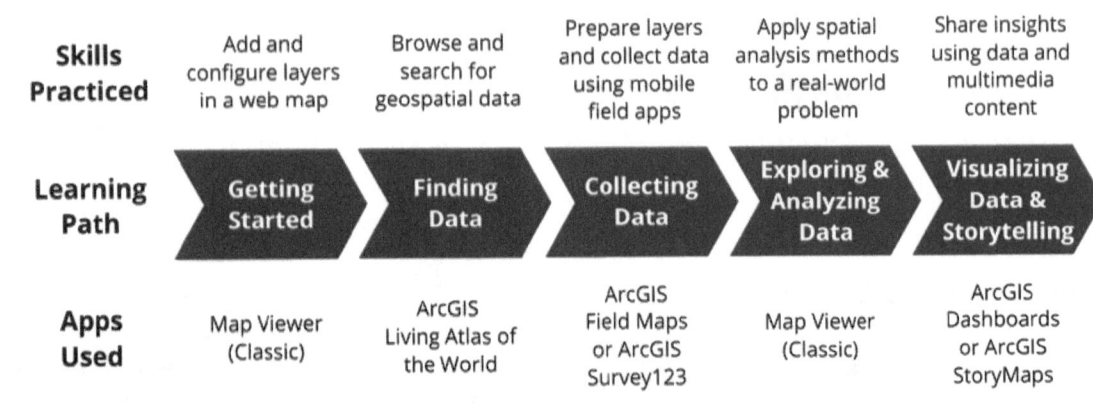

FIGURE 1
Module topics and progression of the ArcGIS Online Challenge learning path.

level learning objectives. The ArcGIS Online Challenge is delivered through Springshare's LibGuides platform, and our guide is shared in the LibGuides community for reuse by librarians at other institutions (see Additional Resources).

INGREDIENTS

- ArcGIS Online organizational account
- Content—short readings and videos highlighting key concepts and relevant use cases for each module
- Activities—hands-on tutorials providing training on the methods and tools covered in each module
- Brainstorming questions—weekly opportunities for learners to think about how to transfer new knowledge and skills to their own research or teaching
- Content management system (e.g., LibGuides)
- Event registration system
- Communication channel (e.g., e-mail)
- Survey software (e.g., Qualtrics)
- Videoconferencing software (e.g., Zoom; optional)

PREPARATION

1. Review the content and activities included in the learning path to ensure that everything is current and working appropriately. Update content and activities as needed.
2. Add the ArcGIS Online Challenge to your organization's event registration system, and begin promoting it at least three weeks before the start date.

   While the ArcGIS Online Challenge guide can be accessed any time once published, promoting it as an event and collecting registrations allows for easier communication with a cohort of participants. We suggest organizing one challenge each semester.

3. Use your preferred survey software to create two forms with relevant questions to assist in planning a project or assignment using ArcGIS Online (or GIS, more generally).

   After completing the learning path, participants can use these forms to help transfer the knowledge and skills they have learned to their own research or teaching. Consider including a question so participants can indicate if they would like to schedule a consultation for more focused support. Refer to the section "Resources: Using Your New Skills" of our guide under "Resources and Feedback" for examples of project and assignment planning forms (see Additional Resources).

4. (Optional) Create a form so participants can provide feedback about their learning experience.
5. (Optional) Use your preferred videoconferencing software to schedule weekly "Ask Me Anything" sessions, where interested participants can join a meeting to ask questions and pitch project or assignment ideas based on what they are learning through the challenge.

   Be prepared to demonstrate relevant functionality in ArcGIS Online based on the module topic and apps used during that particular week.

INSTRUCTIONS
Module 1: Getting Started
- Send a welcome e-mail to all registrants at the beginning of the first week describing the structure and expectations for the learning path.

- Provide a primary contact for participants to ask questions, such as a shared e-mail accessible by all facilitators.
- Ensure that participants have clear instructions for creating their own ArcGIS Online organizational account to begin the challenge.

Module 2: Finding Data
- Starting with the second module, use your weekly e-mail to participants to draw connections between the concepts and skills being learned from one module to the next. Emphasize the self-paced nature of the challenge, and call attention to any upcoming "Ask Me Anything" sessions (optional).
- Highlight how ArcGIS Living Atlas of the World can be used not only to find content in ArcGIS Online but also to identify the original sources contributing that content for further browsing.
- If you have created your own resource for helping learners find geospatial data (e.g., a separate guide or tutorial), be sure to call attention to that too.

Module 3: Collecting Data
- Highlight the distinction between map-centric and form-centric data collection strategies in ArcGIS Online so that learners can consider what makes the most sense for their own work.
- Inform participants that they will be required to temporarily install a mobile data collection app on their smartphone or tablet to complete the activities.

Module 4: Exploring and Analyzing Data
- Use examples from different subject areas to highlight the value of spatial analysis for solving a wide variety of real-world problems.
- Encourage participants to reach out if none of the example activities seem relevant to their work in case an appropriate alternative might be available.

Module 5: Visualizing Data and Telling Your Story
- Call attention to the potential uses of ArcGIS Dashboards and ArcGIS StoryMaps in an instructional context, such as creating a dashboard as a digital alternative to a research poster or a story map as an alternative to a traditional research paper.

Following Up
- Send a follow-up e-mail to participants after the challenge to encourage their use of the project and assignment planning forms and feedback form (optional).
- Provide information about how they can continue their learning (e.g., additional trainings) or get more focused support (e.g., scheduling a consultation).

ASSESSMENT STRATEGY
Quantitative assessment data include the number of registrants for each ArcGIS Online Challenge cohort and the total number of guide views during the associated event period. Qualitative assessment data include participant feedback provided through a survey at the end of the learning path, as well as any impacts resulting from deeper engagement with participants during or after the challenge (e.g., focused consultations to advance research projects or teaching assignments, invitations for classroom instruction).

ALLERGY WARNING
Esri continually updates the functionality of its ArcGIS Online suite of apps and regularly launches or retires associated training resources. Facilitators using such third-party resources as content or activities in the learning path should be prepared to evaluate and update the resources included on an ongoing basis. While this poses a minor challenge for learning path maintenance, it has the benefit of helping facilitators to remain current with the latest developments and resource offerings associated with ArcGIS Online.

ADAPTING THE RECIPE
- Our ArcGIS Online Challenge guide is shared in the LibGuides community, and the content and structure are available to be reused and adapted to fit other institutional contexts (see Additional Resources). When adapting the guide, update the content and activities to be most relevant for your learners' needs: for example, selecting content about your geographic region or specific subject areas and activities demonstrating methods and tools associated with research questions you commonly receive.
- We designed the learning path to be

largely self-directed and delivered asynchronously in a remote learning environment. However, this learning path could be adapted for use in other instructional settings. For example, a hybrid workshop could be designed to focus on only one of the module topics. Participants could be asked to review the content prior to the workshop, with the synchronous session devoted to completing the activity and discussing the brainstorming questions.
- We focused on ArcGIS Online, but this learning path structure could be adapted for use with other GIS tools and geospatial data topics, as well as data literacy topics in other areas of librarianship.

Esri Documentation, https://doc.arcgis.com.

Esri Newsroom, https://www.esri.com/about/newsroom/overview/.

Learn ArcGIS, https://learn-arcgis-learngis.hub.arcgis.com.

CHEF'S NOTES

The authors wish to acknowledge Nancy Courtney, whose Research Impact Challenge guide (https://guides.osu.edu/researchimpactchallenge) served as an inspiration as we designed and developed the ArcGIS Online Challenge.

ADDITIONAL RESOURCES

ArcGIS book series, https://learn.arcgis.com/en/arcgis-book-series/.

ArcGIS Living Atlas of the World, https://livingatlas.arcgis.com.

ArcGIS Online Challenge guide, https://guides.osu.edu/agol-challenge.

Esri Academy, https://www.esri.com/training/.

GIS for Success Series
Learning the Basics of QGIS Workshop

Kelly Grove, GIS and Earth Sciences Librarian, Florida State University

NUTRITION INFORMATION

This is an introductory workshop and serves as part one of a series titled "GIS for Success." This series teaches students about using QGIS, a free and open-source desktop application, through which students can learn the basics of geospatial data structure, create visual representations of real-world data, and gain hands-on experience with GIS work. The recipe discussed in the chapter "GIS for Success Series: Let's Make a Map Workshop" builds upon this workshop to teach more in-depth operations and lead learners to make a printable map. The sequencing of these two chapters allows for greater exploration of the QGIS software without being overwhelming.

TARGET AUDIENCE AND NUMBER SERVED

Target audience: Anyone interested in GIS
Number served: Can be scaled for small groups to large classes

LEARNING OUTCOMES

Participants will be able to
- identify the essential icons in the QGIS toolbar
- perform the series of steps to add and view multiple data layers from different sources
- use spatial data attribute tables to
  - filter data to find specific data points
  - interpret data held in columns and rows in attribute tables
  - perform basic calculations on attribute tables in QGIS

COOKING TIME

Workshops are designed to take one hour. Writing the lesson plan will take a few hours.

DIETARY GUIDELINES

After students leave their institution, they may not have access to proprietary programs such as ArcGIS. Lack of access can hinder their ability to apply skills learned during their degree program and hamper their desire to make research and learning part of their life after school. To encourage the idea of research and learning as a lifelong practice that ACRL's *Framework for Information Literacy for Higher Education* strives to embolden students with once they leave college, the two "GIS for Success" workshops are designed to teach attendees how to find open data sources and use freely available tools used to run analyses and illustrate stories.

INGREDIENTS

Instructors can make substitutions based on audience and subject disciplines. For example, a lesson plan can be written using data sources for an environmental science audience, or the data sources can be swapped out for health data for a public health audience with a few tweaks to the base lesson plan.

Basic equipment and technology:
- Computers
- Internet access
- Ability to screen share or project an image
- QGIS
- Google Folders
- Google Docs
- Open Science Framework (OSF) project

Data Sources
- Shapefiles describing local surroundings provided by state agencies. Example: Florida's Geospatial Open Data (https://geodata.floridagio.gov)
  - Florida State Park Points of Interest
  - Florida State Park Trails
  - Florida County Lines
  - Florida
- Shapefiles provided by commercial companies. Example: ArcGIS (https://www.arcgis.com/home/search.html?q=florida county boundaries) Florida County Boundaries

Note: To personalize the workshop, search state and local government websites for GIS repositories to obtain data unique to the area.

Section 6. Geospatial Data

PREPARATION

Each workshop in the series has accompanying documentation written in lab-style format, meaning each step is listed with details and accompanied by annotated screenshots. This documentation helps participants follow along with the instructor, aids participants if they get lost during a step, and allows them to keep it as a reference document. This style of writing documentation is involved and can be overwhelming at first.

Preheating the oven: Starting with a blank lesson plan is not required. Rather, preheat the process by using the *QGIS Training Manual* found at https://docs.qgis.org/3.22/en/docs/training_manual/index.html. The *QGIS Training Manual* is a wonderful guide for getting started, and it supplies inspiration for where to add screenshots with captions, ideas for phrasing instructions, and how to order steps that build upon each other to complete a goal. This manual aids in the process of learning new versions of the software and serves as a reminder of the small nuances that are required to complete a task.

Gathering ingredients: This recipe can be any flavor; while the basic ingredients are the same (QGIS, internet, computers), it can be spiced up by using different data sets and shapefiles. Consider making the workshop specific to the institution, state, or workshop attendees by using GIS data from the area. When an attendee works with a place with which they are familiar, it piques their interest and helps keep them focused on the workshop. The QGIS manual offers prearranged data from the surrounding areas of Swellendam in South Africa. Looking at the attributes of these data provides an idea about the type of shapefiles and underlying attribute tables that would be best for each part of the lesson.

Mise en place: Prior to cooking (teaching) the lesson plan for these workshops, it is necessary to ensure everything is in place. Start with writing the learning objectives for each workshop and create a scenario that frames the learning objectives with a common goal. For example, the following scenario has been used in the original GIS for Success workshops: "As part of a group dedicated to Florida outdoor activities, we were asked to make a map that would help people identify many of the trails and points of interest located within the Wekiwa Springs State Park." This scenario helps narrow down the area the data sets define, provides an end goal for participants to work toward, is loose enough to allow participants to develop their own interpretations of map creation, and allows for exploration of the QGIS software and the data.

Help workshop participants prepare with a few pre-workshop activities listed at the top of the shared lab documentation. Include pre-workshop download instructions for the QGIS software from https://qgis.org/en/site/forusers/download.html and the data sets. Make sure to note the latest software versions available for the various operating systems and ensure the lab instructions are updated. An example of these pre-workshop instructions can be found in the documentation listed under Additional Resources.

Cleanup tips:
1. Reread the lesson plan (from the point of the view of the participants) and follow the written instructions. Ask: Are there any assumptions made about the participants' knowledge of the software? Were any small steps left out when performing a function? Is it clear when a section is complete? If you have difficulty getting to the next step, rewrite the instructions or add an annotated screenshot.
2. Writing the lab-style documentation for participants takes time, but when completed it gives instructors a detailed lesson plan that they can annotate with their own reminders and notes. Add reminders to stop and ask if there are questions and notes about timing to stay on pace.

INSTRUCTIONS

Combine dry ingredients: This step introduces participants to the interface of QGIS.
1. Give a tour of the software interface; point out the different windowpanes and toolbars. This step may feel unnecessary, but if the participants are new to GIS tools, elements such as the layers list or the map canvas may be foreign to them.
2. Review essential toolbar icons that will be used during the workshop. Referencing the *QGIS Training Manual* for this step can help with naming and defining the function of each icon. Explain the process of uploading data. Depending on the number of shapefiles, the participants can

download each individually from the data source's web page, or the instructor can create a shared folder on a Google Drive so that participants can download all the files at once. After all participants have downloaded the data, walk them through the process of uploading files using the Data Source Manager.
3. Note: Depending on size of files, including instructions about unzipping files may be necessary.

Combine wet ingredients: This section of the recipe is about getting the hands dirty digging into the data attribute tables by learning how to read attribute tables and manipulate the tables to answer simple questions. Ask participants to navigate to one of the attribute tables that describes a layer in a moderate amount of detail. With the table open, explain the organization of the data.
1. This is a good time to discuss what questions these tables could answer. For example, in a large attribute table, ask the participants how many of the data points meet a certain criterion. When an answer is given or if everyone gives up because the number is too great to count, this leads right into the next step of this section.
2. Demonstrate a few quick and easy tasks that can be accomplished in the attribute table. Demonstrate how to sort the table in ascending and descending order. Show how a feature on the map canvas is highlighted when the corresponding feature is selected in the attribute table. As a simple demonstration of the power of QGIS, use the Select by Expression tool to answer the question posed earlier.
3. Open a new attribute table that has at least one column containing attribute data similar to that in the previous table. Discuss how these tables relate to each other and how data from one table might enhance the data in the other table.

Mix: With the dry and wet ingredients processed, build upon those skills and concentrate on performing the more complicated tasks of joining tables and running functions. Depending on the data chosen, the recipe's flavor will change because the data will influence the type of functions and queries run.
1. Using two layers identified as having common attribute features in the section above, walk the participants through joining the tables. Run a more complex query in the Select by Expression tool that uses a mix of the two tables.
2. Demonstrate how to use a tool in QGIS to answer a question. Choose a tool based on the type of data and the type of question asked. I suggest using one of the Vector Analysis Tools as they can perform quick analysis of data that would be easy for new GIS researchers to understand. Demonstrating these types of tools will help participants imagine the type of analysis that is possible with GIS data.

Bake: The final stage of this workshop is a preview of a skill needed to complete the subsequent workshop described in "GIS for Success Series: Let's Make a Map Workshop." It gives participants a chance to explore QGIS.

1. Have the participants navigate to the Properties menu of a layer and work within the Symbology tab. Demonstrate how to change the color, fill, size, or shape of the chosen layer. Participants can choose their own symbols and see how it affects the map.
2. As the participants are experimenting, explain that in the next workshop, "Let's Make a Map," a discussion about map design and a deeper exploration into the functionality of the symbology menu will occur.

REVIEWS/ASSESSMENT STRATEGY
A simple way to assess the hands-on workshop is to see if participants were able to successfully produce the end product that the written instructions guide them to create.

CHEF'S NOTES
This recipe lays the foundation for the "GIS for Success Series: Let's Make a Map in QGIS Workshop" recipe.

ADDITIONAL RESOURCES
Downloadable copies of the QGIS lesson plans and more can be found in the Open Science Framework https://osf.io/q84up/?view_only=33959329d6fd4f1491299bb8cea45cf3.

GIS for Success Series
Let's Make a Map in QGIS Workshop
Kelly Grove, GIS and Earth Sciences Librarian, Florida State University

NUTRITION INFORMATION

This recipe is a sequel to the chapter "GIS for Success Series: Learning the Basics of QGIS Workshop." The focus of this workshop is on map creation. Elements from the "GIS for Success Series: Learning the Basics of QGIS Workshop" are reinforced and expanded upon. If detailed instructions and explanations about downloading software, data, and a brief explanation of the QGIS interface are included at the top of the lab-style documents (given to the participants at the start of the workshop), it is not necessary for participants to attend the "Learning the Basics of QGIS Workshop: first. However, if participants do not attend the "Learning the Basics of QGIS Workshop" first, they will miss out on learning about attribute tables and working with various QGIS tools and functions.

TARGET AUDIENCE AND NUMBER SERVED

Target audience: Anyone new to or interested in GIS.

Number served: Workshops can be scaled to fit small groups to large classes.

LEARNING OUTCOMES

At the end of the session students will
- identify the icons used in the New Print Layout window in QGIS
- identify appropriate predefined symbols for points, lines, and polygons based on an object's attributes
- design their own printable map from a set of data files

COOKING TIME

Workshops are designed to take one hour each.

The writing of the lesson plan will take a few hours.

DIETARY GUIDELINES

By creating a map, the students must make choices about how they represent objects and determine the best way to analyze data that fit with the end goal. These choices, which are informed by their point of view, reinforce the idea that the map creation process impacts the purpose and potential perceptions of their research products. The process of creating a map ties directly into ACRL's *Framework for Information Literacy for Higher Education* through the frames Scholarship as a Conversation and Information Creation as a Process.

INGREDIENTS

Make substitutions based on the audience and subject disciplines.

Basic equipment and technology:
- Computers
- Internet access
- Ability to screen share or project an image
- QGIS
- Google Folders
- Google Docs
- Open Science Framework (OSF) project

Data sources:
- Shapefiles describing local surroundings provided by state agencies. Example: Florida's Geospatial Open Data (https://geodata.floridagio.gov)
  - Florida State Park Points of Interest
  - Florida State Park Trails
  - Florida County Lines
  - Florida
- Shapefiles provided by commercial companies. Example: ArcGIS (https://www.arcgis.com/home/search.html?q=florida county boundaries) Florida County Boundaries

Note: To personalize the workshop, search state and local government websites for GIS repositories to obtain data unique to the area.

PREPARATION

The preparation for this workshop is like the preparation steps used in the recipe "GIS for

Success Series: Learning the Basics of QGIS Workshop" with a few minor alterations or suggestions noted below.

Preheating the oven: Having written the lesson plan and lab document for "Learning the Basics of QGIS Workshop" means there is no need to start with a blank page. Using a similar structure is highly encouraged for consistency between the two workshops. Refer to the *QGIS Training Manual* (https://docs.qgis.org/3.22/en/docs/training_manual/index.html) for help creating the order of steps and providing some insight into the capabilities of QGIS.

Mise en place: Using the same scenario as "Learning the Basics of QGIS Workshop" is beneficial because it makes it easier for the instructor to transition from one lesson to the next and helps solidify the two "GIS for Success" lessons. The scenario used in the original series is, "As part of a group dedicated to Florida outdoor activities, we were asked to make a map that would help people identify many of the trails located within the Wekiwa Springs State Park." In this chapter, participants will fully complete the task laid out in the scenario. Include the same pre-workshop instructions used in "Learning the Basics of QGIS Workshop" as they will help new participants catch up to the rest of the group.

INSTRUCTIONS

With the preparation steps in place, it is now time to assemble the lesson. Just as in baking, it is important to do this process in steps and use the ingredients in the proper order.

Combine dry ingredients: This serves as a brief reminder to participants about how to get data into QGIS; it also ensures that any newcomers are properly set up.

Start the workshop by providing a quick demonstration of how to upload a shapefile to the program. If there is a problem, refer the participants to the lab documentation used in the chapter "GIS for Success Series: Learning the Basics of QGIS Workshop" for the uploading steps.

Combine wet ingredients: In this portion of the recipe, the goal is for the participants to learn and experiment with the variety of options available to them in the symbology of the data.

1. Ask everyone to examine the types of symbols available for polygon layers. Using the Symbology tab in the Layer Properties window, explore the color and fill options. Repeat this with point layers and line layers, while allowing participants to choose their own symbols. For a real crowd-pleaser, demonstrate how to change a point layer to categorized symbology so that different points will be different colors based on an attribute.

Ask if anyone can clearly read their map. Did they use too many colors? Are they able to tell the difference in the colors of the points to know which color represents which attribute? Can they tell the difference between the line layers and the lines in the polygon layer? Start a conversation about basic cartographic design. A suggested topic to start with is visual contrast and the color wheel.

2. For a simple introduction to the principles of cartography design, check out the Esri post "Design Principles for Cartography" (https://www.esri.com/arcgis-blog/products/product/mapping/design-principles-for-cartography/) from October 2011. This post discusses basic principles with illustrated examples and provides a short list of recommended cartography textbooks for more in-depth reading.

3. Ask participants to change their symbols to fit with the design principles discussed. Make suggestions about which colors and fills to use. Ask participants to remove points from the point layer that was categorized. Discuss where maps may differ based on the person creating the map. For example, one map maker may choose to include bathroom locations, while another does not deem that necessary to meet the goals of the scenario.

Walk the participants through the process of creating labels for a layer of the map. Create labels for a layer that contains multiple polygons, as these elements tend to be larger and manipulation of the label around the polygon is easier then when working with a point layer. Demonstrate how to edit the labels by changing font size, adding a buffer, and so on.

4. Talk about map legibility. This is especially important when producing paper maps as the viewers cannot zoom in to the map to read the labels.

Mix: It is time to introduce the participants to the Print Layout window of QGIS. This step

Section 6. Geospatial Data

will allow participants to do more experimenting to get them ready to create the final map.

1. Help the participants navigate to the Print Layout window. Spend a few moments explaining the new icons used to add map elements to the workspace.

   Demonstrate how to add a new map to the workspace. As this is the main component to create a map, spend some time showing everyone how to move the map image around, zooming in and out to different levels, and how to best situate the large map on to the page.

2. Tip: Mention leaving a little room around the margins, or else when the map is printed, it will get cut off during printing.

3. Have the participants add the legend to map and point out the editing features. More time will be spent on this step when creating the actual map. This is a time to get participants familiar with the new window and the tools.

Bake: With the batter made, it is time to make the actual map. This step will combine everything learned in this workshop and incorporate a little bit from the recipe "Learning the Basics of QGIS Workshop."

1. Bring everyone back to the main QGIS workspace. Help the participants use the attribute tables to zoom the map canvas to the correct location. Tell them they can double-check this by using the information mouse icon.
2. With the map zoomed to a better scale, you will likely need to modify the symbols. Check the line thickness of line layers, remove any unnecessary symbols from line and point layers that do not help meet the goal of the map.
3. If desired, participants can add additional labels to line or point features.
4. Have the participants open a new print layout window to add map elements. Have them find the right position of the area and where to place the legend. Working with the legend elements will take the most time.
5. At some point, a label or symbol will not be right for a print map. Demonstrate how to move between the print layout window and the map canvas to make changes.
6. Allow the participants to add map elements and create symbology as they see fit. While they are working, remind them about basic cartography principles that were discussed earlier.
7. Finally show participants how to export or print their map. Ask volunteers to show their maps so the group can compare the different design choices and how they affect the message the map presents.

CHEF'S NOTES
Additional notes about the background and theory of this workshop, cleanup, and a possible assessment strategy can be found in the chapter "GIS for Success Series: Learning the Basics of QGIS Workshop."

ADDITIONAL RESOURCES
Downloadable copies of the QGIS lesson plans and more can be found in the Open Science Framework: https://osf.io/q84up/?view_only=33959329d6fd4f1491299bb8cea45cf3.

Statistical and Geospatial Literacy for Integrative Genetics

Jay Forrest, Data Scientist Librarian and subject specialist, Georgia Institute of Technology, jay.forrest@library.gatech.edu;
Chrissy Spencer, Principal Academic Professional, Georgia Institute of Technology, chrissy.spencer@biology.gatech.edu

NUTRITION INFORMATION

In this baklava-inspired recipe, we introduce students to R and RStudio (the phyllo dough), geospatial literacy (our walnut filling), and statistical techniques (the honey syrup that saturates each layer). As they create different layers of phyllo, filling, and topping, students apply raster and vector data to use in their analysis of urbanization's impacts on European honey bee *Apis mellifera* colony collapse. Colony Collapse Disorder is a global phenomenon that has dramatically reduced population sizes of the common pollinator *Apis mellifera* in recent decades.[1] Common hypotheses for factors that influence loss of bee colonies include industrial agricultural practices, anthropogenic landscape impacts, and the spread of parasites and viruses that weaken individual bees and spread quickly throughout a colony.[2]

Over a 15-week semester, students complete a course-based undergraduate research experience (CURE) to examine the potential for urban, human-built environmental features to affect the number and quantity of RNA viruses, called the viral load, infecting the colony's workers. As described in Spencer and Harrison, students read background material and develop a research question and hypothesis.[3] Background material on bees includes the USDA *Bee Basics* book, primary literature, and papers the students identify on their own (see Additional Resources for more recommended reading). The course website at https://ingeneticslab.biosci.gatech.edu describes the detail and flow of student progress across the 15-week semester, by week, with links to references and resources. Aside from molecular genetics benchwork, the students select metrics of urbanization, defined as human impact on the landscape, that they hypothesize would detrimentally affect colony health. Typical urbanization metrics, such as percent impervious surface, use data from the geospatial analysis. The geospatial data provide the independent variable for additional data analysis using regression, t-test, or ANOVA models.

TARGET AUDIENCE AND NUMBER SERVED

Ideas presented here are relevant for undergraduate researchers at any level conducting research with applied geospatial or GIS variables. Students were enrolled in a sophomore-level genetics lab course, and 85 students participated over the past two years.

LEARNING OBJECTIVES

After completing these lessons, students will be able to
- conceptualize geospatial information and apply geospatial data to define the independent variables (categorical or continuous) of a research study using spatial data.
- analyze data using the geospatially defined independent variable and a differentially described dependent variable using basic statistical analysis
- interpret and evaluate results
- cite relevant primary literature

COOKING TIME

Prep time: 2 sessions of 2 hours each using R
Bake time: Semester-length one-credit-hour laboratory course

DIETARY GUIDELINES

The recipe serves to expose students in the biological sciences to geospatial and statistical techniques using the R programming language.

INGREDIENTS

- Cooking utensils: R and RStudio.
- Cooking techniques: spatial overlays, ANOVA, t-test, regression.
- 15 sheets of R packages for data transformation and spatial analysis (raster, rgdal, rgeos, tidyr).
- ¾ cup point feature data—latitude/longitude of bee colonies: While the actual bee colony data are protected to prevent theft, you can create your own list

of 20–30 locations. The source .csv file should contain two columns: a numeric ID and a comma-separated coordinates pair in decimal degrees. To match the other geographic data, we recommend selecting latitudes between 33.73 and 34.73 and longitudes between −83.6 and −84.6.
- ½ tbsp linear feature data—Georgia roads (see Additional Resources).
- ¼ cup polygon feature data—Georgia county boundaries/census tract data (see Additional Resources).
- ½ cup raster feature data—Georgia impervious surfaces (see Additional Resources).
- 1 cup of knowledge of honey bees (see Additional Resources for recommended reading on bee basics).
- ½ cup of application knowledge of statistics.
- Students will customize their recipes through research and incorporation of additional ingredients (data) into their data analysis, leveraging R and geospatial techniques learned and results obtained during two RStudio 2.5-hour sessions.

PREPARATION

Mise en place: We have found that introducing students to RStudio and geospatial data and analysis works best within a cooking studio where each student has a "cooking station" with R/RStudio and required spatial data packages preinstalled, as some of the packages can take time to install. Students complete RStudio and package installation on their own machines to continue working after class.

INSTRUCTIONS

Preheat the oven: Provide instructions and an interactive session to help the students become familiar with the RStudio interface and the four-pane layout (Source, Console, Environment, Files/Plots/Packages/Help).

Prepare the phyllo dough: Teach students basic R data structures and syntax. Then have the students practice creating data objects, loading external data, and developing simple linear regression or between-group comparisons such as t-test and analysis of variance.

Mix the filling: Give the students a basic overview of geospatial literacy, including coordinate systems, map projections, and raster versus vector, and a theoretical overview of data joins, spatial joins, and spatial overlays.

Layer the dough and filling: Provide code to guide the students to apply layers of geospatial information to the colony location information.
- *Layer 1—raster data:* We begin with the largest file type, a statewide raster file of impervious surfaces using raster(). (See Additional Resources, "Raster feature data")
- *Layer 2—polygon data:* In layer two, students use shape files of Georgia counties using readOGR. Students use spTransform() to match the projection of the original raster file. Students use list-based subsetting (%in%) to select specific counties of interest and intersect() to isolate spatial data to those counties. Optionally, students can be introduced to other polygon features, such as census tracts.
- *Layer 3—data merge:* Students use sp::merge() to join features from external data on either the GEOID10 and FIPS codes of their polygon data.
- *Layer 4—linear features:* Students import Tigerline files of Georgia roads subset by road type, re-project the data to match other data layers, and then intersect() on county boundaries.
- *Layer 5—point features:* Bee colony locations are brought in from a .csv file with lat/long coordinates (see Additional Resources). Students load the file, add a coordinate system (coordinates(df) <- ~lon + lat), project the data (proj4string(df) <- CRS("+proj=longlat +ellps=WGS84 +datum=WGS84")), and then re-project to match other data layers.

Bake: The output is a data frame with each observed colony in rows and variables created from a geospatial analysis in columns: distance to the nearest road using gDistance(), average impervious surface within a defined buffer using raster::extract(), and other characteristics of the containing county with sp::over() and cbind().

Prepare and pour honey mixture over the top: Next, students use the colony geospatial information to inform a continuous or categorical independent variable for one of the following parametric analyses: linear regres-

sion lm(y~x), t-test t.test(x,y), or analysis of variance aov(y~x) with pairwise comparisons using Tukey's Highly Significant Difference test TukeyHSD(). In each case, students use their calculation of viral load from the pooled genetic data collected by the entire class as the dependent or response variable. To complete their analysis, students create a data set with bee colony as the unit of replication and organize their independent and dependent variables in a data table appropriate for their selected analysis. First, each student applies the R code to complete each analysis using mock data to learn how the code works; then they organize their own data by imitation and complete a parametric data analysis.

Slice and serve: Informed by their inferential statistical analysis, students interpret the results to support or refute their scientific hypothesis. The results are incorporated into a written lab report of their work over the semester. They draw conclusions grounded in the evidence to address their research question.

ASSESSMENT STRATEGY

We assess student work related to the geospatial and inferential statistics component of the course as their active participation in the practice of R coding, their responses to pre-lab questions to download R and to interpret their data, the completeness and accuracy of their lab notebook entry on the process of coding inputs and outputs, and finally their lab report methodology, result, and discussion that puts the coding work and statistical analysis in the context of their scientific research question.

CHEF'S NOTES

While there are many map projections available to use, for simplicity of processing information in R, we bring the raster layer in first and then re-project additional vector layers into the same projection system.

While the recipe will present bee colonies as point features, roads as linear features, and county boundaries as polygon features, alternate geographies can be used depending on the focus of the course.

ADDITIONAL RESOURCES

Course website containing links to research articles and inferential statistics R code: https://ingeneticslab.biosci.gatech.edu.

Spencer, Chrissy C., and Colin D. Harrison. "Designing Course-Based Undergraduate Research Experience (CURE) Labs with Citizen Science and Service Learning." *Proceedings of the Association for Biology Laboratory Education* 39 (2018), article 17. https://www.ableweb.org/biologylabs/wp-content/uploads/volumes/vol-39/Spencer.pdf.

Resources on Bee Basics

Moissett, Beatriz, and Steve Buchanan. *Bee Basics: An Introduction to Our Native Bees*. USDA Forest Service and Pollinator Partnership, 2010. https://efotg.sc.egov.usda.gov/references/public/SC/Bee_Basics_North_American_Bee_ID.pdf.

Sguazza, Guillermo Hernán, Francisco José Reynaldi, Cecilia Mónica Galosi, and Marcelo Ricardo Pecoraro. "Simultaneous Detection of Bee Viruses by Multiplex PCR." *Journal of Virological Methods* 194, no. 1–2 (December 2013): 102–6. https://doi.org/10.1016/j.jviromet.2013.08.003.

Steinhauer, Nathalie A., Karen Rennich, Michael E. Wilson, Dewey M. Caron, Eugene J. Lengerich, Jeff S. Pettis, Robyn Rose, et al. "A National Survey of Managed Honey Bee 2012–2013 Annual Colony Losses in the USA: Results from the Bee Informed Partnership." *Journal of Apicultural Research*, 53, no.1 (2014): 1–18. https://doi.org/10.3896/IBRA.1.53.1.01.

Youngsteadt, Elsa, R. Holden Appler, Margarita M. López-Uribe, David R. Tarpy, and Steven D. Frank. "Urbanization Increases Pathogen Pressure on Feral and Managed Honey Bees." *PLOS ONE* 10, no. 11 (2015). https://doi.org/10.1371/journal.pone.0142031.

R Language and RStudio Development Environment

R Core Team. *R: A Language and Environment for Statistical Computing*. Vienna, Austria: R Foundation for Statistical Computing, 2021. https://www.R-project.org.

R language is open source and is available for Linux, MacOS, and Windows at The Comprehensive R Archive Network, https://cran.r-project.org.

The RStudio IDE is open source and is available for Linux, MacOS and Windows at https://www.rstudio.com/products/rstudio/download/.

Geographic Data

Raster Feature Data

University of Georgia Natural Resources Spatial Analysis Laboratory, "Ga. Land Use Trends Impervious Surface Cover 2008." https://data.georgiaspatial.org/index.asp?body=preview&dataId=43979. Free registration required to access.

Polygon Feature Data

Georgia County shapefile, https://arc-garc.opendata.arcgis.com/datasets/dc20713282734a73abe990995de40497_68.

Linear Feature Data

Georgia Road shapefile, https://www2.census.gov/geo/tiger/TIGER2014/PRISECROADS/.

NOTES

1. Nathalie A. Steinhauer, Karen Rennich, Michael E. Wilson, Dewey M. Caron, Eugene J. Lengerich, Jeff S. Pettis, Robyn Rose, et al., "A National Survey of Managed Honey Bee 2012–2013 Annual Colony Losses in the USA: Results from the Bee Informed Partnership," *Journal of Apicultural Research* 53, no. 1 (2014): 1–18, https://doi.org/10.3896/IBRA.1.53.1.01.
2. Elsa Youngsteadt, R. Holden Appler, Margarita M. López-Uribe, David R. Tarpy, and Steven D. Frank, "Urbanization Increases Pathogen Pressure on Feral and Managed Honey Bees," *PLOS ONE* 10, no. 11 (2015), https://doi.org/10.1371/journal.pone.0142031.
3. Chrissy C. Spencer and Colin D Harrison, "Designing Course-Based Undergraduate Research Experience (CURE) Labs with Citizen Science and Service Learning," *Proceedings of the Association for Biology Laboratory Education* 39 (2018), article 17, https://www.ableweb.org/biologylabs/wp-content/uploads/volumes/vol-39/Spencer.pdf.

Web Map Layer Cake
Teaching Web Mapping Skills with Leaflet for R

Sarah Zhang, Librarian for Geography, GIS, & Maps, Simon Fraser University, s_zhang@sfu.ca; Julie Jones, Research Commons Librarian, Simon Fraser University, julie_jones@sfu.ca

NUTRITION INFORMATION

Web maps, or interactive maps, are tools for both data exploration and communication. Leaflet is a popular JavaScript library for interactive maps; however, to master it, users need basic knowledge of HTML, CSS, and JavaScript. In this session, students learn Leaflet for R, an R package that makes it easy to integrate and control Leaflet maps in R without knowledge of JavaScript. They learn how to create an interactive web map and, equally importantly, are encouraged to think critically about the decisions they make that determine the way a web map is designed, displayed, and communicated. As a result of this process, they will choose a final format that suits their purposes: for example, an R Markdown file or an HTML file.

TARGET AUDIENCE AND NUMBER SERVED

Graduate students across disciplines who are familiar with the R environment and interested in creating web maps to explore data and communicate their research. Number served: about 20.

LEARNING OUTCOMES

After this workshop, students will be able to
- describe introductory concepts of web mapping, including map tiles and pop-ups
- create a simple interactive web map using Leaflet for R
- create an R Markdown file and publish it as a GitHub page (see figure 1)
- identify and defend standard decisions

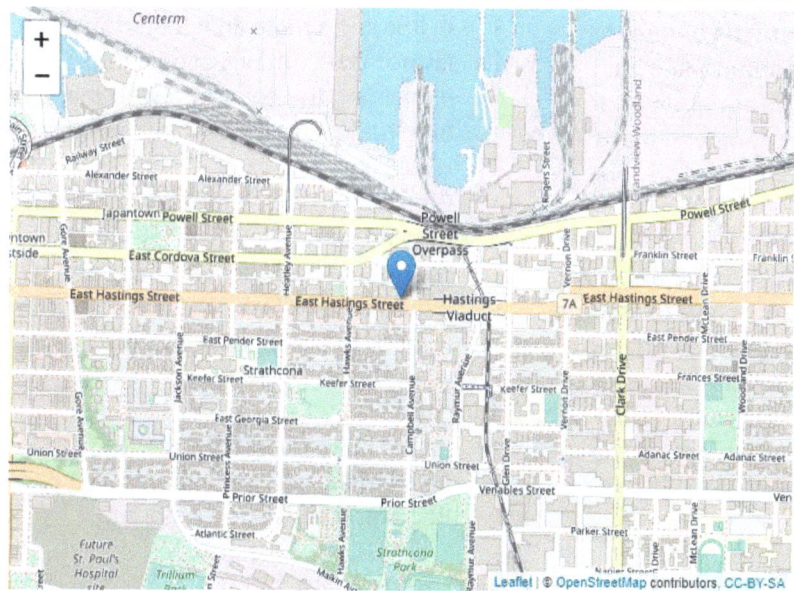

FIGURE 1
Screenshot of the finished GitHub page (converted from a R Markdown file and hosted on GitHub)

Section 6. Geospatial Data

made when publishing a web map, e.g. who the target audience is and what the most suitable format for communicating the map to the target audience is

COOKING TIME
Prep time: 120 minutes for instructor, 10 minutes for students
Cook time: 120 minutes
Total time: 250 minutes

DIETARY GUIDELINES
The session ties into the frames Information Creation as a Process and Scholarship as Conversation from ACRL's *Framework for Information Literacy for Higher Education*, with necessary adaptation to the geospatial data context. Information Creation as a Process recognizes the evolving landscape of information creation and formats of research delivery. Leaflet, a web mapping API, is a new(ish) way of creating information and sharing research in the form of web maps. This workshop supports many of the concepts within this frame.

Leaflet also allows students to step away from being information consumers and to assume the role of creators and to learn by doing. This also allows students to "see themselves as contributors to scholarship rather than only consumers of it,"[1] which is one of the dispositions of the frame Scholarship as Conversation.

INGREDIENTS
- Presentation slides
- Access to R and RStudio on lab or students' personal computers
- An R Markdown file (.Rmd) that contains code chunks, text explaining the code, and web maps
- A GitHub repository that contains an R Markdown file and an HTML file that the .Rmd file is converted to, shapefiles or other data sets
- A web page published as a GitHub page (same contents as the R Markdown file)

PREPARATION
Instructor:
- Find a shapefile from an open data portal (e.g., https://opendata.vancouver.ca/pages/home/, which is also linked from the GitHub page included in this chapter).
- In RStudio, create an R Markdown file (.Rmd) that contains code chunks, text explaining the code, and web maps.
- In RStudio, knit the R Markdown file to a HTML file.
- Create a GitHub repository (repo), upload the .Rmd file, HTML file, and shapefile or other data sets to the repo.
- Rename the HTML file as index.html.
- Publish the repo as a GitHub page.

Students:
- Install RStudio and R.
- Install the package Leaflet for R.
- Register for a GitHub account.

INSTRUCTIONS
1. Provide a short theoretical introduction to basic concepts: web maps and how they differ from other maps, map tiles and geojson, Leaflet and Leaflet for R.
2. Working in RStudio, demonstrate the following and have the students complete each step along with you:
    » how to write code
    » how to initiate a Leaflet Map in R
    » how to tweak Map Attributes (including zoom levels, pop-ups, and basemap) (See figure 2)
    » how to change a marker
    » how to add data (WMS and shapefile)
3. Ask students to create an R Markdown file in RStudio and add code chunks, text, and headings.
4. Knit the R Markdown file to an HTML file and have the students do the same.
5. Lead the students through the following: Create a GitHub repository (repo), upload the .Rmd file, HTML file, and the shapefile or other data sets to the repo; rename the HTML file as index.html; publish the repo as a GitHub page.
6. Mix in the following active learning activity to explore the decisions a researcher makes when creating a web map:

Think-Pair-Share
- Instruct students to *think* about the following for a few minutes on their own, writing down some notes:

    Think about the target audience for your web map. Is this a map for the general public or for other researchers who are experts in your subject area? How does the

Tweaking Map Attributes

Let's talk about zoom levels on a tile. For Open Street Maps, these levels run from 0 to 20

```
t1 <- leaflet(options = leafletOptions(minZoom = 7, maxZoom = 18)) %>%
  addTiles() %>%
  addMarkers(lng=-123.085269, lat=49.281178, popup="Prototype Cafe; on the want-to-try list")
t1
```

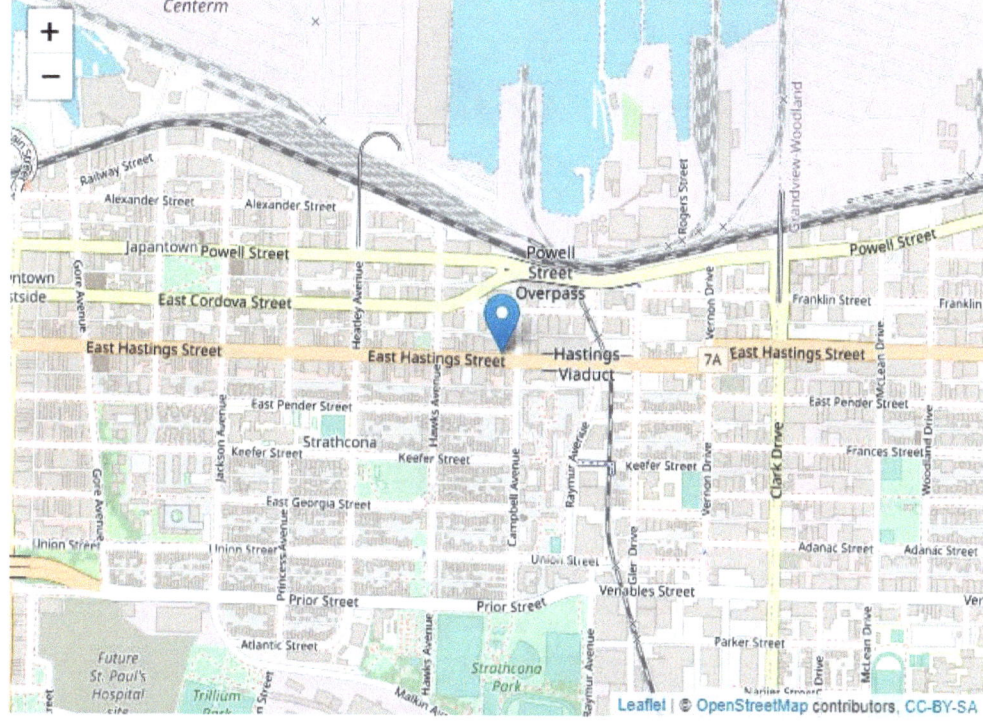

FIGURE 2
Screenshot of the finished GitHub page: how to tweak Map Attributes (zoom levels)

intended audience impact how you design, display, and communicate your web map?

- *Pair*. Have the students pair up and share and discuss their thoughts with a partner.

- *Share*. Facilitate a mini-discussion where pairs share what they discussed.

If conversations stall, add prompting questions to the mix:

– How can an R Markdown file (or a web page when knitted to HTML and hosted on GitHub) facilitate communication with other researchers?
– Would the above approach suit the purpose of communicating with the general public? If not, how would you use your web map?
– What other elements should you consider in designing a web map? For example, how much information in a pop-up should you include? Would you include working data?

REVIEWS/ASSESSMENT STRATEGY

Have the students access a Google Form at the end of the session with the following questions:

Think about today's session...
1. What is one thing you learned today?
2. Is there anything you are confused about? If yes, what?
3. What is one thing you want to learn more about?

CHEF'S NOTES

For each chunk of demoing and playing around with code, leave enough time for students to try out the code and ask questions. The session is designed in a way that makes sure that students with minimal R experience can succeed.

The recipe preparation for this workshop—creating an R Markdown file, knitting it to HTML, and hosting it as a GitHub page—serves two purposes:
1. It provides all the code so that any stu-

dents struggling to have a chunk of code yield expected results can copy and paste the code into RStudio and actually create something without giving up.
2. It provides an example of the end product of the workshop, demonstrating how to use an R Markdown file containing web maps for communication.

Please refer to "Web Mapping with R" by Aateka Shashank for reference.

ADAPTING THE RECIPE
This workshop can be delivered in person or online.

ADDITIONAL RESOURCES
Shashank, Aateka "Web Mapping with R." November 27, 2020. https://aticup.github.io/webmapr/.

NOTE
1. Association of College and Research Libraries, *Framework for Information Literacy for Higher Education*, (Chicago: Association of College and Research Libraries, 2016), https://www.ala.org/acrl/standards/ilframework.

Section 7.
Data in the Disciplines

167   [[Ch43]]**Data in Context: How Data Fit into the Scholarly Conversation**
Theresa Burress

171   [[Ch44]]**Let the Dough Rise! Integrating Library Instruction in a Digital Humanities Course**
René Duplain and Chantal Ripp

175   [[Ch45]]**Ethics and Biodiversity Data**
Rebecca Hill Renirie

179   [[Ch46]]**Data Decisions and the Research Process in the Sciences and Social Sciences**
Nicole Helregel

182   [[Ch47]]**Financial Data for Economics Students**
Jennifer Yao Weinraub

183   [[Ch48]]**Stuffed Shiny App with Business Intelligence**
Yun Dai and Fan Luo

189   [[Ch49]]**Fast Casual Marketing Strategies**
Juliann Couture, Halley Todd, and Natalia Tingle Dolan

192   [[Ch50]]**When and Where: A Framework for Finding and Evaluating Social Science Data for Reuse**
Ari Gofman

197   [[Ch51]]**Data Literacy Layered Lasagna for Preservice Teachers**
Brad Dennis and Allison Hart-Young

Data in Context
How Data Fit into the Scholarly Conversation

Theresa Burress, Assistant Director of Research and Instruction, Nelson Poynter Memorial Library, University of South Florida, tburress@usf.edu, *https://works.bepress.com/theresa-burress/*

NUTRITION INFORMATION
This activity uses peer-reviewed research articles to introduce students to the concept of scholarly conversation, with a focus on how data are collected, analyzed, and presented in research.

TARGET AUDIENCE AND NUMBER SERVED
- Undergraduate or graduate students, any level
- 3–30 students

LEARNING OUTCOMES
The learning outcomes for this activity tie into the following frames from ACRL's *Framework for Information Literacy for Higher Education*:
- Scholarship as Conversation
  - Recognize that local faculty and researchers are both authors of a research study and part of a scholarly community.
  - Interpret and evaluate whether and how effectively the format of the data visualizations in a research study supports the stated research objectives.
- Research as Inquiry
  - Interpret and critically evaluate the use of data sources in the context of original research, considering research design factors such as qualitative vs. quantitative, experimental vs. compiled, etc.
  - Communicate data-based insights of the research study, considering such factors as reproducibility, etc.
- Information [and Data] Creation as a Process
  - Differentiate between different types of data sources as represented by data visualizations in the scholarly literature of a discipline (e.g., experimental data sets, compiled data sets, etc.).
  - Interpret and evaluate data visualizations in a research study in the context of best data presentation practices.

COOKING TIME
- Librarian preparation time: 1–4 hours to identify research articles appropriate to the discipline and authored by researchers or faculty who are part of the campus or local research community.
- Student work time:
  - Reading/pre-assignment: 60 minutes
  - Synchronous discussion: 45–90 minutes (virtual or in person)

DIETARY GUIDELINES
Faculty working in data-intensive disciplines expect students to have a range of data-related skills when they begin graduate programs.[1] In addition, faculty across disciplines are incorporating data literacy competencies into undergraduate courses to help students prepare for a data-driven workforce.[2] Data literacy competencies such as interpreting or evaluating data often require an understanding of the broader context of the data sources, including how and why the data were collected. Thus these data literacy competencies are suited to be embedded within library instruction that focuses on the frames Scholarship as Conversation, Research as Inquiry, and Information Creation as a Process. Prado and Marzal situate data literacy competencies along a continuum that includes information literacy.[3] This framework serves as a practical and flexible road map for librarians to successfully incorporate data literacy into their information literacy instruction.

INGREDIENTS
- 3–6 full-text, original research articles on topics that fall within the scope of the course.
- Pre-assignment questions (see Data Literacy Teaching Toolkit website for example).
- 10-minute presentation introducing how data, and the research articles reporting these data, fit into the research process (see figures 1a and 1b for an example of the research process).

Section 7. Data in the Disciplines

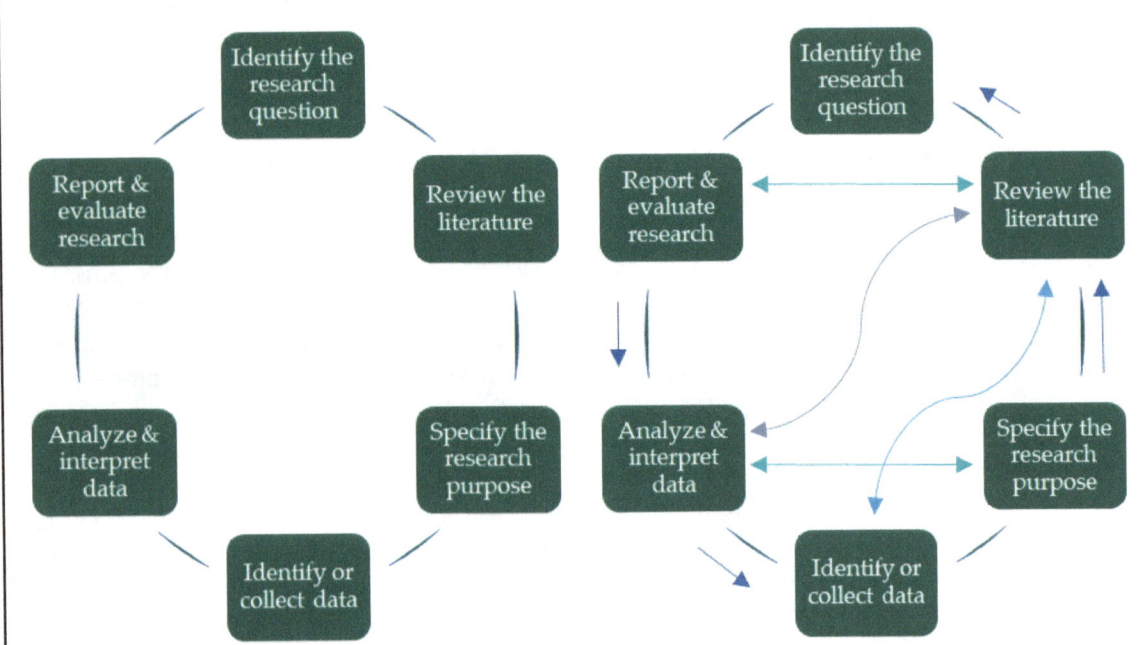

FIGURE 1A
Diagram showing the recursive nature of a research project, which often begins with a research question and ends with a new research question. This "tidy" example of a research cycle is applicable across a range of disciplines.

FIGURE 1B
Diagram showing the "messy reality" of the research cycle, which often requires revisiting the literature at multiple points during the process. It can be helpful for students to visualize the messiness of the research process, which isn't as tidy or linear as depicted in figure 1a.

session. The Data Literacy Teaching Toolkit hosted by the University of South Florida Libraries[4] contains an example pre-assignment worksheet (https://lib.stpetersburg.usf.edu/ld.php?content_id=59861334) that includes a checklist of best data presentation practices for tables and graphs adapted from Duquia and colleagues[5] and a series of questions about the research study and how the data support the research objectives.

- Provide the full-text articles and pre-assignment worksheet to the instructor, who then directs each student to choose an article to read and complete the related assignment. The students bring their responses to a synchronous virtual or in-person session for discussion.

- 3–30 students divided into groups of 2–3 students based on the article they read.

PREPARATION

- Consult with instructor regarding the scope of the assignment and discussion, and select 3–6 research articles based on these criteria:
  - subject matter pertinent to the course
  - authored by researchers/faculty who are part of the local research community
  - published in peer-reviewed journals (1 or more open-access examples)
  - demonstrate varied examples of research design, data collection, compilation, sources, visualizations
- Customize the pre-assignment worksheet as needed to accommodate the level and learning objectives of the

INSTRUCTIONS

- Begin the session with a brief presentation about how peer-reviewed articles and the corresponding data reported within the articles fit into the research process. Review the worksheet that students used to complete the pre-assignment. This worksheet provides the structure for the discussion of each assigned article.
- Gather the students into groups organized by article, and ask them to spend 5 minutes discussing the article and comparing their individual responses to the worksheet questions. Then have them choose a spokesperson from their group to lead discussion of their article.
- Ask the students to reconvene as a

single group. A spokesperson from each group leads the class discussion about their article, providing an overview of the research objective and aspects of the data sources, methods, visualizations, and how well the data support the research findings. The students describe at least one figure with regard to best data presentation practices. Students consider the effectiveness of the data presentation and how they might present the data.

- As the discussion proceeds, highlight various points of interest that reinforce information and data literacy concepts such as provenance of data; the relationships between articles, if any; the broader context of the research; and relationship to the local research community. Careful selection of varied examples creates opportunities to compare and contrast disciplinary approaches, research design and methods, data sources, and aspects of data visualization.
- Depending on course goals and time constraints, conclude with time for students to work independently on classroom computers, using the article to (1) find related articles, (2) shape a research topic, or (3) identify opportunities for internships, independent study, or research mentors.
- At the conclusion of the workshop, share information about how to make appointments for individual research consultations with librarians.

REVIEWS/ASSESSMENT STRATEGY

- *Formative:* As the discussion proceeds, the librarian can assess student comfort levels with academic literature in the relevant topic and fill gaps, offer to provide supplemental material to the instructor, etc.
- *Summative:* Student responses on the worksheets can be reviewed; customized feedback or grades based on a rubric can be provided.

CHEF'S NOTES

- When cooking for the first time in a college workshop or a freshman honors seminar, the discussion may focus on getting to know faculty as part of a research community. The presentation introduces the concept of original research based on data collection and differentiates between faculty as teachers and faculty as authors with research agendas. The discussion also encourages students to seek out faculty whose research aligns with the student's interests for the purpose of research mentorship or employment as research assistants.
- To spice up a senior or graduate seminar, the overview presentation may address the audience as authors of research studies, focusing more deeply on research design methods, the data sources and visualizations, and research reproducibility. You could also highlight the ways in which bibliographies reflect the overarching history of the various subdisciplines. For example, you could contrast microbiological or genetic studies that primarily use very recent literature due to the advent of technological advances with behavioral studies on observation of fauna like sharks, which may cite studies authored by Charles Darwin.

ADDITIONAL RESOURCES

Burress, Theresa, and Emily Mann. "Data Literacy Teaching Toolkit." Nelson Poynter Memorial Library, University of South Florida St. Petersburg, 2019. https://lib.stpetersburg.usf.edu/dataliteracy.

Burress, Theresa, Emily Mann, Susan Montgomery, and Rachel Walton. "Data Literacy in Undergraduate Education: Faculty Perspectives and Pedagogical Approaches." In *Data Literacy in Academic Libraries: Teaching Critical Thinking with Numbers*, edited by Julia Bauder, 1-22. Chicago: American Library Association, 2021.

Carlson, Jake R, Michael Fosmire, Chris Miller, and Megan R. Sapp Nelson. "Determining Data Information Literacy Needs: A Study of Students and Research Faculty," preprint. Libraries Faculty and Staff Scholarship and Research, Purdue University e-Pubs, paper 23. 2011. https://docs.lib.purdue.edu/lib_fsdocs/23.

Carlson, Jake, and Lisa R. Johnston, eds. *Data Information Literacy: Librarians, Data, and the Education of a New Generation of Researchers*. West Lafayette, IN: Purdue University Press, 2015. http://www.datainfolit.org/dilguide/.

Duquia, Rodrigo Pereira, João Luiz Bastos,

Section 7. Data in the Disciplines

Renan Rangel Bonamigo, David Alejandro González-Chica, and Jeovany Martínez-Mesa. "Presenting Data in Tables and Charts." *Anais Brasileiros de Dermatologia* 89, no. 2 (April 2014): 280–85. https://doi.org/10.1590/abd1806-4841.20143388.

Prado, Javier Calzada, and Miguel Ángel Marzal. "Incorporating Data Literacy into Information Literacy Programs: Core Competencies and Contents." *Libri* 63, no. 2 (2013): 123–34. https://doi.org/10.1515/libri-2013-0010.

NOTES

1. Jake Carlson and Lisa R. Johnston, eds., *Data Information Literacy* (West Lafayette, IN: Purdue University Press, 2015), http://www.datainfolit.org/dilguide/.
2. Theresa Burress, Emily Mann, Susan Montgomery, and Rachel Walton, "Data Literacy in Undergraduate Education: Faculty Perspectives and Pedagogical Approaches," in *Data Literacy in Academic Libraries: Teaching Critical Thinking with Numbers*, ed. Julia Bauder (Chicago: American Library Association, 2021), 1-22.
3. Javier Calzada Prado and Miguel Ángel Marzal, "Incorporating Data Literacy into Information Literacy Programs: Core Competencies and Contents," *Libri* 63, no. 2 (2013): 123–34, https://doi.org/10.1515/libri-2013-0010.
4. Theresa Burress and Emily Mann, "Data Literacy Teaching Toolkit," Nelson Poynter Memorial Library, University of South Florida St. Petersburg, 2019, https://lib.stpetersburg.usf.edu/dataliteracy.
5. Rodrigo Pereira Duquia, João Luiz Bastos, Renan Rangel Bonamigo, David Alejandro González-Chica, and Jeovany Martínez-Mesa, "Presenting Data in Tables and Charts," *Anais Brasileiros de Dermatologia* 89, no. 2 (April 2014): 280–85, https://doi.org/10.1590/abd1806-4841.20143388

Let the Dough Rise!
Integrating Library Instruction in a Digital Humanities Course

René Duplain, Research Librarian (GIS), University of Ottawa Library, rene.duplain@uottawa.ca; Chantal Ripp, Research Librarian, University of Ottawa, Chantal.Ripp@uottawa.ca

NUTRITION INFORMATION
These instruction sessions support a fourth-year undergraduate digital humanities capstone project. Digital humanities are a method of research and discovery that uses digital tools to answer questions that have traditionally belonged to the humanities. Students collaborate on a group project and explore the cultural development of the city of Ottawa as it relates to its immigrant population. ArcGIS StoryMaps was chosen as a tool in this project for its ability to create a compelling narrative by leveraging text, images, and maps. An emphasis is placed on integrating data literacy principles into the sessions, particularly since not all students have an extensive background working with data.

TARGET AUDIENCE AND NUMBER SERVED
The sessions have been delivered to a fourth-year digital humanities course with seven students. However, the content and model can be adapted to other disciplines. We recommend a small class size (e.g., 6–12) to facilitate group discussion. Alternatively, the sessions can be completed in smaller breakout groups for larger class sizes. The advanced concepts, including data and geospatial literacy, are most appropriate for third- or fourth-year undergraduate students and possibly even graduate students.

Note: Given the advanced nature of this instruction, we recommend having two presenters so that one may provide a supporting role while the other is actively leading the session. We also suggest that at least one instructor have some GIS expertise, ideally experience working with ArcGIS Online.

LEARNING OBJECTIVES
Students will
- assess the quality and suitability of data sources
- access, extract, and prepare data
- create data visualizations using ArcGIS Online
- become familiar with creating a basic ArcGIS StoryMap

COOKING TIME
Instruction session 1: Preparing the ingredients (60 minutes)
Instruction session 2: Kneading the dough (90 minutes)
Instruction session 3: Baking (60 minutes)

DIETARY GUIDELINES
These instruction sessions relate to the following frames from ACRL's *Framework for Information Literacy for Higher Education*:
- Research as Inquiry: Students engage in an iterative research process when selecting their data sources to respond to their research question.
- Authority Is Constructed and Contextual: Students investigate data from the census as a reflection of the world in which the data were defined, collected, and disseminated.
- Information Creation as a Process: Students explore data visualizations and creating a narrative around data using ArcGIS StoryMaps.

INGREDIENTS
- Computer lab with internet access
- Access to Esri ArcGIS Online, preferably with an institutional license
- Access to MS Excel
- Two enthusiastic library instructors

PREPARATION
- Reserve computer lab.
- Develop the presentation materials (i.e., handout)

INSTRUCTIONS
This instruction could be offered in a single extended cooking session. However, for best results, it is recommended to deliver the con-

tent over three separate sessions (see Chef's Notes).

Instruction Session 1: Preparing the Ingredients (60 minutes)

1. Gather students in a classroom or computer lab.
2. In this introductory session, you will go over the ingredients and preparation instructions. It is always recommended to read a recipe in full before jumping in.
3. Explore various data sources to align with the students' data needs. Explain the importance of finding good quality ingredients—data sets with rich metadata from reputable sources—as the base for their recipe.
4. Allow for 15–20 minutes to engage in an open discussion with the students. Have them reflect on what parameters meet the scope of their project to help identify preferred data products.
   a. When students are exploring data sources, explain the range in complexity of data products (e.g., number of variables) that are available for various levels of geography (e.g., national, provincial, municipal). Remind them that there is often a trade-off between data available at finer levels of geography and the number of variables in a table.
   b. The Census of Canada was identified as the preferred source for this project, given that it provides a rich set of aggregate data about populations at relatively finer levels of geography than typically available from other data sources.
5. Explain inherent challenges in studying changes over time. After all, recipes can be altered, so it is important to ensure that your ingredients and measurements match your recipe.
   a. For instance, census geographic boundaries can evolve. Similarly, how concepts are defined and captured in the census can also change over time.
6. Show participants where to find additional resources to learn more about their selected data source, such as questionnaires and data dictionaries.

Instruction Session 2: Kneading the Dough (90 minutes)

1. Gather students in a computer lab equipped with workstations that have MS Excel and internet access.
   a. Students can work individually or in small groups depending on class size.
   b. As this is a hands-on session, consider offering a recipe card with snapshots and detailed step-by-step instructions. See figure 1.
2. Demonstrate how to extract Statistics Canada census data.

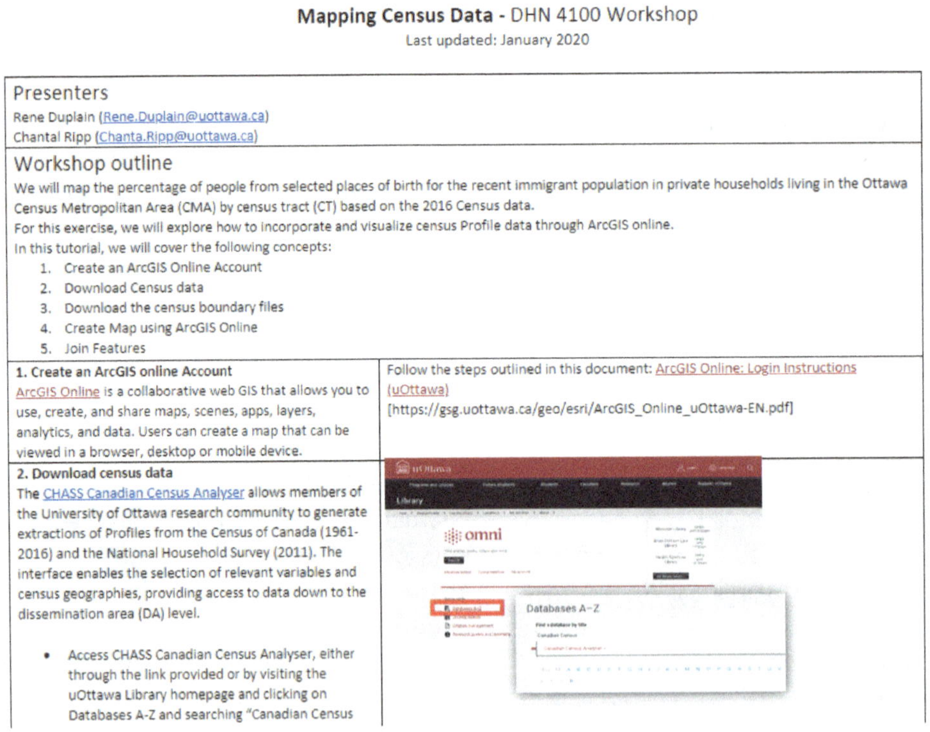

Mapping Census Data - DHN 4100 Workshop
Last updated: January 2020

Presenters
Rene Duplain (Rene.Duplain@uottawa.ca)
Chantal Ripp (Chanta.Ripp@uottawa.ca)

Workshop outline
We will map the percentage of people from selected places of birth for the recent immigrant population in private households living in the Ottawa Census Metropolitan Area (CMA) by census tract (CT) based on the 2016 Census data.
For this exercise, we will explore how to incorporate and visualize census Profile data through ArcGIS online.
In this tutorial, we will cover the following concepts:
1. Create an ArcGIS Online Account
2. Download Census data
3. Download the census boundary files
4. Create Map using ArcGIS Online
5. Join Features

1. Create an ArcGIS online Account
ArcGIS Online is a collaborative web GIS that allows you to use, create, and share maps, scenes, apps, layers, analytics, and data. Users can create a map that can be viewed in a browser, desktop or mobile device.

Follow the steps outlined in this document: ArcGIS Online: Login Instructions (uOttawa)
[https://gsg.uottawa.ca/geo/esri/ArcGIS_Online_uOttawa-EN.pdf]

2. Download census data
The CHASS Canadian Census Analyser allows members of the University of Ottawa research community to generate extractions of Profiles from the Census of Canada (1961-2016) and the National Household Survey (2011). The interface enables the selection of relevant variables and census geographies, providing access to data down to the dissemination area (DA) level.

- Access CHASS Canadian Census Analyser, either through the link provided or by visiting the uOttawa Library homepage and clicking on Databases A-Z and searching "Canadian Census

FIGURE 1
Sample recipe card.

a. Explain the importance of working with interoperable data formats (e.g., comma-separated values) so that these can be imported and manipulated in a broad range of tools and platforms.
   b. For tastier results, sprinkle in some salt by incorporating best practices in data management, such as using recommended file naming conventions.
3. Have participants knead the dough—massage the data—prior to baking. Time spent may vary based on previous experience.
   a. Show the students how to import the extracted data set into Excel.
   b. Manipulate the data set to make it GIS-compatible. This could include operations such as renaming variables, defining formats for each column, removing unwanted records, and recoding blank records.
4. While the dough rises, give a general overview of ArcGIS Online.
   a. Have the students log in to a—or create a new—ArcGIS Online account.
   b. Download the associated boundary file for the desired geography.
   c. Import the boundary file in ArcGIS Online and add to a new Web Map.
   d. Repeat for the massaged data set and add it to the same Web Map.
   e. Perform a Join Features analysis between the massaged data set and boundary file.
5. Take the time to go over proper cartographic principles. After all, it takes patience to make the perfect bread, and it should not be rushed!
   a. Explain the importance of showing the right type of symbology to reflect the variables that are being showcased (e.g., use of color schemes).
   b. Guide students to create a choropleth map. Choropleth maps use the Counts and Amounts (Color) smart mapping symbol type to show normalized data as shaded points, lines, or areas. Choropleth maps help answer questions about your data, such as "How do rates or percentages compare by geographic feature?"
   c. Explain that for them to be able to compare regions within the city, data should be normalized using ratios, percentages, or weights. To normalize data in ArcGIS Online, select a variable to map (numerator) and a variable to scale against (denominator). In this case, you could use the "place of birth" field as the numerator and the "total population" field as the denominator.
6. Have the students work on the steps above for their own data sets individually or in groups. Provide advice and troubleshooting, as needed.
7. Now that the dough is ready to go in the oven, have the students begin to reflect on how they want to share their final product with others.

Instruction Session 3: Baking (60 minutes)
1. Gather students in a computer lab with internet access.
2. Give an overview of ArcGIS StoryMaps.
   a. Have the students explore a selection of StoryMaps to get a sense of what is feasible. For a sample of StoryMaps, please reference the page "Gallery" by Esri listed in the Additional Resources section.
   b. Demonstrate how to incorporate the map created in session 2 using ArcGIS Online into a new StoryMap.
3. Encourage the students reflect on their target audience and use this reflection to inform the style, outline, and relevant sections to create a compelling story. A bread without a nice crispy crust is not a bread that is worth your time!
4. Now that it is all coming together, have the students explore what maps, images, or videos could be incorporated into the StoryMap to communicate the key message.
5. Take time to check the temperature on how the class project is progressing.

   Are they experiencing technical issues with the platform (e.g., uploading data, editing map properties, displaying features)?

6. A perfect bread is mastered after lots of attention and practice!

REVIEWS/ASSESSMENT STRATEGY

We recommend using an interactive polling application to assess knowledge and comfort with the tool at the beginning and end of

each session. This will potentially allow you to adjust your instruction based on your audience's level of familiarity with the tools and concepts.

Sample question: *How familiar are you with finding, evaluating, and preparing data to support a project? Rate your response from 1 to 5, 1 being least familiar and 5 being very familiar.*

CHEF'S NOTES

By having multiple sessions with the same group, we can develop better relationships and trust so that learners can feel comfortable reaching out to us with additional questions. We are also more familiar with the specific needs of the class and can adapt lessons accordingly. Having the professor's buy-in also gives a level of legitimacy to the experience for students, when they understand that our instruction is directly tied to a project that will count toward their grade.

Instructors can learn more about the ArcGIS Online platform through Esri's help guide: https://doc.arcgis.com/en/arcgis-online/get-started/get-started.htm.

ADDITIONAL RESOURCES

Esri. "Gallery." ArcGIS StoryMaps. Accessed on June 1, 2021 https://www.esri.com/en-us/arcgis/products/arcgis-storymaps/stories.

Uyanze, Candice, Lori Antranikian, Xuanqi Dai, Sébastien Sanscartier, Aretha Kalonji, Liam Baker, Yasmine Robinson, and Jada Watson. "International Seasonings." ArcGIS Online Stories, April 27, 2020. https://storymaps.arcgis.com/stories/abcedf57495745d5b824c-e419cd8c0c9.

Ethics and Biodiversity Data

Rebecca Hill Renirie, Medical Librarian, Research and Instruction Librarian, and Assistant Professor, Central Michigan University, rebecca.renirie@cmich.edu

NUTRITION INFORMATION

The explosion of data in the ecological sciences in the past several decades has led to a need for researchers to possess data information literacy skills. One core skill is data ethics. There are many aspects of using data ethically, including proper citation of data, licensing and legal uses of data, and protecting sensitive data. When working with biodiversity data, the last may refer to sensitive species information, particularly geolocations, if those species are protected, threatened, or trafficked.[1] Steps taken to protect or generalize these data are typically documented in metadata, a concept with which undergraduate students may not be familiar.

TARGET AUDIENCE AND NUMBER SERVED

This activity introduces undergraduate students to the ethical considerations of working with biodiversity data. It could be used in ecology, zoology, botany, or other life sciences courses. It is suitable for individuals or groups and it serves between 5 and 40 students.

LEARNING OUTCOMES

Upon completion of this activity, each student should be able to
- search, locate, and access a relevant data set and associated files in an open-access repository
- identify ethical uses of a data set as stated in a repository's terms of service (TOS) or in a data set's metadata and associated files
- apply aspects of Creative Commons or other licensing associated with data sets
- compose a data citation that includes a specific identifier such as a digital object identifier (DOI)

COOKING TIME

This lesson fits into a 45-to-60-minute class period. It includes
- a presentation to students on data ethics in general and with biodiversity data (15–20 minutes)
- the activity including brief instructions (15–20 minutes)
- wrap-up discussion (15–20 minutes)

DIETARY GUIDELINES

The frames Scholarship as Conversation and Information Has Value from ACRL's *Framework for Information Literacy for Higher Education* touch on appropriate citation of information and copyright concerns. The frame Research as Inquiry addresses using information ethically and legally.

As with bibliographic instruction, data information literacy instruction should take the students' disciplines into account;[2] pairing the data ethics content with what students are already learning in class makes for a more authentic learning experience.[3] Using a disciplinary data repository such as Dryad supports this experience. Students can look at licensing information in the repository (such as Creative Commons licensing) to discover how to use the data legally and ethically. Students can also examine metadata or README files, or other associated documentation, to see notes about any protections the data have received. All of these skills can be introduced using an activity wherein students search a data repository for an existing data set; note the license, terms of service or use, or other documentation that describes legal and ethical use of the data; and write their own data citation.

INGREDIENTS

- Background presentation on data ethics.
- Handout on data ethics.
- A relevant open data repository. Suggested here: Dryad (https://datadryad.org/stash). Choose a repository of biodiversity data relevant to student work
- Relevant handouts. Suggested here: Data ONE, "Lesson 8: Data Citation" (https://dataoneorg.github.io/Education/lessons/08_citation/index.html), and "Lesson 10: Legal and Policy Issues" (https://dataoneorg.github.io/Educa-

Section 7. Data in the Disciplines

tion/lessons/10_policy/L10_LegalPolicy_Handout.pdf).

- *Activity with clear instructions:* To give students enough time to search for and find a relevant data set and examine information about ethical use of those data, provide clear instructions for how students should proceed. This can be in the form of a handout, the information on the last slide of the presentation, or both.

PREPARATION

Get instructor buy-in and collaboration: Instructor buy-in is key to ensuring the activity ties in with existing learning outcomes for the course. To maximize the impact of this activity, it should be directly tied to students' assignments or research topics. When speaking with the instructor, inquire about upcoming laboratory topics, species of interest, or key assignments and research papers. Discover which model organisms are being used in either the course or its laboratory. If none, discuss an example organism that might be used to search the data repository. Specific learning objectives for the data ethics activity should also be identified. Those listed here can be a guide for beginning the discussion.

Plan an outline: Map an outline that leaves enough time for the presentation, the activity, and the discussion.

- *Prepare the background presentation on data ethics:* Before examining data sets, the students need some context for why ethical considerations matter with biodiversity data. This explanation can take the form of a presentation that should cover data ethics in general, and then specific ethical concerns with biodiversity data.

  The presentation should also detail examples of ethical data use, such as checking requirements from the author and database in the terms of service or on the data record page; checking requirements in README files; avoiding the sharing of any sensitive information, including species locations and personal contact information; and providing a proper citation. It should also give examples of data use that is unethical, such as distributing the information without any context or credit to the author, sharing sensitive or protected information, or failing to provide a citation.

- *Prepare the handout:* Be sure to print enough copies of handouts for the students.
- *Practice the activity, and prepare clear instructions:* Familiarize yourself with the data repository and the location of its licensing and terms of service or use information. Practice searching a selected organism with different keywords, such as scientific name (e.g., *Rudbeckia hirta*), common name (Black-Eyed Susan), or family (asters). Check which data sets are coming first in your searches and investigate the files. Do they include metadata? README documents? Be prepared for the data sets the students are likely to use and any questions they may have.

INSTRUCTIONS

- Distribute the handouts and welcome the students to the presentation.
- Deliver the presentation, making sure to connect students' course or lab work with the information about data ethics.
- Distribute the instructions for the activity, explain the instructions (below), and allow students to start searching the data repository for their species of interest.

Example student instructions for the activity:

1. Go to Dryad: https://datadryad.org/stash. Go to the very bottom of the page to Dryad's Terms of Service, 3.2 Content Distribution.
2. Explain who is responsible for the use of the data sets that can be viewed or downloaded. Is it Dryad? The data creators? Those who downloaded the data?
3. Search Dryad for [species, genus, etc.] and choose a resulting data set. Write a citation for the data set, including the DOI.
4. Scroll to the bottom of the page. What is the license given for this data set? Briefly describe what the license allows.
5. Beneath the button to download the data set, open the files associated with the most recent date
6. Are there README files, or any other files associated with the data? If so, open one. What does it contribute to the data files?
7. Open a data file. Do you see notes or a codebook? Describe what this

contributes to the data file.
8. Choose a member of your group to report your findings to the class.
- Walk the room as students are working; check in on every student or group and be available to answer questions as they occur.
- *Post-activity discussion:* Ask students to report their findings and the ethical considerations they've identified from working with biodiversity data.

REVIEWS/ASSESSMENT STRATEGY

Once the students or groups have had time to examine the ethical and legal uses of their data set, bring them back together to discuss the activity. Depending on time, each group or selected volunteers can report on what they found. Allow time to discuss trouble points, interesting findings, or anything the students found especially noteworthy.

CHEF'S NOTES

The first version of this activity was created as an in-class exercise and then adapted to be an educational module for the NSF-funded Biodiversity Literacy in Undergraduate Education (BLUE) project.[4] You can read more about this initiative to build biodiversity data literacy in undergraduates by visiting https://www.biodiversityliteracy.com/ and see version 1 of this module in QUBES.[5]

ADAPTING THE RECIPE

More modules for data information literacy in the life sciences include those from DataONE and the New England Collaborative Data Management Curriculum (NECDMC; see Additional Resources).

The activity could focus on other disciplines and disciplinary data repositories by allowing students to use the Registry of Research Data Repositories (https://www.re3data.org) to find a repository on a topic that interests them. However the recipe is adapted, be sure students are able to find ethical data use indicators such as terms of service or terms of use, licensing information, metadata, or README files, and to cite data sets appropriately. Librarians could also flip the classroom with this activity, which would allow a more in-depth wrap-up activity or assessment.

ADDITIONAL RESOURCES

Australian Research Data Commons, "Publishing and Sharing Sensitive Data," https://ardc.edu.au/wp-content/uploads/2020/05/Publishing-and-sharing-sensitive-data-flowchart.pdf.

Chapman, Arthur D. "Current Best Practices for Generalizing Primary Species-Occurrence Data," Version Master. Global Biodiversity Information Facility (GBIF), November 16, 2020. https://docs.gbif.org/sensitive-species-best-practices/master/en/.

DataONE, "Data Citation," handout. Data Management Skillbuilding Hub. https://dataone-org.github.io/Education/lessons/08_citation/index.html.

DataONE, "Legal and Policy Issues," handout. Data Management Skillbuilding Hub. https://dataoneorg.github.io/Education/lessons/10_policy/index.html.

Dryad Digital Repository, https://datadryad.org/stash.

Lamar Soutter Library, University of Massachusetts Medical School. "Module 5: Legal and Ethical Considerations for Research Data." New England Collaborative Data Management Curriculum, https://library.umassmed.edu/resources/necdmc/modules.

Peterson, A. Towsend. "*Biodiversity Informatics* Training Curriculum, version 1.2," *Biodiversity Informatics* 11, no. 1 (March 1, 2016), https://doi.org/10.17161/bi.v11i0.5008.

Full Bibliography: Renirie, Rebecca. "Ethics in Biodiversity Data: Chapter Bibliography: Home." Central Michigan University Libraries, Research Guides. https://libguides.cmich.edu/Renirie_Ethics_Biodiversity_Data.

NOTES

1. Robin Rice and John Southall, "Dealing with Sensitive Data," in *The Data Librarian's Handbook* (London: Facet Publishing, 2016), 121.
2. Jacob Carlson, Michael Fosmire, C. C. Miller, and Megan Sapp Nelson, "Determining Data Information Literacy Needs: A Study of Students and Research Faculty," *portal: Libraries and the Academy* 11, no. 2 (2011): 654, https://doi.org/10.1353/pla.2011.0022.
3. Catherine M. O'Reilly, Rebekka D. Gougis, Jennifer L. Klug, Cayelan C. Carey, David C. Richardson, Nicholas E. Bader, Dax C. Soule, et al., "Using Large Data Sets for Open-Ended Inquiry in Undergraduate Science Classrooms,"

BioScience 67, no. 12 (December 2017): 1060, https://doi.org/10.1093/biosci/bix118.
4. Biodiversity Literacy in Undergraduate Education Data Initiative, "BLUE Data Network," accessed March 23, 2021, https://www.biodiversityliteracy.com.
5. Rebecca Hill Renirie, "Data Ethics," Biodiversity Literacy in Undergraduate Education, QUBES Educational Resources, version 1.0, March 19, 2020, https://doi.org/10.25334/ZKY1-6C12.

Data Decisions and the Research Process in the Sciences and Social Sciences

Nicole Helregel, Student Success Librarian, Purchase College (SUNY), nhelregel@gmail.com

NUTRITION INFORMATION

The purpose of this workshop is to teach upper-level science or social science undergraduates some basic principles of data literacy within the context of their field. A short introductory lecture explores the processes of data collection and visualization and the deliberate choices therein. A group activity offers students the chance to identify and critically evaluate choices made throughout the research process through the examination of a scholarly article. A summative large-group discussion helps students develop a richer understanding of the role that data and data visualizations play in the narratives of scientific scholarly communication.

TARGET AUDIENCE AND NUMBER SERVED

The primary audience for this session is mid- to upper-level undergraduates in STEM or the social sciences, though it could be modified for graduate students. The setting for the session would ideally be within the context of a mid- to upper-level STEM or social science undergraduate course with assignments tied to reading and reviewing scholarly literature. The students should already be familiar with scholarly articles as a type of source. The session could also serve as a stand-alone workshop. The session works best with a class of 30 students or fewer.

LEARNING OUTCOMES

Students will
- identify and critique choices made throughout the research process (e.g., data collected, methods used, scope of the project, etc.), as illustrated in a scholarly research article in their field
- analyze and evaluate data visualizations within scholarly research articles in their field

COOKING TIME

Minimum: 1 hour if covering everything in a live class session—see Adapting the Recipe for other options.

DIETARY GUIDELINES

This recipe's learning outcomes and lesson plan draw from ACRL's *Framework for Information Literacy for Higher Education*, most heavily from Information Creation as a Process and Research as Inquiry. Students examine a research article to determine the methods of research inquiry, data-related elements, and major research findings. The class then engages in discussion about the processes and decisions that make up the research process, including initial inquiry, data collection and analysis, visualization and writing, and publication.

INGREDIENTS

- Peer-reviewed research article (link if digital; 6–7 copies, if print)
- Worksheets (link if digital; 6–7 copies, if print)

PREPARATION

- Choose a research article that is relevant to the course, contains original research (rather than a review article), and contains graphs or charts.
- Read the article for yourself, and note a few possible answers to the activity questions that are open-ended. This will help you guide the class discussion.
- If your article or worksheets are digital, make sure that the setting of your session is such that all participants have access to a computer (e.g., a computer lab; a class in which all students are required to bring their own computer; an online class in which everyone is participating via a computer). Also, get a permalink for the article and consider shortening the link for ease and simplicity.
- Some settings may necessitate conferring with the course instructor to confirm the technical setup; be sure to do so

before the session.

INSTRUCTIONS

1. *Introduction and context* (15 minutes): Introduce (or reinforce, depending on the context within the course) these concepts: research as a process; data visualization as a process. Both processes are determined by several subjective decisions; for example, decisions about the scope of the research project, decisions about what data to collect and how, decisions about how to analyze the data, decisions about which data to highlight in visualizations, and decisions about which chart or graph type to use. For more background information see the Additional Resources section.

   Challenge the idea that data visualizations exist in a stark "good vs. bad" binary; just like information sources, data visualizations should be analyzed with a critical eye. This is part of what you offer as a librarian—helping students develop nuanced ways of analyzing and critiquing data visualizations. Such critiques should be based on data visualization best practices: transparency, clarity, highlighting a specific result, and including appropriate contextual information. Show examples of charts that illustrate these best practices (find examples in the works listed in the Additional Resources section). Draw connections to research that the students will engage in as part of their major, such as research term papers or projects, undergraduate theses, or other research opportunities. This helps drive home the relevance of being able to both evaluate and create data visualizations.

2. *Conduct the small group activity* (minimum 20 minutes): Divide students into small groups (3–5 students) and give each group a copy of the research article and a copy of the worksheet. See appendix A: "Small-Group Worksheet."

3. *Conduct the large-group discussion* (minimum 20 minutes): Talk through some of what students found. Have some answers prepared for questions 3 and 5 to go over in case the discussion is sparse. Let students steer the conversation as much as possible, and try to address some ideas about research methodology choices, data visualization choices, and research reproducibility.

REVIEWS/ASSESSMENT STRATEGY

Use the group worksheets to assess how well students achieved the learning outcomes of the session. A basic rubric is helpful for this—feel free to adapt to suit the needs of your specific context. See appendix B: "Instructor Worksheet."

If the session is within the context of a course, assess the learning outcomes within a larger applicable assignment (e.g., an article critique assignment or annotated bibliography). Confer with the instructor to see if they have a rubric for the assignment.

ADAPTING THE RECIPE

The times given in Instructions are minimums. Each section could easily be expanded. Work with the instructor to determine how much time is appropriate and whether or not some content should be flipped or front-loaded.

The lesson plan can be flipped or front-loaded with the context section being shared via a prerecorded video and the small-group activity being completed as a group homework assignment. This would allow the entire live session to be a discussion.

The lesson plan can be adapted for graduate students in the context of a graduate-level course or a research group. Choose an article (or multiple articles) specific to the course or research group focus (try not to choose an article written by a faculty member at your institution; this may inhibit students from critically evaluating the data decisions). Weave in more connections to the students' own research endeavors and the decisions they are or will soon be making regarding their own data collection and visualization.

ADDITIONAL RESOURCES

Gough, Phillip. "From the Analytical to the Artistic: A Review of Literature on Information Visualization." *Leonardo* 50, no. 1 (February 2017): 47–52. https://doi.org/10.1162/LEON_a_00959.

Heer, Jeffrey, Michael Bostock, and Vadim Ogievetsky. "A Tour Through the Visualization Zoo: A Survey of Powerful Visualization Techniques, from the Obvious to the Obscure." *Queue* 8, no. 5 (May 2010): 20–30. https://doi.org/10.1145/1794514.1805128.

Huff, Darrell. *How to Lie with Statistics*. New York: Norton, 1993.

Seeing Data. "Developing Visualisation Literacy." Accessed June 11, 2021. http://seeing-data.org/developing-visualisation-literacy/.

Tufte, Edward R. *The Visual Display of Quantitative Information*. Cheshire, CT: Graphics Press, 1983.

APPENDIX A: SMALL-GROUP WORKSHEET

1. What research question(s) did the authors set out to answer? Paraphrase the research question:

2. What kind of data did the authors collect? How did they collect the data? Summarize some of the methods for data collection:

3. Do these data collection methods make sense for answering this research question? Would any other data or method of data collection have been helpful for answering this research question?

4. Find figure X [specific figure number]. What specific data/result (variable) does this figure depict? Paraphrase the specific data/result figure X depicts:

5. Is there any other type of graph or chart that would also work to display figure X's data? Why do you think the authors chose to visualize the data in this way?

6. Read the conclusion. Summarize in 1–2 sentences why this research is important.

APPENDIX B. INSTRUCTOR WORKSHEET

	No/Limited Engagement (0)	Baseline Engagement (1)	Moderate Engagement (2)	Complex Engagement (3)
Question 1	Incomplete or incorrect	Copies/pastes thesis	Paraphrases thesis	N/A—moderate is max
Question 2	Incomplete or incorrect	Copies/pastes methods	Summarizes methods	N/A—moderate is max
Question 3	Incomplete or incorrect	No alternatives suggested	One alternative suggested	Two or more alternatives suggested
Question 4	Incomplete or incorrect	Copies/pastes title and/or axes labels	Paraphrases data displayed	N/A—moderate is max
Question 5	Incomplete or incorrect	No alternatives suggested; author's choice based on objectivity/without much explanation	One alternative suggested; explanation of possible decisions behind author's choice	Two or more alternatives suggested; explanation of possible decisions behind author's choice with nuance/multiple factors
Question 6	Incomplete or incorrect	Copies/pastes from conclusion	Summarizes conclusion	N/A—moderate is max

Financial Data for Economics Students

Jennifer Yao Weinraub, Librarian, The New School Libraries, weinraubj@newschool.edu

NUTRITION INFORMATION

Economic and financial data are used in a wide range of economics scholarship. Students learn statistical software programs such as R and EViews in undergraduate and graduate econometrics courses, but the curriculum does not necessarily cover identifying, finding, and assessing data for economic analysis.

There are many sources of economic data, including governments, agencies, and intergovernmental organizations. With financial data, however, the researcher quickly finds that free sources such as Yahoo! Finance have limited content and functionality. There are numerous commercial financial data providers, including Bloomberg, Refinitiv, S&P Global, Global Financial Data, and Wharton Research Data Services (WRDS). These provide deep and broad access to financial data, as well as advanced tools that allow researchers to search for, organize, and download large data sets.

Libraries seeking to support economics programs by providing access to financial data face multiple hurdles, including restrictive user agreements that limit access and sharing with consortial arrangements and interlibrary loan, rapidly evolving products, faculty unfamiliarity with products, limited availability of user testing trials, and high cost. Finding financial data might be especially challenging at institutions where there is no graduate business program or specialized business library.

This recipe calls for inviting financial data providers to demonstrate their products and speak directly to students and faculty. While librarians may be accustomed to providing database demonstrations, it is advantageous to have financial data companies provide instruction. This type of demonstration allows students to become acquainted with financial data, learn how experts use the platform, preview new features and tools, and gain insight into the knowledge creation process.

TARGET AUDIENCE AND NUMBER SERVED

Undergraduate and graduate economics and business students interested in financial data for statistical analysis
5 to 20 participants (or more)

LEARNING OBJECTIVES

- Gain awareness of commercial financial data resources and the data points they provide. These may include data on stocks, bonds, and commodities; indices; structure and ownership data; executive compensation figures; interest rates; exchange rates; macroeconomic data; financial news; and more.
- Understand how financial data are used by both academics and financial professionals.
- Learn about tools, protocols, and software used to search for and download data.
- Meet professionals with knowledge of the financial information creation process.
- Critically evaluate and compare economic data platforms.

COOKING TIME

2 to 3 one-hour-long meetings or webinars

DIETARY GUIDELINES

This activity relates to the following frames from ACRL's *Framework for Information Literacy for Higher Education*:

- *Information Creation as a Process*: Students gain insight into how financial data is collected, organized, and presented in financial data platforms.
- *Information Has Value*: Students gain awareness of the vast financial data industry that provides financial data that are not available on the open web.

INGREDIENTS

For in-person presentations: space with a computer, large-format display screen, and internet access

For online webinars: student computers and internet access

PREPARATION

Gauge interest in financial data platforms. Assess current access to financial data throughout the institution, including at the library, data center, and academic departments.

Review library database lists, LibGuides, or professional literature (see Additional Resources) to determine the financial data information providers.

Prepare for scheduling the meeting. Contact department administrators or advisors, and view class schedules, to determine which days of the week and times of day students and faculty are typically available. View the academic calendar, taking care to avoid holidays, breaks, exam periods, and important events.

INSTRUCTIONS

Invite financial data vendor representatives to present on campus or online. Among those who might attend are sales representatives, analysts, economists, data specialists, and education specialists. Commercial financial data platforms are aware of how data are increasingly being used for data analysis, statistical inference, and economic modeling and can speak to these purposes.

Visits should be spaced strategically—at least a day apart, but within a few weeks of each other so that students can compare resources but not be in meetings all day. Advertise the event to students and faculty.

Communicate expectations to vendors as to topics and features that should or should not be discussed. Prepare questions for presenters relating to the learning outcomes (e.g., how to search for and download data) in case they are not answered during the course of the presentation. Schedule an optional, short follow-up meeting after the demonstration with library staff only, in order to learn more about contracts, pricing, and access.

Host the event. Allow students and faculty time to ask questions, or ask questions you've prepared in advance. Inquire whether students and faculty may contact commercial data providers directly; many of these services provide extensive support to customers and will gladly accept questions. Inquire whether a trial is possible. Obtain names and e-mail addresses of attendees so that you can follow up with them. If trial access is available, coordinate trials for students and faculty. For some services, ask for a faculty member to serve as a point person; help coordinate this if any faculty members are interested.

REVIEWS/ASSESSMENT STRATEGY

Survey attendees at the end of the presentations (and trials, if they were given). Ask faculty and students to rank the financial data resources, share what they liked and didn't like about each platform, and describe how their preferred financial data platform would support their teaching, learning, research, and career objectives.

In assessing financial data services, libraries may want to consider these factors in addition to cost-related metrics (such as cost per use, cost per student, cost of technical support or hosting):

- Scale and scope of data: countries, financial instruments (stocks, bonds, commodities, derivatives), industries and sectors, private/public companies, etc.
- Coverage span, periodicity, and frequency of updates of data
- Other content, including news, reports, histories, and biographies
- Educational content such as course support materials
- Unique, proprietary content vs. publicly available content
- Ease of searching, browsing, and accessing data
- Tools for screening, graphing, downloading, and comparing data
- Account setup and access
- Support provided

ADAPTING THE RECIPE

There are opportunities for collaboration at schools that have a data science program, business program, data lab, or business library.

ALLERGY WARNING

Subscriptions to financial data platforms are costly and ultimately might not be feasible. If the library faces insistent demands for a subscription, it might need to explore creative ways to fund it, such as sponsorship or cost sharing with departments.

ADDITIONAL RESOURCES

Bloomberg, https://www.bloomberg.com.

Global Financial Data, https://globalfinancial-data.com.

Refinitiv, https://www.refinitiv.com.

Ross, Celia. "Online Resources for Business Research." *CHOICE* 53, no. 12 (August 2016): 1715–32. https://doi.org/10.5860/CHOICE.53.12.1715.

S&P Global, https://www.spglobal.com.

Wharton Research Data Services (WRDS), https://wrds-www.wharton.upenn.edu.

Stuffed Shiny App with Business Intelligence

Yun Dai, Data Services Librarian, Library, New York University Shanghai, yun.dai@nyu.edu, http://shanghai.hosting.nyu.edu/data/;
Fan Luo, Digital Scholarship Technologist, Library, New York University Shanghai, fan.luo@nyu.edu

NUTRITION INFORMATION

Not every student will stay in academia after graduation. This recipe is prepared for those students who plan to be data analysts in the business world. In this recipe, we present a project-based technical workshop where attendees solve a practical business problem.

The audience of the workshop will build a Shiny app that is a prototype of a one-stop-shop solution for business intelligence needs. The application bridges the gaps along operational lines from the back-end database to the business side and handles database query, analytics, visualization, and other ad hoc requests.[1] Shiny is an R package that enables building interactive web apps, which can be hosted on a web page or embedded in R Markdown documents or in dashboards.[2]

This is one of the three recipes on weaving data literacy competencies into a project-based technical workshop. See also "Text Mining Charcuterie Board" in the Data Manipulation and Transformation section and "Web-Interfacing Data Visualization in a Rainbow Layer Cake" in the Data Visualization section.

TARGET AUDIENCE AND NUMBER SERVED

Serves a small group of 5–10 attendees. The target audience is those seeking solutions, with Shiny or other tools, to the business intelligence problems in their own work or internship. Knowledge of Shiny is helpful but not required.

LEARNING OUTCOMES

Attendees of the workshop should be able to
- articulate the project needs
- identify the key concepts, technical terms, and processes that describe the needs
- break the project needs down to manageable tasks and construct strategies to solve the problems
- retrieve business data stored in an SQLite database in the environment of a Shiny app
- direct user inputs through graphical widgets and display outputs from analytics and visualizations in an interactive web application to communicate insights from business data

COOKING TIME

2 hours

DIETARY GUIDELINES

This recipe uses the frame Research as Inquiry in ACRL's *Framework for Information Literacy for Higher Education*. It uses the following performance indicators in ACRL's *Information Literacy Competency Standards for Higher Education*: Standard 1, Outcome 1.e, identifying key concepts and terms describing the information needs; Standard 4, Outcome 1.a, organizing the content to support the purposes and format of the product; Standard 4, Outcome 1.d, manipulating data to transfer them from their original locations and formats to a new context; and Standard 4, Performance Indicator 3, communicating the product effectively to others, considering the communication medium, information technology, principles of design and communication and style.

INGREDIENTS

- A sample database
- Link to the Shiny app that attendees will build (hosted on https://www.shinyapps.io)
- Scripts of stand-alone demo Shiny apps that are parts of the finished Shiny app with incomplete code
- R script with (1) sample code for demonstration and (2) snippets of code for attendees to grab and fill in the stand-alone apps in the exercises
- Slides that outline the workshop with the conceptual model of the Shiny product and the workflow of creating it
- Sequenced short video tutorials and

a step-by-step written tutorial shared with the audience for reference after the workshop[3]
- One instructor and one helper (recommended for an online session and optional in person)

PREPARATION

1. *Create a sample database with SQLite.* The database consists of 5 tables that store data of order items, order details, product information, customer information, and retail store information. Order information is generated randomly; store and product information are partially based on a company's public data. Each table contains 3 to 7 columns. The tables that record order and customer information contain around 10,000 to 20,000 rows, which would approximate the volume of monthly transactions in real life, but which would not be burdens to attendees' personal computers. The store table contains around 200 rows; the product table contains around 50 rows. These two tables ensure some variety and details in visualizing geographical distribution and displaying product categories.
2. *Build a Shiny app that connects with the sample database.* Stuff it with business intelligence. Add the tables, visualizations, analytics, and widgets to the Shiny app.
3. *Dissolve the finished Shiny app into stand-alone demo Shiny apps.* See figure 1. Each delves into one aspect of the business intelligence app or a combination of several of its chief features: UI layout, inputs (select input, date range input, action button), outputs (data table, plots, metrics, prediction), database connection, parameterized query, and updating database.
4. *Scoop the commands that are fundamental to each demo app into a separate R script.* The R script contains bits of code that attendees can copy, paste, and tweak in the exercises.
5. Prepare a skeleton script that contains the higher-level structures of the finished Shiny app.
6. *Record a short video tutorial for each demo app.* Generate a step-by-step written tutorial with R Markdown.

INSTRUCTIONS

1. *Set the workshop agenda around a business problem.* Explain the roles of a data analyst and business intelligence in enterprise management. Prompt:

A technology company, PineApple Inc., has an executive team meeting. The business managers are reviewing the budget, as well as the costs and revenue of the past few months in order to make budgets for the second half of the fiscal year. They start to ask questions like: How many orders have been placed in the last quarter? What are the order details? What is the revenue of the last month? What do sales look like across the country? What will

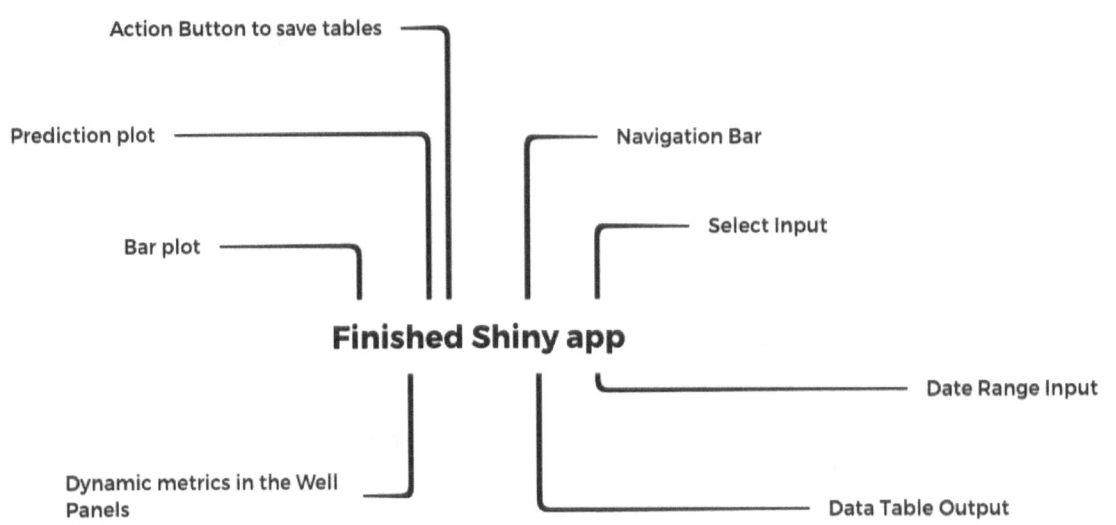

FIGURE 1
Decomposing finished Shiny app into stand-alone demo apps. Created by Fan Luo and Yun Dai under CC BY-NC 4.0.

the estimated sales revenue be for the next six months?

The managers decide to assign data analysts to the project to build a system that answers those questions.

2. *Analyze the project needs from the perspective of a data analyst.* Solution: The system can be a one-stop shop that connects a back-end database and a front-end dashboard. It grabs data from the database, then processes the extracted data to generate business analytics and incorporates the outputs in a decorated dashboard for reports.
3. Give an overview of relational databases and the sample SQLite database.
4. Describe how Shiny functions conceptually.
5. *Illustrate with code the three essential steps of working with an SQLite database in a Shiny environment.* Each step is built upon the previous one: (1) compiling database queries with SQLite; (2) retrieving business data stored in an SQLite database in an R environment; and (3) embedding the queries in a Shiny web application.
6. *Decompose the project needs into parts to address them one by one in the stand-alone demo apps.* The core components of these apps have already been removed prior to the workshop. Attendees need to fill in the missing code as they follow the instructions or on their own in the exercises.
   a. *Navigation bar.* It is a fairly simple element, and therefore quickly gets the audience started in the Shiny environment. Attendees learn to set up the UI and server functions and to disconnect from the database when exiting the app.
   b. *Select input.* Attendees learn to connect to the database and construct the Select Input UI to read user inputs. In this step, there is no communication with the server, and no output is expected.
   c. *Date range input (exercise).* Attendees apply what they have learned from the select input example to make the date range input work. The mechanisms behind it are identical.
   d. *Data table output.* Attendees see how the three key steps are implemented in this small concrete example. Adding the output to the input piece, attendees learn to use a parameterized query to Insert a value from *one* user input into a database query. A parameterized query can be thought of as a pre-compiled SQL statement that an application wants to execute and that can be customized using parameters. In this Shiny app, that would be user inputs of product category, starting date, and ending date.
   e. *Dynamic metrics in Well Panels (exercise).* Attendees test their knowledge of input and output in Shiny by associating the input with the correct output, when there are multiple outputs (number of orders, sales revenue, and average order value). For instance, in order to find out the sales revenue for a product, attendees would associate the input of product type with the output of sales revenue.
   f. *Bar plot.* This demo app details how to render an output utilizing a parameterized query with values from *two* user inputs. Therefore, attendees would be able to retrieve a result such as the sales revenue for a product (product type) within the last two weeks (date range).
   g. *Prediction plot and map (exercise).* Attendees apply what they have learned from the bar plot example to create two plots of sales revenue: a forecast plot and a thematic map. Code to generate plots are provided in a separate R script, so attendees can concentrate on what makes a Shiny app work.
   h. *Action button to save tables to the database.* The instructor introduces a new input UI element, the action button. This demo app also shows passing signals to the server to trigger an event when talking to the database, compared with asking something from the database.
7. *Deploying the Shiny app.* Now each demo app is fully functional and intact. Attendees gather the code segments from the demo apps into the skeleton Shiny app script to complete the one-stop-shop app. Get it stuffed with business intelligence!

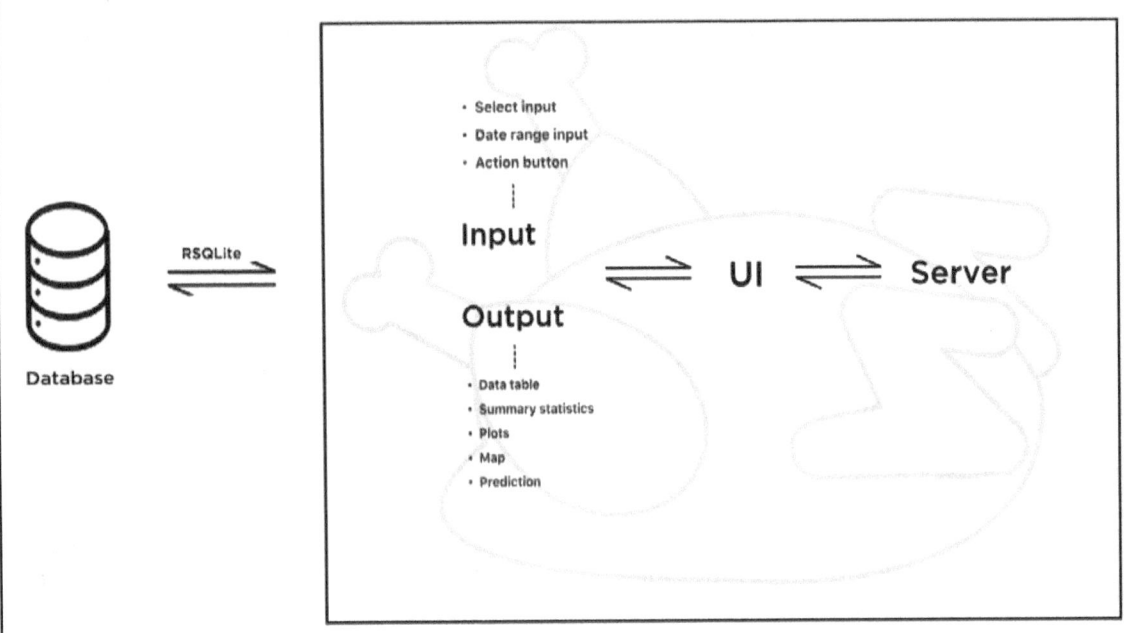

FIGURE 2
Shiny app stuffed with business intelligence. Created by Fan Luo and Yun Dai under CC BY-NC 4.0. Database by Shmidt Sergey from the Noun Project (2020), https://thenounproject.com/term/database/691819/.

NOTES
1. Yun Dai and Yujie Xiang, "Connecting SQLite Database and Shiny App for Business Intelligence," November 2020, https://shanghai.hosting.nyu.edu/data/r/case-3-sql-shiny.html.
2. RStudio. Shiny home page. 2020. https://shiny.rstudio.com.
3. Dai and Xiang, "Connecting SQLite Database."

REVIEWS/ASSESSMENT STRATEGY
At the beginning of the workshop, invite attendees to share their business intelligence problems or stories from their work or internship. What brings one to the workshop? What are the missing pieces in one's puzzle? Learning effectiveness is assessed in the hands-on exercises. Following the exercises, attendees and the instructor sit down in a circle to review the teaching effectiveness in terms of how helpful the workshop is to answer their questions or to solve their own business intelligence problems.

ALLERGY WARNING
A project-based workshop covers many facets of data literacy, so attendees have lots to digest. It may also have more technical barriers than an introductory or topical session. Instructors should share the class materials beforehand, both for attendees to preview and to manage their expectations.

CHEF'S NOTES
You may replace Shiny with other tools (e.g., Python Dash) that offer a framework for building web applications.

Fast Casual Marketing Strategies

Juliann Couture, Head of William M. White Business Library and Associate Professor, University of Colorado Boulder, juliann.couture@colorado.edu; *Halley Todd, formerly Business Research and Instruction Librarian and Assistant Professor, University of Colorado Boulder,* halley.todd@gmail.com; *Natalia Tingle Dolan, Business Collections and Reference Librarian and Associate Professor, University of Colorado Boulder,* natalia.dolan@colorado.edu

NUTRITION INFORMATION
Business students often have limited or no access to library subscription resources post-graduation. This one-shot instruction session is designed to help shift students away from reliance on subscription resources to freely available sources. This activity introduces students to freely available information sources and asks them to consider how this information could be used in their future workplace to target a particular demographic in a given location.

TARGET AUDIENCE
This recipe is adaptable and meant for those with advanced palates at the upper undergraduate or graduate level. It can serve any number of students as long as they can form smaller groups of 2 to 3 people. Classes with fewer than 50 students work best.

LEARNING OUTCOMES
- Identify government and local sources of demographic data
- Articulate how to use this data for marketing purposes
- Compare and contrast data collected from sources at the state and local levels
- Practice using advanced Google search strategies to increase the effectiveness of their queries

COOKING TIME
45–50 minutes

DIETARY GUIDELINES
The lesson plan and related activities were created around ACRL's *Framework for Information Literacy for Higher Education* frames of Information Has Value, Authority Is Constructed and Contextual, and Searching as Strategic Exploration. We don't expect students to remember each unique information source we show them. Instead, the focus is on guiding students to consider who or what entity collects data, the value of the data, the purpose the data have been collected for, and other critical thinking strategies that translate across resources and contexts. Our intent is to empower students to use these skills in their future careers and daily lives.

INGREDIENTS
- Internet access.
- Device to access the internet.
- Projection device for sharing the instructor's screen.
- Access to PowerPoint or Google Slides.
- A mix of data websites. Options include US Census Bureau (https://data.census.gov/cedsci/), state demography office (e.g., Colorado State Web Portal, https://demography.dola.colorado.gov), and local chamber of commerce, (e.g., Boulder Chamber of Commerce, https://boulderchamber.com) or other similar sources, such as a local economic development association (e.g., Boulder Economic Council, https://bouldereconomiccouncil.org).
- Zip codes for comparing locations.
- Prompts and questions for exploring the zip codes and assigned data source.
- Slides that show how to search Google using file type and site limiters.

PREPARATION
- Create slips of paper (at least one for each student) with zip codes and names of government resources to distribute to student groups (e.g., 80305 and Boulder Economic Council)
- Create a slide with prompts for students to consider to help frame post-activity discussion
- Optional: Ask the course instructor to post this message in the learning man-

agement system or by e-mail to the class: "Prior to class, please visit https://data.census.gov/cedsci/. Search your current zip code and the zip code of the area where you grew up (alternatively, use Boulder [80305], Longmont [80501], Fort Lupton [80621], Denver [80205] as options). As you explore available information, consider: What can you find out about the people who live there? Look at education attainment—how many have at least a bachelor's degree? How many have the internet in the home? What else do you find useful? What questions do you have about using census data?"

- Optional: Create a LibGuide, or shareable document using a tool such as Google Sheets or Padlet, with links to access the information sources and space for student comments.

INSTRUCTIONS

As part of the introductory framing of the session, discuss what librarians do: help connect people to information, navigate proprietary information resources, and understand how information is created, produced, and disseminated. Librarians are uniquely situated to be authoritative resources for thinking creatively about sources of information, especially in our local communities. Finally, talk about the purpose of this assignment: What do you do when you do not have a lot of time to gather primary research through surveys and research and no budget to purchase secondary data? What strategies can you use in your future careers to help you find this information quickly?

The first part of the session introduces government information sources. We facilitate a discussion based on the assumptions students have about the Census Bureau and, if assigned, the presession prompts. We invite students to share their observations on the data that they found and how they might be useful, and to share their insights. We address assumptions that census data are limited to the decennial census and discuss the supplementary surveys that are used in the intervening years to understand community changes.

Model a search using the census site (https://data.census.gov/cedsci/) to demonstrate how students can retrieve information at the state, county, city, and zip code level. In exploring the census site, examine the search results to highlight the vast amount of data provided in this resource, not only the decennial census but also the annual American Community Survey (https://www.census.gov/programs-surveys/acs). In less than a minute anyone may freely access historical and current data using one website. We also mention that other federal agencies provide data, such as the Bureau of Economic Analysis (https://www.bea.gov) and the Bureau of Labor Statistics (https://www.bls.gov/). The US government is a great source for free demographic data because it systematically collects and disseminates them.

Next, explore government information sources at the state and local city levels. Using our city and state as our example, we introduce the Colorado Demography Office and the Boulder Economic Council sites. We briefly model how to explore those two websites by commenting on keywords to look for such as *demographic profiles*, *interactive data*, or *publications*. These sites often have prepared tables and reports that are easy to access and read. We discuss how and why these two entities collect, publish, and highlight data on their websites. These offices provide data to encourage the development of new programs and initiatives based on community profiles, to attract industries to the area, and to support business endeavors.

Next, we ask students to form small groups (using breakout rooms in Zoom, or nearby classmates in person) to explore the sites further using the assigned zip codes we distributed (in paper in person, or through a Google Sheet on Zoom). We ask them to respond to these prompts via a Google Sheet:
- What kind of information can you find here?
- When might it be useful? Why might it be useful?
- What are some things to know about searching here?
- Anything interesting that you found?
- Can you export?
- Are there advanced search options?

In person, we ask students to explore the census data site, as well as one of the local sites, with an assigned zip code. We suggest 15–20 minutes to explore these sites, and then bring the class back together for 10 minutes of discussion. In the Zoom version, we send students into breakout rooms for 10

190

minutes, followed by 5–10 minutes of large group discussion. The discussion is focused on what information was found using these sources and how to apply that information in a marketing context. As teaching librarians we focus our attention on explaining how to find and access information, whereas the instructor of record may prompt students to consider how to use that information to complete their assignments.

The final part of the session is on Google search strategies (site limiting, file type searching) to fill in the gaps in what they previously found. We suggest thinking through who cares enough about this information to keep track of it. Often an industry association collects information. Some of this information may be for members only, but some of it may be free. It depends on the association. Additionally, we encourage students to examine the subscription market research reports provided by the library to current students and to note the sources they find: industry associations and government agencies alongside commercial or proprietary surveys.

REVIEWS/ASSESSMENT STRATEGY
In order to assess student learning, we use a Think-Pair-Share formative assessment activity. The instructors walk around while students explore their assigned sources and zip codes. We listen to conversations and respond to questions as students work in groups. When we conduct the session online, students go into preassigned breakout rooms via Zoom to complete the activity. We use the responses the student groups' input on the Google Sheet to gain insight into student comprehension and inform learning needs. After the allotted working time, students share observations with the class about their sources during an all-class discussion.

ADDITIONAL RESOURCES
American Community Survey, https://www.census.gov/programs-surveys/acs/

US Bureau of Economic Analysis, https://www.bea.gov

US Bureau of Labor Statistics, https://www.bls.gov

US Census Data website, https://data.census.gov/cedsci/

Examples of Google power searching:
- Site limiter: This searches a specific website type of site, or domain.
  - Example: demographics Boulder, CO site:.gov
  - Example: statistics site:restaurant.org
  - Example: trends site:vendingtimes.com
  - Example: statistics site:nrf.com
- File type limiter: This searches for a specific type of file (PDF, xls, etc.).
  - Example: demographics Boulder, CO filetype:pdf
  - Example: consumer survey filetype:pdf
  - Example: colorado demographics filetype:xlsx

When and Where
A Framework for Finding and Evaluating Social Science Data for Reuse

Ari Gofman, Social Science Data Librarian, Tisch Library, Tufts University, Ari.Gofman@tufts.edu

NUTRITION INFORMATION
This framework introduces undergraduate and graduate learners to questions to ask to successfully find and evaluate social science data for reuse. A graphic is presented on the board to support learners who prefer visual information, and more detail is provided verbally with multiple prepared opportunities for collaborative formative assessment. This modular lesson is usually combined with an interactive activity introducing a data source appropriate to a course's topics such as PolicyMap (a subscription tool), IPUMS (https://www.ipums.org), or Data.gov.

TARGET AUDIENCE AND NUMBER SERVED
This recipe serves advanced undergraduate students (juniors and seniors) and first-year graduate students working on a project that involves finding quantitative social science data for reuse, such as finding census data to use in a research paper. It also works well with medium-size classes (10–30 students) that can be separated into 3–4 breakout groups with different tutorials; each group is asked to teach back data sources using the *Finding Data Framework* covered in this chapter.

LEARNING OUTCOMES
Students will

- identify and select appropriate data for a given inquiry (microdata or aggregate data)
- identify how the creation process impacts the way the information can be used
- use brainstorming and other techniques when searching, including flexibility with proxying concepts for variables
- match information need with search strategy to represent the concepts in a research question

COOKING TIME
10–15 minutes for the basic activity (or as an asynchronous video); 45–75 minutes total for the theory presentation and additional resource exploration with teach-back activity.

DIETARY GUIDELINES
This framework provides an easy-to-understand theoretical guide to evaluating social science data for reuse and gives practical recommendations that improve the searching experience. It touches on the frames Searching as Strategic Exploration, and Information Creation as a Process from ACRL's *Framework for Information Literacy for Higher Education*.

INGREDIENTS
- Instructor presentation (can be PowerPoint or Google Slides).
- Student computers to search and access data sources.
- (optional) Worksheets for resource exploration and teach-back activity. Can be uploaded to Box or Google Drive for remote synchronous activity.

PREPARATION
Align the session with the broader academic course: discuss learning objectives and any relevant assignments with the professor. Discuss what sort of data students will be expected to find and the general types of analyses. If planning a longer session, collaborate to identify 3–4 resources for additional exploration. These can include exclusively data sources or a mix of data sources and secondary literature sources such as PsycInfo. If the session is in person, print enough worksheets for all students, plus two additional copies of each exercise for the librarian and professor. Walk through the instructions of each exercise to ensure the interface and links are effective. A sample worksheet is provided in the appendix.

INSTRUCTIONS
Welcome students to the session in whatever way you are accustomed to. For example, you could introduce yourself and make a pitch for scheduling a research appointment.

Outline the objective of the lesson. This module will introduce you to different types of data, identify important questions to ask when searching for and evaluating data, and provide an introduction to the resources you can find in your online research guide.

If the session is synchronous, ask students what they think of when they hear the word *data*. Most responses will be in the general theme of "numbers" or "Excel spreadsheets." After students have shared, offer a definition that data are units of information that can be analyzed and synthesized into new information. These sorts of quantitative, structured data make up a substantial part of research in the social sciences but are not the only form of data. Other forms of data include qualitative data in the form of words and text, such as interviews; images; video; audio; art; and more. However, this session will focus on structured quantitative data.

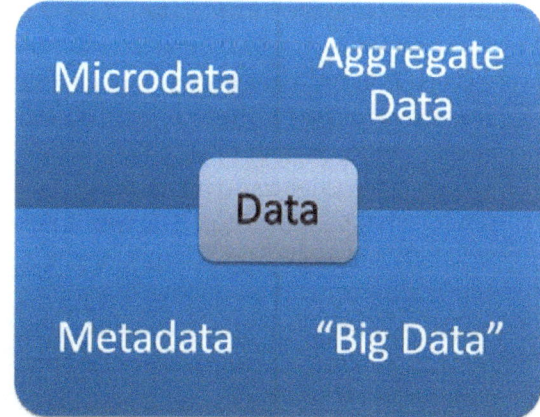

FIGURE 1
Data Types

Forms of Data

Proceed to introduce the framework for different forms of data offered in figure 1, which is one of many ways of breaking down types of data. We will use the US census as our example data set.

Microdata. Every ten years, the US Census Bureau attempts to count every individual in the United States, regardless of their immigration status. Every household receives a postcard asking them to respond to some basic demographic questions. Most governments do some version of a census to understand the makeup of the population for core functions such as taxation, military, education, welfare, and more. An individual person submits answers—which together in the United States will become approximately 330 million rows of data representing 330 million individuals. These are known as microdata—individual-level responses. They are very useful for statistics and running cross-tabulations. However, imagine Maria Gonzalez at 123 North Street, who just shared her income, address, the names and ages of her children, and even more private information. How would she feel about that information being publicly accessible? There can be significant privacy concerns in releasing microdata that include location, age, and other identifiable information, which means that most microdata are released only in an anonymized form.

Aggregate data. Most data are released in aggregate form. Instead of sharing all individual data, they are aggregated, or combined, into geographic or demographic categories. For example, the census may release a table identifying the percentage of people in Massachusetts who are in the 60–64 years age range. These data pose fewer risks to individuals and are often sufficient to answer many research questions.

Big data. The data we've talked about involve many observations, but you can engage with them using typical statistical techniques and readily accessible software and computational power. Big data are known for having high volume, variety, and velocity; we are not focusing on big data in this module.

Metadata. If you've ever downloaded a data set, you may have wondered: What do those variable names mean? What is the sample size, or sampling method? Have the data been cleaned? The answers to these questions and many more can be found in the metadata, commonly described as the data about the data. Metadata is similar to bibliographic information about a book. You search for and choose a book before you actually start reading it. First you look at its title, author, subject, publication date, publisher, genre, etc. Similarly, data sets have titles, authors, subjects, funding, geographic coverage, methodology, and more. This information can usually be found in a document called a codebook, or technical documentation.

After answering any questions about the previous section, continue to the next theoretical framework, finding data.

Finding Data

Where?
Geography
Location
Unit size: tract, county, state

What?
What variables could to answer the research question?
Check the codebook!

When?
Do you want a snapshot or longitudinal data?

Snapshot: one time
Longitudinal: covers multiple years for identifying changes

Why?
Why would someone collect this data? Where will you be able to find it, and what biases do their goals introduce?

Who?
Demographics!
Age, gender, socioeconomic status, etc.
Protected/vulnerable groups: children, prisoners, cognitively impaired people, etc.

How?
How is this type of data gathered?
Surveys, financial data, census, self-reported, government information

FIGURE 2
Finding Data Framework

Finding Data

By reordering the journalistic questions shown in figure 2, learners are guided by the instructor to ask six categories of questions.

1. *Where?* The biggest limiter in most social science data is geographic. What location are you researching? Is it the United States, Ethiopia, or Brookline Village? The second way to ask where refers to the unit size of location: an entire country, or broken down by state, county, census tract, etc.? Generally, the smaller the unit of geography, the more restricted the data will be.
2. *When?* The next question is temporal: When do you want data from? The year 1851, or the most recent available? You'll also consider the type of time you need. Do you want a snapshot (for example, population in 1851) or do you require longitudinal data that you can compare over time, such as how the unemployment rate changed each month from 2007 to 2012? Make sure that the same question is being asked of the same representative population. This generally means you want to find longitudinal data in a single data set to ensure comparability.
3. *Who?* Ask about people: Who do you want data about? The entire population? A particular age, gender, socioeconomic class, race, or other demographic group? Are any of the groups you want to research protected or vulnerable, such as children or people who are incarcerated? If so, you may find that access to those data is more restricted to protect people who are less able to advocate for themselves. You will need to be cognizant of your responsibility as a researcher to do no harm.
4. *What?* What variables could answer the research question? This is intentionally asked after the three previous questions because often there are multiple variables that could address your question. Emphasize flexibility and creativity in accordance with Searching as Strategic Exploration to proxy the concepts in a research question. If we want to evaluate farmer incomes, what variables can we look at? Answers

could include self-report survey, tax filings, or proxies such as rainfall. To find out what the variables actually mean, check the codebook, which describes how that question was asked and what it represents. Ask what type of data: Do you need microdata or aggregate data?

5. *Why?* Why would someone collect these data? This question can be used to identify stakeholders who may gather data, which you can then intentionally search for. Where will you be able to find them, and what biases do their goals introduce? If we want to get data about electricity access in Nigeria, who are the stakeholders? Answers include the power company, the government, nongovernmental organizations (NGOs), researchers/academics, and local community groups or newspapers. We discuss strategies for searching for each stakeholder's data in different sources. What are the goals or incentives of each group, and how might that influence or bias their results? For example, is Coca-Cola funding this research on sugar?

6. *How?* How refers to how these data were gathered and how that affects your analysis? I discuss known data quality issues through humor—by observing that everyone says they're taller, richer, and more attractive than they actually are—to introduce methodological considerations of self-reported data, among other discipline-relevant issues.

REVIEWS/ASSESSMENT STRATEGY

A graphic is presented on the board to support learners who prefer visual information, and more detail is provided verbally with multiple prepared opportunities for collaborative formative assessment, as discussed above.

For longer sessions, participants have the opportunity to apply the framework to an exploration of a particular data source using a guided worksheet in a small group and demonstrate their understanding by teaching back to the class.

ADAPTING THE RECIPE

Worksheets can be adapted to different data sources with the same basic structure and questions, with specific instructions on accessing the desired information.

APPENDIX: SAMPLE WORKSHEET

Resource: NYC Open Data Portal

Topic: Start and end locations of taxis in 2019

Adapted from a worksheet by Erica Schattle

Goal: To find out more about the contents of this resource and think about if and when it might be useful for your research.

Deliverable: After completing the steps and considering the questions outlined below, you will have *3 minutes* to present and demonstrate what you learned about this resource to the rest of the class.

Steps
- Navigate to https://data.cityofnewyork.us/. What is it? Who funds it? What does it include and not include?

- Begin searching for data on your topic.
- What geographic regions are covered in this resource? Time periods?

- Identify a data set of interest to you. Is the data set derived from another source, or is it primary? How can you tell?

APPENDIX: SAMPLE WORKSHEET (continued)

- In what file format(s) is this data set available?

- Download the data set for Excel (.csv or .xlsx file formats). Explain the steps you took to download the file(s).

- Create a citation for your data set (hint: https://www.datacite.org/cite-your-data.html has more information).

- Do other municipalities have similar data? Can you find open data portals for Mexico and Cambridge, MA? How detailed are their data? How do you find geospatial data? (Divide your group in half—one half looking at Cambridge and one half looking at Brazil).

Questions to Consider for Your Presentation

- A basic overview of the resource: What type of data does it include? What does it not include? Who runs it, and what are their goals?

- What geographic regions are covered in this resource? Time periods?

- Is the data set derived from another source, or is it primary? How can you tell?

- In what file format(s) is this data set available? Can you get a command file?

- How granular were the variables in the data set you downloaded? Are these microdata or aggregate data?

Data Literacy Layered Lasagna for Preservice Teachers

Brad Dennis, Associate Professor and Education Librarian, Western Michigan University, brad.dennis@wmich.edu;
Allison Hart-Young, Professor of Teaching, Learning and Educational Studies, Western Michigan University, allison.young@wmich.edu

NUTRITION INFORMATION

Like all great library liaison and teaching faculty relationships, this assignment has evolved over the years. This chapter will describe this assignment and its impact on information literacy and data literacy at Western Michigan University (WMU). The assignment was a collaboration between Brad Dennis, associate professor and education librarian, and Allison Hart-Young, professor of teaching, learning, and educational studies at Western Michigan University in Kalamazoo, Michigan. Education students have a series of questions that they must incorporate into a demographic summary of the school they attended before college and its community and a demographic summary of the local school that they attend as part of a service-learning experience.

TARGET AUDIENCE AND NUMBER SERVED

The cooks have tested this recipe with groups of 10–25 preservice teachers enrolled in elementary or secondary education programs. At WMU, most of these students are upper-level undergraduates (sophomores, juniors, and seniors).

LEARNING OUTCOMES

The student will be able to
- conduct research on the school they attended as a youth and the surrounding community and the school that they will attend as a service-learning component of the class
- differentiate between the different data sources and employ good judgment in the selection of those sources
- synthesize information from a variety of sources into a demographic paper and form questions based on the self-identified gaps in their knowledge
- utilize different search strategies effectively within each database
- match search tools and resources to their information needs and determine which source is best suited to provide community and school data
- determine which sources are credible and which are not
- discuss credibility of sources
- credit the work of others through proper attribution and citations

COOKING TIME

75-minute library instruction session followed by one week to complete the assignment (see appendix).

DIETARY GUIDELINES

This recipe best connects with the frames Research as Inquiry, Searching as Strategic Exploration, Authority Is Constructed and Contextual, and Information Has Value from ACRL's *Framework for Information Literacy for Higher Education*.

Teacher preparation at WMU is evaluated and accredited through the Council for the Accreditation of Educator Preparation, which uses model core teaching standards put forward by the Interstate Teacher Assessment and Support Consortium (InTASC; see Additional Resources). This assignment addresses three essential knowledge standards:

- *Learning differences:* "The teacher [candidate] knows how to access information about the values of diverse cultures and communities and how to incorporate learners' experiences, cultures, and community resources into instruction" (InTASC 2(k)).
- *Professional learning and ethical practice:* "The teacher [candidate] is committed to deepening understanding of his/her own frames of reference (e.g., culture, gender, language, abilities, ways of knowing), the potential biases in these frames, and their impact on expectations" (InTASC 9(m)).
- *Leadership and collaboration:* "The teacher [candidate] understands that alignment of family, school, and community spheres of influence enhances student learning and that discontinuity in these spheres of influence interferes with learning" (InTASC 10(m)).

INGREDIENTS

- A subject area instructor/colleague chef
- You, the enthusiastic librarian chef
- Enthralled student sous chefs
- Teacher station with presentation software (PowerPoint, Prezi, etc.).
- PCs or laptops with internet access for all of the students for active learning
- Optional—Social Explorer database (for access to InfoGroup Religion survey)

PREPARATION

- Prior to beginning of the class cooking demonstration, meet with your chef colleague and determine which major data sources to use for the assignment. Familiarize yourself with the data, interface, and access to the data. For example, in the state of Michigan, assessment data are stored at the State of Michigan Center for Educational Performance and Information (CEPI). However, the parent tab interface is designed and simplified for use by parents. Preservice teachers really need to learn how to use the CEPI Combined Annual Reports, the data intended for use by educators, which are available only in the educators tab interface.
- Find a specific school to use as an example when searching at the school, district, and state level. The students will use the school they attended before college.
- Census data are now found at https://data.census.gov. This replaces American FactFinder. The system is very comprehensive, but it can be overwhelming for students if you search using open queries. Therefore, it is best to search by geographic region and browse the tables.
- If you decide to use community religion data, find out where religion data are easily accessible for your class. The Social Explorer database includes two different religion surveys.

INSTRUCTIONS

1. Administer a pretest. We used multiple-choice questions using iClicker, but you can select any polling software available to you.
2. When class begins, note that this assignment translates to different subject areas or even outside of the college of education. This is important because student sous chefs will use the school and community data in order to
   a. write knowledgeably about the schools and community for this assignment,
   b. interview with the schools or any local employer, and
   c. compare areas to one another when making a decision to move or relocate.
3. Discuss goals and objectives of this class and assignment. Cover the importance of contacting a librarian, and distribute contact information.
4. Discuss the importance of authority, and introduce students to four main sources for primary school and community data:
   a. US Census Bureau for community demographic data (https://data.census.gov)
   b. National Center for Education Statistics (NCES) for school demographic data (https://www.nces.ed.gov/)
   c. State of Michigan Center for Educational Performance and Information (CEPI) for standardized test and other state education data (https://www.michigan.gov/cepi/)
   d. Religion Data (InfoGroup) using the Social Explorer database for demographics of religious groups.
5. Use active learning methods by demonstrating a search for the data using a preselected example school while the students search using the school they attended before college. Access the state department of education website that stores your state's educational performance measures (Michigan example: https://www.michigan.gov/cepi/). Locate the annual reports for testing data for the school, school district, and state. Teach the students how to find teacher credentials, completion rate (dropout rate), and postsecondary enrollment data.
6. Demonstrate how to find the individual school and school district enrollment characteristics such as race/ethnicity, gender, special needs, and socioeconomic data, including free and reduced price lunch data. Examine school staff population and fiscal data such as revenue and expenditures. Allow time for students to look for some of these data.

7. The NCES data also link to the Census Bureau and American Community Survey (ACS). Access primary source data through https://data.census.gov for the most common community statistics such as population by race/ethnicity, poverty rate, household income, and educational attainment.
8. Collaboratively discuss how religion affects the community, how religious practice can influence the students, and how teachers should be aware of the community in which they serve. Show how to access religion data using US Religion Data (InfoGroup) through the Social Explorer database. Examine the larger classifications and individual groups by congregation.
9. Utilize the University Libraries Citing Sources LibGuide (https://libguides.wmich.edu/citing) to discuss APA citations and references.
10. Administer the post-class assessment using iClicker or any polling software available to you. Hand out the worksheet to the students (see appendix). The demographics assignment is a progressive assignment that appears in multiple courses in the education program at WMU. Once the table is completed using the data, the student completes a writing assignment based on the data. The student will
    a. identify where reliable, valid demographic information is found
    b. identify and describe their home community using reliable demographic data (personal demographic data)
    c. compare the data to their own perceptions of their home community
    d. identify and describe communities in which they are placed for their clinical experiences (clinical placement demographic data)
    e. compare the data to their own perceptions of the community and schools in which they are placed for pre-internship
    f. compare and contrast their home community and schools with the community and schools of their clinical placements

REVIEWS/ASSESSMENT STRATEGY

Give the students a test of four multiple-choice questions before and after the session:
1. Which resource would you use to find religion demographics for a community (city, town, or village)? (Correct answer: C—Social Explorer)
2. Which resource would you use to find assessment scores for a school if the school was located in Michigan? (Correct answer: A—CEPI)
3. Which resource would you use to find demographics for a community (city, town, or village)? (Correct answer: D—Data.census.gov)
4. Which resource would you use to find demographics for a school? (Correct answer: B—NCES)

Multiple-choice answers
A. CEPI
B. NCES
C. Social Explorer
D. Data.census.gov
E. I don't know.

Also see appendix.

CHEF'S NOTES

Note that the religion survey is optional if your library does not have access to Social Explorer. Pew Research also conducts a religion survey: https://www.pewforum.org/religious-landscape-study/.

Every state has a site like the Michigan CEPI. The site usually contains the annual report data of every school in the state. It is usually maintained by the state department of education.

ADDITIONAL RESOURCES

CCSSO's Interstate Teacher Assessment and Support Consortium (InTASC). *InTASC Model Core Teaching Standards: A Resource for State Dialogue*. Washington, DC: CCSSO Council of Chief State School Officers, April 2011. https://ccsso.org/sites/default/files/2017-11/InTASC_Model_Core_Teaching_Standards_2011.pdf.

InfoGroup. "Religion (InfoGroup)." Prepared by Social Explorer, August 13, 2021. https://www.socialexplorer.com/tables/Religion_InfoUSA10.

Michigan. "Center for Educational Performance and Information." State of Michigan, 2021. https://www.michigan.gov/cepi.

National Center for Educational Statistics. "Search for Public Schools." CCD (Common Core of Data) Public School Data, 1919–2021. https://nces.ed.gov/ccd/schoolsearch/.

US Census Bureau. "Explore Census Data." Accessed August 13, 2021. https://data.census.gov/cedsci/.

APPENDIX: DEMOGRAPHICS COMPARISON

	Database	Your Hometown	Kalamazoo
Community name			
Community			
Geographic Context			
Population			
Racial/Ethnic Breakdown			
Socioeconomic Status (SES)			
Major Employers			
Religious Diversity			
District			
Geographic Context & # of Schools			
Population			
Racial/Ethnic Breakdown			
SES/FRL			
Teacher Educational Attainment			
High School			
Geographic Context & # of Schools			
Population			
Racial/Ethnic Breakdown			
SES/FRL			
Teacher Educational Attainment			
Graduation Rate/Attrition Rate			
Postsecondary Access			

Write a paragraph describing your perceptions of growing up in this community. What did it feel like to grow up in this community?

Write another paragraph about what it was like to go to the schools in this district. What is this district like?

Finally, write a paragraph about what it was like to attend your high school. What kinds of things were offered at your school?

Section 8.
Data Literacy Outreach and Engagement

203   [[Ch52]]**Data Visualization Day: Promoting Data Literacy with Campus Partners**
Wenli Gao

206   [[Ch53]]**Getting Messy Ourselves: An Experiential Learning Curriculum for Subject Librarians to Engage with Data Literacy**
Adrienne Canino

211   [[Ch54]]**Research Data Management Stone Soup: Gauging Team Competencies**
Michelle Armstrong, Megan Davis, Ellie Dworak, Yitzhak "Yitzy" Paul, and Elisabeth Shook

214   [[Ch55]]**Data Literacy Family Style: Full-Day Professional Development**
Molly Ledermann, Emilia Marcyk, Terence O'Neill, and Dianna E. Sachs

217   [[Ch56]]**Everyone Is Welcome at the Table: Outreach for Data Management and Data Literacy in Research Assignment Design**
Shannon Sheridan and Hilary Baribeau

220   [[Ch57]]**Seasoning and Simmering: Cultivating Data Literacy Skills through an Open Data Hackathon**
Peace Ossom-Williamson

223   [[Ch58]]**From Soup to Nuts: Finding Your Way around the Data Services Buffet**
Jane Fry and Chantal Ripp

226   [[Ch59]]**Teaching Data Literacy and Computational Thinking in Educational Technology**
Lesley S. J. Farmer

Data Visualization Day
Promoting Data Literacy with Campus Partners
Wenli Gao, Collections Analytics Coordinator, University of Houston Libraries

NUTRITION INFORMATION

With the growing demands for accessing, analyzing, and evaluating data in academia, there is a need for librarians to promote data literacy in higher education. One way to increase the awareness of data literacy on campus is to host data events with campus partners. Given the level of familiarity and existing campus interest in data literacy, we chose data visualization from the twelve Data Information Literacy Competencies developed by Purdue University Libraries as a foundation on which to host campus data events. From 2018 to 2021, the University of Houston Libraries partnered with the Hewlett Packard Enterprise Data Science Institute (DSI) to cosponsor this annual event. Over the years, the content of the event evolved from student data visualization contests to full-day events with presentations from academics and industries and demonstrations of visualization tools. This recipe is focused on the full-day Data Visualization Day event, with easy alteration to convert to a half-day event or an online environment.

TARGET AUDIENCE AND NUMBER SERVED

The target audience for this event is undergraduate students, graduate students, faculty, and anyone interested in learning more about data visualization. The number served could vary depending on space and budget. The recommended minimal number of people served is 50 to ensure engagement from participants and deepen the impact of the event.

COOKING TIME

40 hours of preparation time are required to host this one-day event.
Preparation includes
- planning meetings
- contacting speakers
- setting up student contests
- promoting the event

The event includes presentations from invited speakers, short student presentations on their visualization contest entries, demonstrations of tools, and training sessions. It is recommended to enlist 3 to 5 speakers and 10 to 15 students to enter the contest.

DIETARY GUIDELINES

Data visualization, one of the Data Information Literacy Competencies developed by Purdue University Libraries (see Additional Resources), served as the guideline for this event.

LEARNING OBJECTIVES

- Proficiently use basic visualization tools
- Avoid misleading or ambiguous representations
- Discern and understand the advantages of different types of visualization; for example, maps, graphs, animations, or videos

INGREDIENTS

- One large room for the event, set up for presentation.
- Training rooms for data visualization trainings with specific software preinstalled on the computers.
- Simple online registration form asking for name, e-mail, and department.
- PowerPoint slides for speaker presentations.
- Compiled slides with student entries for the data visualization contest.
- A slide with a link to an online survey to vote in the student data visualization contest. The survey has only one question, asking people to pick their favorite visualization. Qualtrics, Google Survey, or SurveyMonkey can be used to create the survey.
- Award certificates and prizes for student data visualization contest winners.
- Different visualization tools to display depending on local campus availability.
- Drinks, snacks, and lunch. Based on the

estimated number of attendees, pizza can be provided as lunch.

PREPARATION

The event begins with monthly planning meetings between the library and campus partners about six months ahead of the event date. Potential campus partners can be the office of research, data or statistics centers on campus, computer labs that host data visualization software, or campus departments that use data visualization heavily, such as the computer sciences department. In the initial planning meeting, a draft time line is set, and action items are developed and delegated between the library and the partners. Once a tentative date for the event is selected, working group members brainstorm potential speakers and contact them to check their availability. Working group members should come from both the library and the campus partners. Include library personnel who work with data visualization, such as the data librarian, data visualization specialist, or GIS expert. People in charge of communications from both the library and the campus partners should also be part of the group. They can promote the event through their channels. Three to five invited speakers from various disciplines give the audience a taste of different visualizations used in a variety of settings. Working group members should also contact people who have expertise in certain visualization tools to demonstrate those tools in the afternoon. There may also be an option for the audience to attend hands-on training sessions, so trainers may be contacted as well.

For the student data visualization contest, an announcement and online form need to be created to accept entries. Working group members will create an online survey for audiences to vote for their favorite visualization after all students share their visualizations. They will also seek funding for prizes. Prizes can be a certificate, cash, or some sort of data-themed gift.

The library and the campus partners use their own channels to promote the event through e-mail, websites and social media. An event image used online for promotion is shown in figure 1. One of the benefits of this collaboration is to reach a greater audience than any one entity would reach alone.

Additional arrangements that need to be taken care of before the event are reserving rooms and ordering food for catering.

INSTRUCTIONS

The event begins with a welcome speech from both the library and the partners, followed by the invited presentations on visualization topics in different disciplines. The topics are selected to allow participants to see a variety of visualization types and use cases. The presentations are 30 minutes each, including questions and answers.

During lunch, students who submitted visualizations for the contest present and give a lightning talk to explain their visualization.

FIGURE 1
Event image used online for promotion

TABLE 1
Data Visualization Day example schedule

9:00–9:15 a.m.	Opening Remarks
9:15–9:45 a.m.	Presentation 1:The Future of Data Visualization
9:45–10:15 a.m.	Presentation 2: Training Astronauts Using Hardware In-the-Loop Simulations and VR
10:15–10:30 a.m.	Break
10:30–11:00 a.m.	Presentation 3: The Human Body Project and the Anatomage Table
11:00–11:30 a.m.	Presentation 4: Interpretation of Machine Learning with Visualization and HPE AI Solutions
11:30 a.m.–12:00 p.m.	Presentation 5: The Sampled City—Visualizing Granularity and Connection in Health
12:00–1:30 p.m.	Lunch and Student Visualization Contest
1:30–3:30 p.m.	Option 1: Demonstration of Visualization Tools: Anatomage Table, VR goggles, iMotions biometric tools
1:30–3:30 p.m.	Option 2: Training: Introduction to Tableau

Working group members, speakers, and the audiences are the judges. A link to the voting form will be posted on the screen for people to vote using their laptops or a phone. Results are posted immediately after the contest. If there is a tie, working group members and speakers will have a final discussion to decide the winner.

In the afternoon, participants have two options. One is to play with some visualization tools. Those tools are small and can be moved to the event room. The selection of tools is based on your campus availability. In our one-day event, we looked at the Anatomage Table, virtual reality goggles, and iMotions biometrics tool. The Anatomage Table is a movable touchscreen table where you can see images of your organs and move around to see details. VR goggles allow you to experience virtual reality. iMotions is a tool where you look at images and it captures biometrics and generates heat maps of your eye movement. These tools showcase what equipment is available to faculty and students. If no tools are available, a simple version of this activity could be setting up stations to show easy-to-use online data visualization websites. With prepared data sets, you can see how different visualizations can be created. These websites can be Piktochart (https://piktochart.com), Datawrapper (https://www.datawrapper.de), or RAWGraphs (https://rawgraphs.io). The other option is to attend a hands-on Tableau training session in one of the reserved training spaces. Tableau is selected because of its popularity on campus.

A sample schedule for Data Visualization Day is shown in table 1.

REVIEWS/ASSESSMENT STRATEGY

The number of attendees is used to gauge interest in this event. A quick anonymous survey is used to assess the quality of the event. It could be online or on paper. Urge attendees to fill out the survey immediately after the event to ensure a healthy response rate.

ADAPTING THE RECIPE

The one-day event could easily be adjusted to a half-day event, ending at noon after selecting winners for the student visualization contest. With the constraints of COVID-19, this event could also be moved online using virtual platforms such as Zoom or Teams. With virtual events, the demonstration of tools and hands-on training sessions are optional, but presentations and the student visualization could be hosted.

ADDITIONAL RESOURCES

Carlson, Jake R., Michael Fosmire, Chris Miller, and Megan R Sapp Nelson. "Determining Data Information Literacy Needs: A Study of Students and Research Faculty," preprint. Libraries Faculty and Staff Scholarship and Research, Paper 23. 2011. https://docs.lib.purdue.edu/lib_fsdocs/23/.

Getting Messy Ourselves
An Experiential Learning Curriculum for Subject Librarians to Engage with Data Literacy

Adrienne Canino, Data Librarian, Axiom Data Science, adrienne.canino@gmail.com

NUTRITION INFORMATION

This recipe is designed to create a dish of data manipulation with a side of group work. It is based on the definition of digital literacy published by the ALA Digital Literacy Task Force in 2011[1] and a modified data literacy definition from Bhargava and D'Ignazio: "Data literacy includes the ability to read, work with, analyze and argue with data."[2] This definition is the foundation of a three-tier data literacy rubric with skill statements and descriptions for each tier of literacy (see appendix). This rubric is an assessment tool for the workshop.

The participants begin with a pre-workshop marinade of reading assignments that provide an overview of concepts and terms. The recipe includes an introduction and activities for understanding messy data, evaluating variables, proxies, measurement criteria, summary statistics, and graphing/data visualization and dashboards. The recipe is served alongside a final debrief of open questions and comments.

TARGET AUDIENCE AND NUMBER SERVED
- The target diners are librarians and library staff who desire a deeper understanding of working with data sets and who have little to moderate experience. These librarians may go on to conduct similar workshops for students.
- This recipe serves approximately 16 diners, who form 4 groups of 4.

LEARNING OUTCOMES
Participants will
- observe how collecting and storing observations may lead to messy data
- explain one descriptive statistic related to their data set
- assess one of the descriptive statistics' usefulness for analysis
- create a data visualization to relay meaningful information

COOKING TIME
90 minutes

DIETARY GUIDELINES
As data science and research data become more prominent in everyday library work, this curriculum connects the frames Information Creation as a Process and Research as Inquiry from ACRL's *Framework for Information Literacy for Higher Education* to the process of manipulating and evaluating raw data.

INGREDIENTS
- Regular size Post-it Notes
- Easel size Post-it Notes
- Capability to do a digital presentation or paper printouts
- Method for sharing the pre-assignment materials, such as Dropbox or Google Drive
- At least four computers with spreadsheet software (such as Microsoft Excel with the Analysis ToolPak plug-in installed)
- The Marinade:
  – Digital or hard copies of recommended readings: David Herzog's *Data Literacy*, chapters 8 and 11, and the glossary
  – A data set as a comma-separated value file—raw, not cleaned—that is relevant to the participants; for example, collections data or bibliometric data. The data set should include several columns of different kinds of information for best effect (e.g., item title, author, publication date, price, checkout count). Ideally, it will include more than 8 columns and more than 50 rows.

PREPARATION AND SOUS CHEFS

Two Weeks before the Workshop

- Distribute the pre-readings and the data set.
- Share instructions for installing Microsoft Excel Analysis ToolPak, or send documentation on formulas for other spreadsheet software.
- Assign the pre-readings.

Marinade

- Ask participants to install Microsoft Excel and the plug-in.
- Ask participants to download the data set to their local computer.

Optional: Before the workshop, assemble a slide deck to define the terms used for each section of the workshop and the instructions for activities.

The following sous chef roles are recommended: volunteers among the participants for a notetaker, a timekeeper, a helper for assisting during activities, and a parking lot attendant to record follow-up or tangential comments.

Day Of

- On a whiteboard or large Post-it Note, draw a table with two columns—Name and Shirt Color—with space for each participant to complete a row of the table.
- Hang easel-sized Post-it Notes around the workshop space, labeled with the summary statistics from the "Introduction to Summary Statistics" section (below).

INSTRUCTIONS

Understanding Messy Data

Large-group facilitated discussion, 10 minutes.

1. Upon arrival, ask participants to record their answers, one entry per participant, on the Name/Shirt Color table.
2. Begin a discussion about messy data, using the shirt information as a prompt. Ask:
    a. How should we record shirts with patterns or more than one color?
    b. How should we record complex data, like first and last names?
    c. What's not included that we may care about in this fashion survey?
    d. What's the goal of collecting this information?
3. Using this example, build consensus about how we define *data*, and determine methods for analysis.
4. Transition to the comma-separated values data set. Ask what the analysis goals would be. As a large group, discuss which variables are complex, and which variables are more or less important for different audiences (e.g., author information versus catalog number for researchers and librarians).
5. Debrief and takeaways:
    a. Summarize and check that everyone is on the same page when we talk about data and data analysis and how analysis goals influence data collection.
    b. A takeaway is that good metadata is important. Controlled vocabularies are key in the data collection phase. Otherwise, data could be too messy to summarize or work with.
    c. Ask what are the shortcomings of the data set? For example, with only collections data, we don't know how often an item has been cited (or viewed) by students, only how often it's been checked out.

Evaluating Variables, Proxies, Measurement Criteria

Small-group exercise, 4 per group, 20 minutes,

1. Define the terms *variable*, *proxy*, and other relevant terms.
2. Split participants into groups, and make sure at least one participant per group can open the spreadsheet, has Analysis ToolPak or formulas ready, and saves an unaltered copy with "MasterCopy" appended to the file name.
3. *Free exploration* (10 minutes): Each group can use filters or other tools to look through the data, make notes about metadata, and think about questions the data can answer and the answers they want. Decide on criteria for evaluating the data set. The criteria do NOT have to be an exact match for variables in the data. Decide which data points to use and how the analysis or manipulation of the data will contribute to your criteria.
4. Groups can manipulate data for this exercise, but it is not required. Participants can gain a lot of information from filtering, noting calculations, or finding missing values.
5. Regroup and share criteria and variables of interest. Notetaker records.
6. Optional: A participant with Excel can clean the group's data set to hold only

the variables that relate to the evaluation criteria the group picked.
7. Debrief and takeaways (5 minutes):
    a. Are any proxies mentioned by a group? Are there criteria that require two data points used together? For example, in collection data, price can be a proxy for value.
    b. Participants should understand that analysis and decision-making are only as good as the data collected and the measurements used.
    c. Examine the implications of good metadata on reusability: Is the version of record available or important for these variables? What do you want to know about the provenance of the data?

Introduction to Summary Statistics

Presentation, same small groups, 20 minutes.
1. Present slides on summary statistics. Herzog's chapter 8 is the pre-reading. Review:
    a. count and sum
    b. average/mean
    c. mode
    d. median
    e. minimum/maximum
    f. percentiles = percentages, and percentage points
    g. rate of change
2. Introduction to tidy data.
    a. The organization of the spreadsheets affects the visualization and the calculations we can do.
    b. Tidy data is defined as one variable per column, one observation per row, one value per cell. It also usually assumes one header of variable (column) names and consistent data types of the cells.[3]
3. Each group picks a single variable to analyze (e.g., checkouts per year). The next activity builds on this one, and only certain criteria build good visualizations.
4. Each person in the group picks a different summary statistic (from step 1 of this exercise) to calculate for this variable.
5. *Analysis activity:* Using the spreadsheet software and formulas or Analysis ToolPak, the group performs the calculations for each statistic. When the small groups understand the summary statistics of each member, their work is done. For example, one participant will have *average* checkouts per year, one will choose the *minimum/maximum,* and so on.
6. Optional: Ask each group member to explain their statistic and what it means. For example, in collections data, the *highest count* of item location could serve as a proxy for accessibility by patrons if fast access to materials is the criterion.
7. Each participant should record their summary statistic with a little context, such as unit and variable, on a standard size Post-it Note and stick this to the big Post-it Note it matches in the room.
8. Debrief and takeaways (large-group facilitated discussion):
    a. Are participants comfortable with at least one summary statistic and what it represents for the evaluation criteria?
    b. Do these statistics answer the evaluation criteria? Directly or indirectly?

Introduction to Graphing/Data Visualization and Dashboards

Presentation, small groups, 20 minutes
1. Present a slide deck on basic graphs and plots (data visualization). Herzog's chapter 11 is the pre-reading.
2. Review the following:
    a. graph anatomy—title, labels, legend, source, author
    b. bar plots
    c. histograms
    d. scatter plots
    e. line plots
    f. Why the debate over pie charts? (optional)
3. Form new groups of 3–4 based on the summary statistics.
4. Pull the summary statistics Post-it Note off the wall and put up a new easel-size Post-it Note.
5. On the easel-size Post-it Note, design a graph to visualize the selected summary statistic. Calculate more data points if desired.
    a. Draw the axis, and add authors, title, and units or categories.
    b. Record the data points from the standard size Post-it Notes of the summary statistic.
    c. For example, a bar chart of averages would have categories on the x axis, and ascending numbers on y axis. So average checkouts compared to average price may be two variables displayed on this chart.
6. Once the visualization is complete, post it at the front of the room.
7. Debrief and takeaways (facilitated discussion):

a. Each group gives a brief overview of their graph.
b. Celebrate! Together, these statistics show a handmade dashboard of data.

Final Debrief
Facilitated discussion, 10 minutes.
1. Open to questions and comments.
2. Ask for feedback from participants.

REVIEWS/ASSESSMENT STRATEGY

- Get feedback after the session, either during the final debrief or later after digestion.
- Participants self-select via an anonymous pre- and post-survey where they were in the tiers of the data literacy rubric (see appendix).

ALLERGY WARNING

This is a topic where imposter syndrome can be common among librarians. The readings are at an introductory level. Instructors may consider a pre-workshop survey that allows participants to express specific concerns or desired outcomes.

RECIPE VARIATIONS

The workshop can use Google Sheets or OpenOffice Calc by using formulas for the descriptive statistics. If there are different levels of experience with spreadsheet software, small groups can include one person who is experienced with using formulas.

ADDITIONAL RESOURCES

Herzog, David. *Data Literacy: A User's Guide*. Thousand Oaks, CA: Sage, 2015.

Seattle. "Library Collection Inventory." City of Seattle Open Data Portal, 2021. https://data.seattle.gov/Community/Library-Collection-Inventory/6vkj-f5xf.

ACKNOWLEDGEMENTS

Many thanks to the University of Rochester River Campus Libraries team that provided the kitchen testers for this recipe!

APPENDIX: DATA LITERACY SKILLS—TIERS OF UNDERSTANDING

Data literacy includes the ability to read, work with, analyze, and argue with data. A data literate person:			
Explicit skill description	**Tier 1**	**Tier 2**	**Tier 3**
Understands what data are and what aspects of the world they represent	I see data or evidence in use and generally follow the idea it represents.	I am comfortable with data and their use in decisions in my subject specialty or domain.	I always ask for (and read through) technical/ methodological documentation of the data sets I'm working with
Is able to create, acquire, clean, and manage data	I can export or save a file that has my data in it. I don't know if it's messy or not.	I can get or create an interoperable data file, do some version control, and do some cleaning to my data.	I always do version control and thorough data cleaning and encoding. Need a web scrape? OK! API data pull? No problem!
Performs analytic operations on data such as filtering, sorting, aggregating, comparing, and interpreting	I can sort in spreadsheet software. I can compare tables that already show results.	I can sort, filter, do some formulas, maybe a pivot table, in my favorite data software or interface.	I can use an advanced data analysis tool or two (or three+) and manipulate easily. What's the question?
Translates data into representative/accurate visuals	I can make a table showing the basics, maybe a chart from my favorite plug-and-play data viz software.	I feel confident with the basic table, bar graph, pie chart, live chart. I can usually demonstrate a trend I want to show with basic data/analysis.	I can slice and dice and design visual representations that tell a story with the data. Or falsify a story.

Section 8. Data Literacy Outreach and Engagement

APPENDIX: DATA LITERACY SKILLS—TIERS OF UNDERSTANDING (continued)

Data literacy includes the ability to read, work with, analyze, and argue with data. A data literate person:			
Explicit skill description	*Tier 1*	*Tier 2*	*Tier 3*
Differentiates between ethical and unethical uses of data	I know there shouldn't be personal information in the data sets I use.	I can clean out personal private information from my data sets.	I can uncouple bad ID proxies, PPI data, and representations from my data before I begin analysis.
Uses data effectively to support a larger narrative intended to communicate some message to a particular audience	I can find the evidence, primary or secondary, to support a narrative.	I can clean and analyze a data set uniquely to support my narrative, building my own primary conclusions.	I can create, clean, and combine data from multiple sources to build a narrative without overstretching what the data really represent.

NOTES

1. American Library Association, "Digital Literacy," 2011, https://literacy.ala.org/digital-literacy/.
2. Rahul Bhargava and Catherine D'Ignazio, "Designing Tools and Activities for Data Literacy Learners" (paper presented at WebScience: Data Literacy Workshop, Oxford, UK, 2015), https://www.media.mit.edu/publications/designing-tools-and-activities-for-data-literacy-learners/.
3. Hadley Wickham, "Tidy Data," *Journal of Statistical Software* 59, no. 10 (2014): 1–23, https://www.jstatsoft.org/article/view/v059i10.

Research Data Management Stone Soup
Gauging Team Competencies

Michelle Armstrong, Associate Dean, Albertsons Library, Boise State University; Megan Davis, Reference and Instruction Librarian, Albertsons Library, Boise State University; Ellie Dworak, Data Visualization Librarian, Albertsons Library, Boise State University; Yitzhak "Yitzy" Paul, Instruction Librarian, Albertsons Library, Boise State University; Elisabeth Shook, Head of Scholarly Communications and Data Management, Albertsons Library, Boise State University

NUTRITION INFORMATION
This recipe incorporates ingredients from several competency documents designed by an array of library groups to create an exercise that helps to bolster skills and services surrounding research data management (RDM). This assessment allows the library to better understand and visualize the strengths and gaps in knowledge necessary to effectively run an RDM team creating an ever-changing, collaborative "stone soup."

NUMBER SERVED
The recipe scales with the number of library employees working with or interested in research data management work. While the competencies exercise can be used with groups of any size, groups should include a sufficient number to identify areas of strength and weakness.

LEARNING OUTCOMES
- Self-assess strengths and weaknesses of an institution's current RDM team
- identify gaps in knowledge in which professional development should be sought
- build upon the strengths of individuals to form a balanced RDM team in light of the institution's research data management needs

COOKING TIME
It should take approximately one hour for an individual to complete the competencies exercise. Times may vary depending on customization of the recipe. Additional time will be required if you choose to evaluate the strengths and weaknesses of the group as a whole.

DIETARY GUIDELINES
This Stone Soup recipe addresses ACRL's *Framework for Information Literacy for Higher Education* in the following ways:
- Information Has Value: Understanding the strengths and knowledge gaps of individuals and the group can lead to the development of a balanced team, prepared to provide high-level service to researchers.
- Research as Inquiry: As groups responsible for data management may change over time, utilizing the competencies document will allow for ongoing evaluation of individual and group strengths and weaknesses, leading to continued growth in skills and knowledge.
- Scholarship as Conversation: Group members will discuss which data management competencies are applicable and important to themselves and their institution's researchers.

INGREDIENTS
- Research Data Management Self-Assessment Tool Template (see appendix)
- Library staff who are involved or interested in data management, regardless of current skill level
- One or more of the resources from Additional Ingredients to make this recipe your own

PREPARATION
1. Copy the Self-Assessment Template into a text editor such as Word or spreadsheet software such as Google Sheets.
2. Ask the members of your RDM group to review the Additional Ingredients.
3. Review the competencies with your group to discuss what additional ingredients (either from the Additional Ingredients or from scratch) to add to the Self-Assess-

ment Tool Template.
4. Add these additional ingredients, such as data visualization, to taste and distribute the revised Self-Assessment Tool to the group.

INSTRUCTIONS

1. Using the customized Self-Assessment Tool, have each group member rate their level of competence on a sliding scale, from developing to expert, by placing a mark next to each ingredient.
2. Have group members share their results with the rest of the group.
3. Meet to discuss next steps, such as professional development activities. Based on the results, self-study, or group instruction in cases in which more than one or two people must learn or refresh a skill, should be planned.

REVIEWS/ASSESSMENT STRATEGY

After each member rates their level of competence, bring the self-assessments together to identify overlapping or complementary strengths or areas of weaknesses of your group. Compare this information with the assessed needs of your institution and the data services your library would like to provide. By doing this, you will be able to identify either service gaps or opportunities for expansion. The individual competencies list can also be used to identify specific areas for professional development.

ADAPTING THE RECIPE

This recipe can be adapted to include employees outside the library. People outside the library may be needed to fill some of the competency gaps identified the first time this recipe is made. This can be especially useful if you are just starting to investigate research data management at a library or university level.

ALLERGY WARNING

Warning: Sensitivities about an ingredient during assessment can result in false signs of an allergic reaction and feelings of inadequacy. To prevent this, any early signs of allergic reaction should be reported to the group. This should result in a frank and open discussion. The group can then help the individual reorient their perspective.

CHEF'S NOTES

It should be noted that cooks should not discount foundational ingredients. The ability to build meaningful relationships or communicate clearly are examples of important ingredients that create a delicious and nutritious dish.

ADDITIONAL INGREDIENTS

The complete self-assessment instrument is available for use here:

Armstrong, Michelle; Davis, Megan; Ellie Dworak; Yitzhak Paul, and Elisabeth Shook. "Research Data Management Competencies Self-Assessment." ScholarWorks, Boise State University, Data Management Services, paper 8, 2021. https://doi.org/10.18122/dataservices.8.boisestate.

You may wish to use these additional ingredients to customize your recipe.

American Library Association. "Knowledge and Competencies Statements." Accessed March 3, 2021. https://www.ala.org/educationcareers/careers/corecomp/corecompspecial/knowledgecompetencies.

Association of College and Research Libraries. "Roles and Strengths of Teaching Librarians." April 28, 2017. http://www.ala.org/acrl/standards/teachinglibrarians.

Association of Southeastern Research Libraries. "Shaping the Future: ASERL's Competencies for Research Librarians." November 10, 2000. http://www.aserl.org/programs/competencies/.

NASIG. "NASIG Core Competencies for Scholarly Communication Librarians." *NASIG Newsletter* 32, no. 5 (August 2017). https://tigerprints.clemson.edu/nasig/vol32/iss5/1.

Reference and User Services Association. "Guidelines for Behavioral Performance of Reference and Information Service Providers." Latest revisions approved May 28, 2013. https://www.ala.org/rusa/resources/guidelines/guidelinesbehavioral.

Schmidt, Birgit, and Kathleen Shearer. "Librarians' Competencies Profile for Research Data Management." Joint Task Force on Librarians' Competencies in Support of EResearch and Scholarly Communication, June 2016. https://www.coar-repositories.org/files/Competencies-for-RDM_June-2016.pdf.

APPENDIX: RESEARCH DATA MANAGEMENT SELF-ASSESSMENT TOOL TEMPLATE

Name: Date:

I. RDM competencies	Developing	Competent	Proficient	Expert	Notes:
1. Understands data & data management practices					
1.1. Understands & can articulate the research life cycle & where data management fits in					
1.2. Locates, understands, & applies DMP guidance & best practices					
1.3. Understands & applies best practices for file naming & organization					
1.4. Understands & applies best practices for preservation & distribution of research data					
2. Writes data management plans to follow best practices & NSF directorate guidelines as well as meeting researchers' & project needs					
2.1. Identifies & uses guidance & documentation appropriate to the funding agency & project					
2.2. Includes guiding/supporting information specific to project needs					
3. Familiar with research data repositories					
4. Can create organizational structures for researchers & research teams					
5. Communicates clearly verbally & in writing					
5.1. Distills complex concepts for nonexperts					
6. Builds relationships & networks					
7. Sets reasonable boundaries					
8. Has commitment to staying current in a quickly evolving research environment					
9. Manages projects to completion					
10. Understands & models ethical behavior, e.g., regarding research compliance, information security, & intellectual property					
10.1. Respects researcher privacy & maintains confidentiality					

Data Literacy Family Style
Full-Day Professional Development

Molly Ledermann, Faculty Librarian, Washtenaw Community College; Emilia Marcyk, Teaching and Learning Librarian, Michigan State University; Terence O'Neill, Head of Digital Scholarship Services, Michigan State University; Dianna E. Sachs, Health and Human Services Librarian, Western Michigan University

NUTRITION INFORMATION

Improving your data literacy instruction doesn't have to be a lonely task. In this recipe, we'll outline a day-long data literacy boot camp aimed at developing attendees' understanding of data literacy theory and practice while they learn from colleagues around the state. Our state's Information Literacy Interest Group sponsored the event and served as the network to reach our target audience: librarians with a teaching and learning component to their work who are interested in data literacy.

This recipe is a fun, engaging way to build skills and community around data literacy in your professional network. Read on to see how you can make this happen with the expertise, budget, and community that you already have.

TARGET AUDIENCE AND NUMBER SERVED

This family-style recipe is intended for an audience of librarians from two- and four-year colleges and universities. Providing a range of data literacy topics ensures that the menu will have something for everyone, including those working with first-year programs, undergraduates, graduate students, or students in specific disciplines.

Number served: 30 librarians.

An event of this size makes space and food planning easy. It also allows for more discussion and networking opportunities in small groups and as a whole.

LEARNING OBJECTIVES

Participants will explore theoretical frameworks that contextualize data literacy within the larger field of academic librarianship.

Participants will develop practical approaches to data literacy instruction.

Participants will network with librarians interested in data literacy across varying academic contexts.

COOKING TIME

Like the best sourdough, kimchi, and barbecue, this program benefits from substantial preparation and cooking time. Allow at least 6 months to plan and market your program (our program took ten months from inception to completion). If you are planning to request funds or logistical support from an outside organization, such as your state library association or a member library, consider the deadlines of the organization.

Ensure that your program does not coincide with other planned events or conferences that involve your intended audience. Once you have chosen your date, send a save-the-date message to your intended audience and to other organizations, at least two months prior to your event, to avoid any potential conflicts (see figure 1).

FIGURE 1
Event program cover

INGREDIENTS

- 1 planning committee—Source a mix of 4–5 librarians from a variety of institutions. Include large and small public universities, private colleges, and community colleges.
- 1 event space—A room that can hold 30–40 people seated at round tables. Include space for food and beverages. Alternatively, this could be run as a day-long virtual event (without food).
- 1 keynote speaker—Focus on philosophical, big-picture ideas or current complexities in data literacy. Honorarium recommended.
- 3 topic-specific speakers—Focus on practical aspects of data literacy. Include a variety of topics and speakers.
- 1 round of lightning talks—3 speakers, 5 minute each, 15 minutes for Q&A.
- Sprinkles—Top with a generous sprinkling of food and engagement (buttons [see figure 2], stickers, lunchtime activities, etc.).

PREPARATION

For any large meal, preparing your ingredients is key to ensuring service runs smoothly. Here are the primary considerations for a planning committee.

Speakers

Speakers can be recruited from anywhere! Leverage your local networks to identify potential speakers from member libraries, other academic departments, state library organizations, and so on. Review past conference proceedings and publications to find local experts. You can look outside the box and consider data science or journalism experts working in industry in your area. Alternatively, survey potential attendees ahead of time to gain insight on topics of interest. Examples may include ethically using and understanding data from scholarly sources, using and understanding data and data visualizations from popular sources, using and understanding data from government sources, and teaching data literacy.

Logistics

Share best practices for accessible presentations with speakers ahead of time (see Additional Resources). If possible, ask speakers to provide you with their presentation files, as well as any audience handouts before the event. Load presentation files onto a computer already configured for the AV equipment in your venue. Arrange for on-site technical support if needed.

The event space should address accessibility and equity needs, including providing Wi-Fi access for attendees, microphones for speakers and audience Q&A, accessible and all-gender restrooms, locking lactation room, a designated quiet reflection space, sign language interpretation on request, and dietary accommodations on request.

On the day of the event, ensure all volunteers have contact information for on-site technical support, keys or codes to all event spaces, a detailed itinerary, defined responsibilities, and each other's contact information.

We highly recommended that program planners collaborate with a local library or library organization to streamline marketing, process registrations and payments, and coordinate logistics.

Budget

Program planners should identify any financial needs and funding sources as soon as possible. Funding may come from registration, state or local library organizations, or a combination.

It is recommended that speakers be reimbursed for travel expenses and recognized for their contributions with an honorarium. Other items to be budgeted include food and drink, space and equipment rental, printing

FIGURE 2
Buttons distributed during the event

(name tags, schedules, signage), and any prizes or gifts for attendees.

Food

For a full-day event, provide coffee, tea, and snacks throughout the day and a full lunch. We purchased food from local stores and restaurants, which allowed us to save money, but some venues will require you to use their catering services. Be aware of any catering, setup, and cleaning requirements for your venue.

INSTRUCTIONS

The boot camp structure may change based on local needs and what makes sense in your context, but here is as an example agenda:

Data Literacy Boot Camp Agenda

8:00 a.m.
- Volunteers arrive.
- Set up breakfast and coffee.
- Set up registration table and name tags.

8:30 a.m.
- Check-in begins.
- Attendees mingle and eat breakfast.
- Check in with keynote speaker and make sure all sound and projection systems are running.

8:50 a.m.
- Welcome and housekeeping announcements.
- Introduce keynote speaker.

9:00–10:20 a.m.
- Keynote.

10:30–11:30 a.m.
- Interactive session 1.

11:20 a.m.
- Volunteers receive food delivery from restaurant.
- Set up meal.

11:35 a.m.–1:00 p.m.
- Attendees eat lunch and mingle.
- Tour of library spaces relevant to the programming such as a digital scholarship lab.

12:45 p.m.
- Volunteers clean up lunch.
- Box and refrigerate leftovers, if possible. People are excited to take some food home at the end of the day.

1:00–2:00 p.m.
- Interactive session 2.

1:50 p.m.
- Volunteers set up snacks.

2:10–2:40 p.m.
- 5-minute lightning talks.
- Designate a volunteer to introduce talks and a timekeeper.

2:50–3:50 p.m.
- Interactive session 3.

3:50–4:00 p.m.
- Closing remarks and thank-yous to participants, presenters, and sponsors.

4:00 p.m.
- Volunteers break down and clean up space.

REVIEWS/ASSESSMENT STRATEGY

Before planning the event, we scoped the Data Literacy Boot Camp with a survey to our state Information Literacy Interest Group, and solicited input on
- length of activity and how far in future
- general format
- suggestions for speakers
- volunteers for organizing

Following the event, we surveyed participants on whether the event was well organized, the usefulness of the content, the way they heard about the event, the location, the food, and anything else they'd like to share.

ALLERGY WARNING

As with any family-style meal, not all attendees will find that every offering is to their taste, interest, or level of understanding. Make sure to include courses that will appeal to those who are new to data literacy and those with more expertise. Publicize the menu of speakers and topics ahead of time so that attendees will know what to expect.

ADDITIONAL RESOURCES

Michigan Academic Library Association. "Data Literacy Boot Camp." https://www.miala.org/dataliteracybootcamp.php.

University of Washington, Disabilities, Opportunities, Internetworking, and Technology Group. "How Can You Make Your Presentation Accessible?" Accessed May 17, 2021. https://www.washington.edu/doit/how-can-you-make-your-presentation-accessible.

PowerPoint guidelines sent to all presenters.

Everyone Is Welcome at the Table
Outreach for Data Management and Data Literacy in Research Assignment Design

Shannon Sheridan, Research Librarian, Pacific Northwest National Laboratory, shannon.sheridan@pnnl.gov; Hilary Baribeau, Scholarly Communications Librarian, Colby College, hbaribea@colby.edu

NUTRITION INFORMATION

For an instructor to maintain a balanced diet in the classroom, it's important to introduce a healthy portion of data literacy, regardless of the subject area. Data come in all shapes and forms: from bite-sized quantitative snacks to pounds of qualitative banquets. As nutritional experts in data literacy, we looked for outreach opportunities to talk about the importance of integrating data literacy in research assignments.

The University of Wyoming's Ellbogen Center for Teaching and Learning (ECTL) runs an invited presenter series every semester entitled Lunch and Learn. The center seeks session proposals from educators across the university who wish to share teaching strategies that have an impact on student learning.

This recipe is based on a Lunch and Learn session led by librarians to discuss how faculty can design a research assignment that reflects an understanding of data literacy and incorporates data management practices into their research assignments.

TARGET AUDIENCE AND NUMBER SERVED

This recipe serves all instructors and will help to incorporate data literacy and data management into research assignments. Number served is 10–20 for best opportunities for discussion.

LEARNING OUTCOMES

The session is open to instruction faculty and staff from all disciplines and with varying degrees of knowledge of data and data literacy. Therefore, the primary goals of the session are to
- spark conversation
- advertise digital scholarship services
- gauge data service needs

COOKING TIME

Depending on the number of collaborators, chefs will need 3 hours of prep plus a 1 hour instruction session that leaves plenty of time for discussion and digestion of ideas and concepts.

DIETARY GUIDELINES

Much like introducing an unfamiliar fruit with a prickly exterior or a new vegetable to a picky eater, it can be difficult to convince instructors to try a new approach and incorporate data literacy into their course design.

Research assignments today are often situated within a digital communication environment—that is, an environment that requires downloading and managing files in multiple formats, working with other collaborators in an online environment, or extracting large amounts of information at one time to later be synthesized by statistics or textual analysis. By incorporating fundamentals of data literacy such as data quality and documentation, data interpretation and representation, data management, and FAIR data practices and reproducibility, instructors can help students become more familiar with the language of data and manage the data components of their research for a successful project. Librarians are well positioned to guide instructors as they begin to incorporate data literary concepts into their research assignments.

A discussion-based presentation like this offered to a diverse group of instructors not only is an excellent way to discuss incorporating data literacy skills into research assignments across disciplines, but also can function as a form of advertising for specific library services.

INGREDIENTS
- Data literacy expert willing to conduct outreach

- Instruction skills (live, online, synchronous, asynchronous)
- Forum for presentation
- Presentation software, such as PowerPoint or Google Slides
- Examples of disciplinary data literacy instruction, especially those that are large or popular on your campus
- Registration software, such as LibApps
- Software to gauge feedback, such as LibApps or Google Forms

PREPARATION

1. Identify a forum for presentation and any potential collaborators, such as the Center for Teaching and Learning.
2. Strategize outreach and communication to presentation attendees. Utilize resources from any collaborators (mailing lists, event calendars, campus advertising) and capitalize on internal communication resources. Be sure to consider both broad advertising, such as mailing lists, and targeted outreach, such as direct e-mails to faculty with whom you may have a relationship.
3. Determine the knowledge base of your audience. Based on our knowledge of our institution and the traditional attendees of this type of event, we assumed a low level of knowledge of data literacy. We asked attendees during the session about their data and data literacy knowledge, and it was a great conversation starter.
4. Create a welcoming ambiance by inviting all levels of data-literate instructors. Be prepared to discuss broad-level basic concepts, such as identifying data across disciplines and file naming, as well as high-level expert concepts, like reproducibility and metadata.
5. Choose three to four data literacy competencies to highlight, and provide examples. In our presentation, we talked about understanding data sources, interpreting data visualizations, and research data management of both small- and large-scale data. To illustrate these competencies, we pulled examples from well-known news sources where data visualizations were intentionally misleading and talked about how researchers can follow data citations. We also pulled in other real-world scenarios like messy file organization on a researcher's computer as an easy-to-digest way to contextualize research data management for instructors who might be unfamiliar with the concept.
6. Create a presentation that incorporates examples and practical applications, including scaffolding data skills within assignments. You can find a URL for our presentation slides in the Additional Resources section.
7. Practice. We ran through our slides together two times to ensure timing and flow and to anticipate any potential questions from attendees.

INSTRUCTIONS

1. Ask attendees to consider a wide range of research materials as data so that they could contextualize data literacy within their own disciplines.
2. Discuss the three to four data literacy competencies you chose to highlight (see the Preparation section for the list of ours). This is a great opportunity to "pass the dish" and invite others to share how they identify with these competencies or have already incorporated them into their instruction. In the Additional Resources section, you can find sources that provide more complete lists of competencies (Calzada Prado and Marzal 2013).
3. Dig into real-world examples of assignments in different subject areas. Our presentation had one example assignment from the humanities that asked students to create metadata-style tags to describe images, and one STEM assignment that asked students to visualize public health data for general audiences (Cutrara 2019 and Shanks et al. 2017). The articles that provided inspiration for these assignments are cited in the Additional Resources section and in our presentation slides.
4. Ask attendees what data they are already using in their research assignments and how those assignments might be framed more explicitly as data literacy. Instructors may already be incorporating some form of data literacy and not realize it!
5. If a data-savvy instructor is hogging the plate, invite them for a consultation after the presentation. We want everyone to be able to dig into the conversation!
6. Leave your calling card. Make sure to let attendees know that if they would like to continue to discuss ideas, data concepts, data services, and tools the library may offer, they are welcome to reach out.

REVIEWS/ASSESSMENT STRATEGY

After the session, we measured success in three primary ways:
- Interest in the session
- Discussion participation by session attendees
- Follow-up consultations

LibWizard was used to create a feedback form for participants. The session instructors also made sure to connect back with the ECTL Lunch and Learn program facilitators to get their informal feedback on the session and receive useful points of comparison with other Lunch and Learn sessions.

ADAPTING THE RECIPE

Practitioners at other institutions could adapt the general outline of the session presentation to better reflect their capabilities. At institutions with gourmet service capabilities, librarians could incorporate more examples of services they offer, while those with smaller pantries can use such a session to educate instructors on the basics of data literacy and how to incorporate it in research assignment design.

You should also feel free to experiment with the ingredients. Use different techniques, such as Think, Pair, Share," to inspire conversations, or explore different venues for the session. Creativity is rewarded with this recipe!

If you choose to include the option for online attendees, build in the support for moderation and technical troubleshooting.

ALLERGY WARNING

There can be a lot of fear around data terminology. Therefore, it's important to create a welcoming environment for all participants and to give a wide range of examples of data types and data skills across a variety of disciplines.

ADDITIONAL RESOURCES

Original presentation slides available at https://osf.io/ds948/.

Calzada Prado, Javier, and Miguel Ángel Marzal. "Incorporating Data Literacy into Information Literacy Programs: Core Competencies and Contents." *Libri*, 63, no. 2 (May 2013): 123–34. https://doi.org/10.1515/libri-2013-0010

Carlson, Jake, and Lisa Johnston, eds. *Data Information Literacy: Librarians, Data, and the Education of a New Generation of Researchers*. West Lafayette, IN: Purdue University Press, 2015. https://library.oapen.org/handle/20.500.12657/31585.

Cutrara, Samantha. "Metadata Creation/Archival Description: Creating Metadata as a DHSS Assignment." In *Doing Digital Humanities and Social Sciences in Your Classroom*. Toronto, ON: York University Libraries, 2019, https://pressbooks.library.yorku.ca/dhssinstructorsguide/chapter/metadata-creation/.

Shanks, Justin D., Betty Izumi, Christina Sun, Allea Martin, and Carmen Byker Shanks. "Teaching Undergraduate Students to Visualize and Communicate Public Health Data with Infographics." *Frontiers in Public Health* 5 (November 2017): article 315. https://doi.org/10.3389/fpubh.2017.00315.

Seasoning and Simmering
Cultivating Data Literacy Skills through an Open Data Hackathon

Peace Ossom-Williamson, MLS, MS, AHIP, Associate Director of NNLM National Center for Data Services and Associate Curator, NYU Langone Health, peace.williamson@nyulangone.org

NUTRITION INFORMATION

This recipe provides the steps for developing data competitions, such as hackathons, where students practice coding and challenge their thinking around open data and their uses. Hackathons and data competitions foster an environment for applying skills learned in the classroom or elsewhere and build upon them by working on a "project marathon."[1] A hackathon is a project that must be started and completed in a set amount of time and with a given theme. Hackathons may have a lot of structure, including focusing on a specific issue or problem or a particular output, such as an app. However, they often have very little structure, and students are left to develop their own ideas. Open Data Day (https://opendataday.org) is a great opportunity for this event because it is an international celebration of the use and value of open data and the institutions and organizations that support open data. The Open Data Hackathon provides an opportunity for students to challenge themselves to apply or develop new skills, become acquainted with open data sources, and work collaboratively with students outside of their disciplines.

TARGET AUDIENCE AND NUMBER SERVED

15 to 100 participants. (See Chef's Notes regarding larger events.)

LEARNING OUTCOMES

Participants:
- discover how data are formally produced, organized, and disseminated
- explore information sources; identify the purpose and audience of potential resources
- define or modify the information needed to achieve a manageable focus
- define a realistic overall plan and time line to acquire and build from the needed information
- incorporate the data and supporting information into their creation
- describe what they created in order to garner support for winning the competition

COOKING TIME

Preparation time depends upon the size of the hackathon. A hackathon that has around 20 participants may take a month or two to plan, whereas a hackathon with 60–100 participants would take 6 months. Planning for larger hackathons, hackathons that span multiple days, or hackathons that include participants who travel to participate should be planned a year in advance. Regarding the actual event, hackathons can span 2 hours or longer. Common time spans for in-person hackathons are 12 or 24 hours.

DIETARY GUIDELINES

Depending on participants' individual experiences, the hackathon applies at some level concepts from all of the frames of ACRL's *Framework for Information Literacy for Higher Education*; however, it is centered on Information Creation as a Process, Information Has Value, Research as Inquiry, and Searching as Strategic Exploration.

INGREDIENTS

- A location that allows for various working styles, including round tables for group work and small solo tables and couches
- Dry erase boards and markers
- Handouts with schedules and links for submissions
- Name tags
- (optional) T-shirts that are different colors for participants and volunteers
- An abundance of places to plug in (power strips, outlets, etc.) with the expectation of 2–3 outlets per participant
- Wi-Fi access and a network that can accommodate the number of connected

devices and data being transferred
- Enough time for the open data sources to be used (Some APIs take 24 hours to be integrated.)
- Food and drinks, including the meals and snacks to be consumed during the event
- Other amenities as needed for size (See Chef's Notes.)
- Certificates and/or prizes and awards for winners
- (optional) Workshops teaching lessons, such as how to use Git and GitHub and an introduction to machine learning

PREPARATION

Trello or another project management tool is recommended for planning and tracking preparation before the event. The Open Data Hackathon has a theme that designs must incorporate, such as the provision of health information or the incorporation of movement. Technology should be set up for instructions, including (1) a general website with details about the competition, including its rules and the judges' names and information, and (2) a registration form gathering demographic data about participants, including interest in particular workshops and planning information such as dietary guidelines and restrictions and the need for physical accommodations. Recruit volunteers to assist with day-of activities, including registration, handing out name tags and T-shirts, and setting up food, tables, and other ingredients. A place for submissions should be developed and can range from a shared Google Drive folder to using a hackathon platform, such as Devpost (https://devpost.com), which allows for submissions, voting, judging, and contacting participants and winners. Librarians can also develop a LibGuide of resources about available software, open data sources, and files for workshops.

INSTRUCTIONS

Most of the preparation involves marketing the event, developing places for registration and submission, and day-of setup and cleanup. Students from a variety of experience levels and majors are encouraged to attend. This event is especially focused on encouraging students from non-STEM fields to participate. There are many ways to participate in the hackathon, not all of which are coding. Some participants develop the design of their creation, others work on front-end development, some gather data and work with APIs, and others provide content knowledge.

Examples of submissions include teams who have built a COVID-19 tracker, a system that provides students and medical professionals evidence-based information about medicines and illicit drugs and their effects, and a 3D model of a new method of personal identification that works as an all-encompassing ID and payment card. During the workshop, there are typically few questions or needs for assistance. Most of the volunteer efforts occur at mealtimes. However, one hour prior to the conclusion, announcements should be made about concluding documentation, uploading files, and completing submissions. Upon completion, submissions are assessed by three outside judges. One week later, winners receive prizes and certificates.

Table 1 shows a set of review criteria from the UTA Open Data Day Hackathon.

TABLE 1
Review criteria from the UTA Open Data Day Hackathon.

Usefulness	How useful is your project? In what ways does it improve access to or evaluation of information related to the hackathon theme?
Fidelity	Did you use open data sources in your project? Was the use of the data accurate and true to the original data and compliant with existing standards? Did you thoroughly document your project?
Design	Did you put thought into the user experience? How well designed is the interface? Are the purpose and functionality clear for the user?
Technology	How technically impressive was the hack? Did you use a particularly clever technique or many different components? Did the technology involved make people go "Wow"?
Learning	Did you stretch your skills, learn new tricks, or apply new tools?

REVIEWS/ASSESSMENT STRATEGY

Data are collected at registration, including demographic information, which college or department participants are from, (optional) team information if the team is already formed, and other details. This information can be used for creating name tags and for check-in using a digital system. After the hackathon, these data, along with the participants who came and submissions received, can be used to assess the success and growth of the annual event.

CHEF'S NOTES

For large hackathons with over 100 participants who come into town for the event, additional planning will need to be done and addition provisions offered, such as travel buses, hotel and other accommodations, information about the area, security, and authenticating identification of participants. Partnering with the organization Major League Hacking (https://mlh.io) for a large hackathon is recommended because MLH provides promotion on its platform, assists with project management, and assists with resources for a successful large hackathon event.

CLEANUP

A system, typically a rubric, must be in place for judging and selecting winners, and this system should be communicated to participants ahead of the event. Whether there is judging and a presentation at the closing of the competition or whether judges review submissions and hold an award ceremony at a later date, the winners should be recognized with a certificate and/or prizes.

ADDITIONAL RESOURCES

Kaggle. "Competitions." https://www.kaggle.com/competitions.

Open Knowledge Foundation. "Open Data Day." https://opendataday.org.

UTA Open Data Day Hackathon. "Judging Criteria." Devpost. 2020. https://opendatahackathon.devpost.com/.

NOTE

1. Nancy Shin, Peace Ossom Williamson, and Bethany McGowan. "vMLA 2021—On the Edge of Innovation: How Libraries Can Develop Health Datathons and Hackathons for Data Literacy" (immersion session, Medical Library Association 2021 Virtual Conference, May 25, 2021), https://doi.org/10.17605/OSF.IO/SXFHW.

From Soup to Nuts
Finding Your Way around the Data Services Buffet

Jane Fry, Data Services Librarian, MacOdrum Library, Carleton University, jane.fry@carleton.ca; Chantal Ripp, Research Data Management Librarian (Interim), Library, University of Ottawa, chantal.ripp@uottawa.ca

NUTRITION INFORMATION

For many years, the expertise for the provision of data services existed only within larger Canadian universities. As the cost to purchase secondary data decreased and the proliferation of distribution channels increased, the need to provide data services across Canadian academic institutions became apparent. Therefore, in order to maximize the use of data by students and researchers in academic communities, the Canadian library professionals who were members of the Data Liberation Initiative recognized that investments in data literacy skills were required to deliver data services on their campuses. Thus, the DLI-SG (Data Liberation Initiative Survival Guide) was conceived! The importance of upskilling librarians to support data services continues to be paramount because the librarians wore, and continue to wear, more than one hat at most institutions.

The DLI-SG is designed to
- serve as a curriculum map, enabling librarians to effectively fulfil their role as DLI contacts and service providers of data in their local institutional context;
- provide foundational knowledge in data concepts, including the relationship between data, statistics, and information;
- explain the different data products that are released from Statistics Canada, including aggregated data and microdata; and
- demonstrate foundational data literacy concepts, as well as practical skills in working with aggregate and microdata files.

This chapter outlines a scavenger hunt that is intended to be a stand-alone session designed to help librarians become familiar and comfortable with using the DLI-SG. Extensive knowledge of data is not required, merely an interest in knowing more about data. Picture yourself standing in front of a buffet, with many choices available. This training session will help you to figure out which dishes you want to add to your plate of data knowledge.

TARGET AUDIENCE AND NUMBER SERVED

- Librarians at all levels, from beginner to advanced.
- Primarily those supporting data services at an academic institution.
- Secondary audience includes colleagues not familiar with data and other data users.
- Can be done individually or in small groups.

LEARNING OUTCOMES

Librarians will become familiar with the DLI-SG, allowing them to
- match information needs and search strategies to find appropriate resources;
- assess the fit between an information product and a particular information need;
- learn the basics of data literacy; and
- become familiar with the continuum of access to data.

COOKING TIME

From 30 minutes to 1 hour, depending how much time is spent on each section of the DLI-SG, or each dish in the buffet.

It is not recommended that the librarians sample all the dishes at the same time as they would be full halfway through the meal! You should encourage the librarians to pace themselves and go back for seconds or even thirds when they are ready. At that point, they can review what they have already eaten, and perhaps will want to dig into one of the sections to savor it again.

DIETARY GUIDELINES

The learning objectives of this buffet correspond with the following frames in ACRL's *Framework for Information Literacy for Higher*

Section 8. Data Literacy Outreach and Engagement

Education: Authority Is Constructed and Contextual, Information Has Value, Research as Inquiry, and Searching as Strategic Exploration.

INGREDIENTS

Just as you need certain tools to enjoy a buffet, including a plate, cutlery, signage as to the various dishes, and an appetite, there are also some basics needed to conduct this training session. You will need

- Computers with internet access for participants
- Worksheets and pens for participants
- Extra sheets of paper
- A few pads of sticky notes—at least 3 by 5 inches
- Access to the DLI-SG (https://www.statcan.gc.ca/eng/microdata/dli/training-events/dli-survival)

PREPARATION

You should review the DLI-SG before teaching this workshop.

Prep time for you to put together this activity—30 minutes to 1 hour. This is where you will be putting together the scavenger hunt.

- Start putting together a worksheet: "Data Scavenger Hunt Using the DLI Survival Guide."
- Look at each section of the DLI-SG to come up with 2 questions per section. There are 4 sections plus a glossary, so you will have 10 questions for the attendees.
- Put these questions on the worksheet as you come up with them. Some example questions:
  – Why are all concepts asked in a survey questionnaire not reflected in a public use microdata file?
  – What documentation accompanies microdata files?
  – Where can you find presentation and educational materials to support your instruction?
  – How do you cite data?
  – Where can you find a good definition of the difference between microdata and aggregate data?
- Have enough extra pieces of paper to hand out for the participants to submit their own questions. You will ask the librarians to come up with a scavenger hunt question at the end of the exercise. They will return the paper to you with their question at the end of the session. You now have new questions to use the next time you deliver the scavenger hunt. The intention of this practice is to have the participants reflect and truly immerse themselves in the content in order to come up with their own question.
- Suggestion: You might want to try this exercise out on someone else before you do it with the class to make sure that your questions make sense. This is like a taste test of your buffet dishes—always ensure that they are well seasoned.

INSTRUCTIONS

The ideal way to deliver this activity is as an interactive training session, but it can also be provided as a self-directed exercise. Sometimes you need help figuring out the buffet dishes, and other times you are ready to dive in on your own. The times included in parentheses after the different sections are merely suggestions to keep the time under one hour.

Introduce the Lesson (5 minutes)

- Background of DLI-SG.
- Goals and learning outcomes. Explain how the scavenger hunt will work. It is your decision if you want the librarians to work alone or in small groups, followed by a larger group discussion.
- Distribute worksheets, extra pieces of paper, and pens.
- Final instruction for the participant.
  – "Come up with one scavenger hunt question."
  – This question should be written on a separate piece of paper because it will be given back to you at the end of the session.

Work on the Exercise (20 minutes)

- Direct the librarians to work on the exercise either individually or in small groups.
- You should wander around the room to see how everyone is doing.
  – Some may need help getting started and figuring out which dish to eat.
  – Don't do the exercise for them!! If you were at a buffet, you wouldn't eat off their plate!

Hold a Group Discussion (20 minutes)

- Take up answers to each question.
  – Ask a few people their responses.
  – After taking up the answer to each

question, ask if anyone has a different answer. If so, have them share the answer and how they arrived at it, then discuss if their answer is valid.
- Make sure everyone gets a chance to answer.

Summary (10 minutes)
- Go over the learning outcomes. Check to see if they were achieved.
- Summarize the lesson and ask for questions.
- Tell them about the stickies you will be handing out (see the *Assessment Strategy* below)

REVIEWS/ASSESSMENT STRATEGY
Before librarians leave, hand out 2 sticky notes to each person.
- On one sticky note, have them draw a smiley face. Ask, "What is your main takeaway?"
- On the other sticky note, draw a sad face. Ask, "Was there anything missing?"

They should write something on each sticky note and put it on the wall beside the door when they leave. This is similar to filling out a satisfaction survey when you leave a restaurant.

ALLERGY WARNING
The ingredients are all allergy-free; in other words, they can be consumed by anyone. The attendees should be encouraged to feel free to share any of the dishes with their friends to feed their curiosity.

ADAPTING THE RECIPE
- You can adapt this session, intended to be done in person, to an online trivia model.
- This recipe can be easily doubled if you have time for a longer lesson.

ADDITIONAL RESOURCES
Association of College and Research Libraries. *Framework for Information Literacy for Higher Education*. Chicago: Association of College and Research Libraries, 2016. https://www.ala.org/acrl/standards/ilframework.

Statistics Canada. *DLI Survival Guide*. Accessed March 23, 2021. https://www.statcan.gc.ca/eng/microdata/dli/training-events/dli-survival.

Teaching Data Literacy and Computational Thinking in Educational Technology

Dr. Lesley S. J. Farmer, Professor, California State University, Long Beach

NUTRITION INFORMATION

One aspect of data literacy is computational thinking. The 2016 ISTE Standards for students define the goal for computational thinkers this way: "Students develop and employ strategies for understanding and solving problems in ways that leverage the power of technological methods to develop and test solutions." On the other hand, mathematics remains a challenging subject for many people, including preservice librarians and teachers. Furthermore, even fewer preservice librarians and teachers are taught data literacy or computational thinking. Nevertheless, these populations need to know how to integrate those concepts into the curriculum in order to teach K–12 students. This activity ties together instructional design, a core educator function, and these less used literacies by having students develop a WebQuest that involves solving a problem using computational thinking and data literacy.

TARGET AUDIENCE AND NUMBER SERVED

This activity targets preservice school librarians and practitioner K–12 teachers, particularly those who want to integrate educational technology into the curriculum. Ideally, a class of 20–30 students generates enough products so that learners can compare products and see how data literacy and computational thinking occur in a variety of subject domains. Beyond that number, a single instructor might have difficulty connecting with the learners and providing timely, specific feedback.

LEARNING OUTCOMES

- Perform an internet search for educational apps specific to your discipline (or one that you are interested in). These apps should be designed specifically to promote computational thinking, including data literacy.
- Develop a WebQuest that involves solving a problem using computational thinking and data literacy. The WebQuest must include a data set that enables learners to solve the problem.

COOKING TIME

- Online time: 30–60 minutes
- Reading, lesson experience, and reflection: 1–2 hours
- Product development: 2–4 hours

DIETARY GUIDELINES

The search for apps about data literacy and computational thinking addresses several frames from ACRL's *Framework for Information Literacy for Higher Education*:

- Searching as Strategic Exploration
- Research as Inquiry
- Information Has Value

The WebQuest involves computational thinking and data literacy on database resources to solve problems; it addresses several *Framework* frames:

- Information Creation as a Process
- Searching as Strategic Exploration
- Research as Inquiry
- Information Has Value

INGREDIENTS

- "Data Sets for ICT Literacy" (27 resources): https://www.merlot.org/merlot/viewBookmarkCollection.htm?id=1242012
- WebQuest: http://webquest.org
- MERLOT: https://www.merlot.org
- Reading resources about data literacy and computational thinking
- A calculating tool
- A dash of creativity

PREPARATION

- Join MERLOT (https://www.merlot.org) as a free member.
- Explore its Content Builder tool.
- Review the following resources and provide students with access to the following: activity readings, WebQuest

resource, MERLOT and its two identified bookmark collections.
- Provide a means to share learning experiences: via blogs, discussion board, online or face-to-face discussion.

INSTRUCTIONS

1. *Give introduction and overview.* Greet learners, introduce yourself, distribute your contact information, and offer your services to meet privately with learners moving forward. State the activity objective. Note that this learning activity may be done asynchronously or as a self-paced tutorial.
2. *Set context.* Say: "Technology has impacted problem-solving, especially as it helps to gather and analyze data (numerical and more). Computational thinking involves decomposing a problem into manageable parts in order to find patterns that then help to solve that problem and often uses technology to facilitate this process." Explain that this activity helps learners understand how data literacy and computational thinking work together to provide strategies for solving problems.
3. Gain background knowledge. Ask learners to read the following:
    » Burress, Theresa, Emily Mann, Susan Montgomery, and Rachel Walton. "Framework for Data Literacy." Data Literacy Teaching Toolkit. University of South Florida. Last updated May 27, 2021. https://lib.stpetersburg.usf.edu/c.php?g=933381&p=6783477. This website also includes several activities and data literacy research.
    » Steinberg, Luc. "Lies, Damned Lies and Statistics: A Data Literacy Primer." EAVI, 2017. https://eavi.eu/lies-damned-lies-statistics-data-literacy-primer/. This article also lists several data literacy resources.
    » Stephens, Wendy Steadman. "Real World Data Fluency: How to Use Raw Data." Chapter 4 in *Creating Data Literate Students*. Edited by Kristin Fontichiaro, Jo Angela Oehrli, and Amy Lennex. Ann Arbor: Michigan Publishing, University of Michigan, 2019. http://datalit.sites.uofmhosting.net/wp-content/uploads/2016/01/Chapter_4_Stephens.pdf.
    » International Society for Technology in Education. "Operational Definition of Computational Thinking for K–12 Education." Accessed May 24, 2021. https://cdn.iste.org/www-root/ct-documents/computational-thinking-operational-definition-flyer.pdf.
    » BBC. "Introduction to Computational Thinking." Bitesize. Accessed May 24, 2021. https://www.bbc.com/education/guides/zp92mp3/revision/1.
    » Sheldon, Eli. "Computational Thinking Lessons by Subject." Computational Thinking Lessons. Accessed June 24, 2022. http://ctlessons.org.
4. *Link background knowledge to applied lessons.* Ask learners to choose one of the computational thinking lessons on Eli Sheldon's website that includes data (e.g., "Racial Bias in Traffic Stops," "Basketball Motion Analysis," "Mapping Earthquakes to Save the World," "The Enigma Machine"), and do the lesson as a student. Say: "As you are doing the lesson, consider how computational thinking about the data helps you to solve a problem."
5. *Reflect on learning experience.* Ask learners to reflect upon and share their lesson experience, noting how it might inform teaching how to solve problems using computational thinking. This sharing may be done orally or in writing, depending on the shared learning space provided. The reflections might consider the ease or difficulty of choosing a data-centric lesson, or even defining data; the range of applicable topics and types of data; the granularity of computational thinking; how assessment is done.
6. *Apply knowledge to generate a learning activity.* Say, "WebQuests (http://webquest.org) provide a useful structure for developing an engaging, interactive learning activity. They enable you to select appropriate, relevant websites and facilitate problem-based learning that meets content standards. To that end, you will develop a WebQuest that incorporates data literacy and computational thinking.

"In this activity, you will generate a WebQuest that includes the following elements: title, introduction, task, process, resources, evaluation using a rubric you create, and conclusion. Structure your WebQuest as a process where the learner uses data to solve a problem using computational thinking. For your resources, choose a data set from the MERLOT's "Data Sets for ICT Literacy" bookmark collection (https://www.merlot.org/merlot/viewBookmarkCollection.

htm?id=1242012).

"Because this is a WebQuest, this learning activity should be created in a web-accessible format. MERLOT (https://www.merlot.org) includes a Content Builder tool, which is a very simple website creation tool. Please join MERLOT; it takes less than a minute to join, and membership is free to educators, including librarians, and students. Create your learning activity in Content Builder, which will generate a unique URL."

7. Assess learning. Ask learners to pair up and do each other's WebQuest. Use the WebQuest rubric (http://webquest.org/sdsu/webquestrubric.html) to assess the quality of the WebQuest.

REVIEWS/ASSESSMENT STRATEGY

Assess learners in terms of their knowledge of computational thinking and data literacy as they: (1) choose and reflect upon an appropriate lesson; (2) develop a learning activity that fosters computational thinking and data literacy through an authentic task and relevant data resource; and (3) peer-review a WebQuest using a rubric (http://webquest.org/sdsu/webquestrubric.html) to identify and describe the criteria.

CHEF'S NOTES

- Data literacy and computational thinking are sometimes labeled as STEM concepts, but they can and should be applied in the social sciences as well.
- An effective strategy for integrating data literacy and computational thinking is problem-based learning.
- Building data literacy and computational thinking confidence takes time and requires repeated practice, just like learning to cook.

ADDITIONAL RESOURCES

Computational Thinking Initiatives. Computational Thinking Resources for Teaching. Accessed May 24, 2021. https://www.computationinitiative.org/resources/teaching/.

Farmer, Lesley. "Data Literacy Bookmark Collection." MERLOT. July 14, 2016. https://www.merlot.org/merlot/viewBookmarkCollection.htm?id=1188590.

Fontichiaro, Kristin, Jo Angela Oehrli, and Amy Lennex, eds. *Creating Data Literate Students*. Ann Arbor: Michigan Publishing, University of Michigan, 2019. http://datalit.sites.uofmhosting.net/books/book/#toc.

Google. "Exploring Computational Thinking." Google for Education. Accessed May 24, 2021. https://edu.google.com/resources/programs/exploring-computational-thinking/.

Gorman, Michael. "Part 2: Computational Thinking: Over 50 Resources to Teach CT across the Curriculum." *21st Century Educational Technology and Learning* (blog), July 25, 2018. https://21centuryedtech.wordpress.com/2018/07/25/part-2-computational-thinking-over-50-resources-to-teach-ct-across-the-curriculum/.

Herzog, David. *Data Literacy: A User's Guide*. Los Angeles: Sage, 2016.

ISTE. Search Results: Exploring Computational Thinking Repositories. Accessed May 22, 2021. https://learn.iste.org/d2l/lor/search/search_results.d2l?ou=6606&lrepos=1006.

Ridsdale, Chantel, James Rothwell, Mike Smit, Michael Bliemel, Dean Irvine, Dan E. Kelley, Stan S. Matwin, Brad Wuetherick, and Hossam Ali-Hassan. *Strategies and Best Practices for Data Literacy Education*. Halifax, Nova Scotia: Dalhousie University, 2015. https://doi.org/10.13140/RG.2.1.1922.5044.

Wright, Sarah J., Jake Carlson, Jon Jeffryes, Camille Andrews, Marianne Bracke, Michael Fosmire, Lisa R. Johnston, et al. "Developing Data Information Library Programs: A Guide for Academic Librarians." Chapter 9 in *Data Information Literacy: Librarians, Data, and the Education of a New Generation of Researchers*, ed. Jake Carlson and Lisa R. Johnston. West Lafayette, IN: Purdue University Press, 2015. http://www.datainfolit.org/dilguide.

Section 9.
Data Literacy Programs and Curricula

231 [[Ch60]]Cooking Up a Data Literacy Course
Claire Nickerson

238 [[Ch61]]Baking a Data Layer Cake: Scaffolding Data Skills through Video Vignettes
Shannon Sheridan

241 [[Ch62]]Building Data Literacy through Scaffolded Workshops: Experiences and Challenges
Jiebei Luo and Yaqing (Allison) Xu

244 [[Ch63]]Data Literacy Appetizers: LibGuide Data Instruction Modules for Undergraduates
Beth Hillemann and Aaron Albertson

247 [[Ch64]]Data as Curation: Framing Data Creation as a Critical Practice through Collections-Based Research Inquiry
Gesina A. Phillips, Tyrica Terry Kapral, Matthew J. Lavin, and Aaron Brenner

253 [[Ch65]]Quantitative Data Skills for Undergraduates: A Seminar Series for Social Science Students
Whitney Kramer and Amelia Kallaher

Cooking Up a Data Literacy Course

Claire Nickerson, Learning Initiatives and Open Educational Resources Librarian, Forsyth Library, Fort Hays State University, cenickerson@fhsu.edu

NUTRITION INFORMATION

This asynchronous online course, Interdisciplinary Studies 815: Introduction to Data, was developed for graduate students in the information analysis and communication concentration of the Fort Hays State University master of liberal studies degree. The course is designed for professionals who need to make data-driven decisions such as educators, policy makers, and nonprofit employees. It is a survey course, so it does not go into great depth on any of the topics covered but rather provides a basic grounding for developing further data literacy skills. It exclusively uses zero-cost resources, including openly licensed content, library-licensed e-books and articles, and free online content. The course uses weekly mini-assessments as well as three major assignments: a data memo, a data visualization, and a final video presentation.

TARGET AUDIENCE AND NUMBER SERVED

The cap is 15 students. A class size of at least 5 students allows for more peer interaction and feedback. The course was designed for graduate students, but the content would also be appropriate for undergraduates.

LEARNING OBJECTIVES

Students will
- identify different types and sources of data, design a research question that can be answered using data, and determine what type of data is appropriate to answer a given research question
- conduct basic data analysis and visualization and be aware of more advanced tools used in working with data
- articulate the narrative behind data findings and communicate insights about data in various formats to targeted audiences

COOKING TIME

The course has been taught as a 3-credit-hour course over a 15-week semester, but it could be condensed into a trimester or short course.

DIETARY GUIDELINES

The course ties closely into ACRL's *Framework for Information Literacy for Higher Education*. Students learn about Searching as Strategic Exploration and that Authority Is Constructed and Contextual through examining different sources of existing data, such as census and other polling data, and processes for collecting their own data. They participate in Information Creation as a Process and Research as Inquiry by choosing a research question, conducting a survey, and gradually adding background information and audiovisual elements to the presentation of their results throughout the course. Students also read news articles providing case studies about the value of data in the real world.

INGREDIENTS

Reading and video topics by week (see Instructions).

PREPARATION

1. Set up rubrics for grading all assignments. Since the course is asynchronous, scores and comments on rubrics are the primary means by which students receive feedback, so providing detailed comments on each section of the rubric is advised.
2. Review and update news articles and other case studies used as course materials.
3. For each reading, provide reading time estimates using the Niram Read-O-Meter (https://niram.org/read/) and textual complexity estimates using the Lexile Analyzer (https://hub.lexile.com/analyzer).
4. Ask students to complete an intake survey asking the following questions (all optional):
   a. Is English your native language? Would you like feedback on grammar and syntax in your coursework, or only on content?
   b. Are you willing to grant copyright permissions for the instructor to

use your work in specific contexts (e.g., as an example for future classes)?
c. Is there anything you would like the instructor to know about you as a student?

INSTRUCTIONS

Students complete the following readings and assessments/class activities asynchronously. Students receive feedback from the instructor in discussion board postings, via comments on assignment grading rubrics, and during scheduled virtual office hours. Modules are posted at the beginning of the semester, so students can look ahead at assignments and instructions.

Week 1

Topic: What are data? The rise of data capital as a business asset, defining data, and data types.
Readings:
Andrus, Calvin, Jon Cook, and Suresh Sood. "Definitions of Data." Chapter 3 in *Data Science: An Introduction*. Wikibooks, 2017. https://en.wikibooks.org/wiki/Data_Science:_An_Introduction/Definitions_of_Data.

Minitab Blog Editor. "Understanding Qualitative, Quantitative, Attribute, Discrete, and Continuous Data Types." *Minitab Blog*, April 28, 2017. http://blog.minitab.com/blog/understanding-statistics/understanding-qualitative-quantitative-attribute-discrete-and-continuous-data-types.

MIT Technology Review. *The Rise of Data Capital*. White paper, MIT Technology Review Custom + Oracle. March 21, 2016. http://files.technologyreview.com/whitepapers/MIT_Oracle+Report-The_Rise_of_Data_Capital.pdf.

Assessment: "What Are Data?" reflection video. Create a video explaining in their own words what data are and why they are valuable. (Students may use any tools they like, but VidGrid is recommended for students who have not created videos before.) Post the video on the discussion board.

Week 2

Topic: Why use data? Data-driven decision-making, news articles about companies making poor decisions based on poor data.
Readings:
Harvard Business Review Analytic Services. *The Evolution of Decision Making: How Leading Organizations Are Adopting a Data-Driven Culture*. Cambridge, MA: Harvard Business School Publishing and SAS, 2012. https://hbr.org/resources/pdfs/tools/17568_HBR_SAS%20Report_webview.pdf.

Pleven, Liam. "Pecan Buyers Shelled by Bad Data." *Wall Street Journal*, April 6, 2012. https://www.wsj.com/articles/SB10001424052702304072004577326002427369224.

Soble, Jonathan. "Fear of More Bad Data Adds to Woes of Kobe Steel." *New York Times*, October 13, 2017. https://www.nytimes.com/2017/10/12/business/kobe-steel-japan-trains.html.

Class activity: Data-driven decision-making discussion.
Assessment: Make a discussion board post about a time when they needed more data to make a decision.

Week 3

Topic: Data ethics. Research bias, guidelines for research with human subjects, HIPPA, and FERPA.
Readings:
Office for Civil Rights. "HIPAA for Professionals." HHS.gov. Last modified May 17, 2021. https://www.hhs.gov/hipaa/for-professionals/index.html.

Rosser, Sue V. "Bias, Research." In *Encyclopedia of Bioethics*, edited by S.G. Post, 1:273–78. New York: Macmillan Reference, November 30, 2003.

Steneck, Nicholas H. "Data Management Practices." In *Introduction to the Responsible Conduct of Research*, 87–102. Washington, DC: Department of Health and Human Services, 2007.

University of Texas at San Antonio Libraries. "What Is FERPA?" YouTube video, 1:22. University of Texas at San Antonio Libraries, May 26, 2011. https://youtu.be/_5XpRGd8O44.

Class activity: Data privacy scenario discussion.
Assessment: Scholarly article bias assessment. Analyze an empirical scholarly article for potential biases. Write a scenario about a researcher, health-care professional, or educator who violates data privacy rules. Then discuss these scenarios with other students using the discussion board.

Week 4

Topic: Searching for data. Overview of census data tools, where to find polling data and public statistics, how to select a research topic, and how to write a research question.

Readings:
New Literacies Alliance. "Ask the Right Questions." January 25, 2017. https://online.newliteraciesalliance.org/courses/course-v1:NLA+IL1013+20_21/about

Nickerson, Claire. "Finding Data," 2018. https://fhsuguides.fhsu.edu/journalism/data.

University of Michigan–Flint. "Select a Topic." 2022. Thompson Library. https://libguides.umflint.edu/research/topic

US Census Bureau. "Data Tools and Apps." Accessed August 16, 2021. https://www.census.gov/data/data-tools.html.

Class activity: Research question.
Assessment: Data set search. Write a research question and locate a data set that provides useful background information related to the topic of the question. (Students usually choose questions related to their personal interests, local communities, or professional lives. Instructor feedback on questions often relates to broadening or narrowing the question and ensuring that it is feasible to collect data that will answer the question.)

Week 5
Topic: Gathering data. Data collection methods, with an emphasis on collecting survey data
Readings:
Blackstone, Amy. "Survey Research: A Quantitative Technique." Chapter 8 in *Sociological Inquiry Principles: Qualitative and Quantitative Methods*. Orono, ME: 2012 Book Archive. https://2012books.lardbucket.org/books/sociological-inquiry-principles-qualitative-and-quantitative-methods/s11-survey-research-a-quantitative.html.

Blackstone, Amy. "Other Methods of Data Collection and Analysis." Chapter 12 in *Sociological Inquiry Principles: Qualitative and Quantitative Methods*. Orono, ME: 2012 Book Archive. https://2012books.lardbucket.org/books/sociological-inquiry-principles-qualitative-and-quantitative-methods/s15-other-methods-of-data-collecti.html.

Kaplowitz, Rella. "How to Collect Data." Data Playbook. Accessed August 16, 2021. https://www.schusterman.org/playbooks/data/how-to-collect-data/.

Assessment: Data collection plan. Write a data collection plan for conducting a survey to answer the research question.

Week 6
Topic: Basic tech skills. Microsoft Excel basics, statistical software options.
Readings:
Microsoft Office. "Excel Video Training." Microsoft Office. Last modified 2022. https://support.office.com/en-us/article/excel-for-windows-training-9bc05390-e94c-46af-a5b3-d7c22f6990bb.

Goldwater, Eva. "Using Excel for Data Analysis—Caveats." University of Massachusetts Amherst. Last modified February 2007. http://people.umass.edu/evagold/excel.html.

Rubin, Denis. "Which Statistical Software to Use?" Quantitative Analysis Guide. NYU Libraries. Last modified May 25, 2022. https://guides.nyu.edu/quant/statsoft.

Assessment: Excel feature video. Pick one tool or feature in Excel and make a video explaining how it works and why it is useful.

Week 7
Topic: Storytelling. How to write about data, examples of news stories centered around data.
Readings:
Gallo, Carmine. "Jeff Bezos Banned PowerPoint in Meetings. His Replacement Is Brilliant." Inc.com. Last modified April 25, 2018. https://www.inc.com/carmine-gallo/jeff-bezos-bans-powerpoint-in-meetings-his-replacement-is-brilliant.html.

Klass, Gary M. "Tabulating the Data and Writing about the Numbers." In *Just Plain Data Analysis: Finding, Presenting, and Interpreting Social Science Data*, 2nd ed., 61–78. Lanham, MD: Rowman & Littlefield, 2012.

Outside in America team. "Bussed Out: How America Moves Thousands of Homeless People around the Country." *Guardian*, December 20, 2017. https://www.theguardian.com/us-news/ng-interactive/2017/dec/20/bussed-out-america-moves-homeless-people-country-study.

Schwirtz, Michael, Michael Winerip, and Robert Gebeloff. "The Scourge of Racial Bias in New York State's Prisons." *New York Times*, December 3, 2016. https://www.nytimes.com/2016/12/03/nyregion/new-york-state-prisons-inmates-racial-bias.html.

United Nations Economic Commission for Europe. *Making Data Meaningful, Part 1: A Guide to Writing Stories about Numbers*. New York

and Geneva: United Nations, 2009. https://unece.org/fileadmin/DAM/stats/documents/writing/MDM_Part1_English.pdf.

Assessment: News article evaluation, Data Memo (Midterm 1). Choose and evaluate a news article for data storytelling skills (details below).

Data Memo

1. Consider how you could answer the research question you wrote last week using survey data.
2. Make a data collection plan for gathering the survey data you need to answer your question. Your data collection plan should be approximately two double-spaced pages and address the following issues:
    a. What resources you will need to implement your plan.
    b. How you will administer your survey and an explanation of why you picked that method (for instance, a survey could be administered verbally, on paper, or using an online tool such as Google Forms).
    c. Who you will include in your survey.
    d. How the data you plan to collect will answer your research question.
3. Implement your data collection plan. Survey data will be due in week 7. If your data collection plan involves the internet, please *include a link to your survey* in your data collection plan submission.

Week 8
Topic: Types of visualizations. Principles of data visualization design and tips for choosing appropriate types of data visualizations.

Readings:
Nelson, Stephen L. "Ten Tips for Visually Analyzing and Presenting Data." Chapter 15 in *Excel Data Analysis for Dummies*, 2nd ed. New York: John Wiley & Son, 2014.

Pierson, Lillian, and Jake Porway. "Following the Principles of Data Visualization Design." Chapter 9 in *Data Science for Dummies*. Somerset, NJ: John Wiley & Sons, 2016.

Assessment: Data visualization scenarios. Given a series of scenarios, describe what kind of data visualization would be most effective for communicating specific data to a specific audience in order to achieve a specific goal and why.

Week 9
Topic: Creating visualizations. When and how to include figures and tables in your writing, tutorials on creating charts in Excel.

Readings:
UNC Chapel Hill Writing Center. "Figures and Charts." Last modified 2018. https://writingcenter.unc.edu/tips-and-tools/figures-and-charts/.

Infobase Cloud Learning. "Excel Advanced: Exploring Charts and Data and Creating Pivot Tables." 2021. https://learningcloud.infobase.com/5972/learnit/144452 (Subscription required)

Microsoft 365. "Insert Excel Data in PowerPoint." Accessed August 16, 2021. https://support.microsoft.com/en-us/office/insert-excel-data-in-powerpoint-19767daf-672c-43bc-bda1-330b242c57c9.

Assessment: Data visualization creation. Choose a data set and create four different types of data visualizations based on that data set, explaining the advantages and disadvantages of each.

Week 10
Topic: Preparing data. Data quality and how to normalize, anonymize, and clean data.

Readings:
Borgatti, Steve. "Normalizing Variables." BA 762: Research Methods. Last modified 2010. http://www.analytictech.com/ba762/handouts/normalization.htm.

Loshin, David. "Data Quality and Information Compliance." In *Business Intelligence: The Savvy Manager's Guide*. San Francisco: Elsevier Science & Technology, 2003.

Nelson, Stephen L. "Scrub-a-Dub-Dub: Cleaning Data." Chapter 3 in *Excel Data Analysis for Dummies*, 2nd ed. New York: John Wiley & Son, 2015.

Roark, Kendall. "Sensitive Research Data Management: Approaches/Techniques for Sharing Sensitive Data." Research guide. Purdue University Libraries. Last modified 2019. https://guides.lib.purdue.edu/c.php?g=352785&p=2377926.

Assessment: Data set evaluation. Evaluate data cleaning and preparation of a user-contributed data set in a public repository.

Week 11
Topic: Types of data analysis. Types of data analysis and what types of questions data can be used to answer

Readings:

Leek, Jeff. "The Data Analytic Question." Chapter 2 in *The Elements of Data Analytic Style: A Guide for People Who Want to Analyze Data*, 3–9. Leanpub, 2015. https://worldpece.org/sites/default/files/datastyle.pdf.

Peng, Roger D., and Elizabeth Matsui. "Types of Questions." Section 3.1 in *The Art of Data Science*. Bookdown, 2017. https://bookdown.org/rdpeng/artofdatascience/types-of-questions.html.

Assessment: Data analysis types, data visualization (Midterm 2; details below).
Describe what type of data analysis they will use to answer their research question and why.

Data Visualization

1. Create one or more data visualizations to accompany your Data Memo from Week 7. Your data visualizations should:
   a. be directed at the same audience to whom you wrote your data memo; and
   b. use the data visualization creation principles you learned in weeks 8 and 9.
2. Write an explanation of why you made the choices you did in creating your data visualizations. Be sure to refer back to the readings.
3. Submit your data visualizations and explanation through Blackboard.

Week 12–13

Topic: Storing and managing data. The research data life cycle, metadata, and metadata standards. How to write a data management plan, version control, and promoting reuse of data.
Readings:
Gilliland, Anne J. "Introduction to Metadata," v. 2.0. Getty Center. Accessed August 16, 2021. https://www.getty.edu/research/publications/electronic_publications/intrometadata/setting.html.

Jones, Sarah. "How to Develop a Data Management and Sharing Plan." Data Curation Centre. September 8, 2011. https://www.dcc.ac.uk/guidance/how-guides/develop-data-plan.

Riley, Jenn, and Davin Becker. "Seeing Standards: A Visualization of the Metadata Universe." Scholars Portal Dataverse. Last modified 2018. https://doi.org/10.5683/SP2/UOHPVH.

UK Data Service. "Data Backup." Last modified 2021. https://www.ukdataservice.ac.uk/manage-data/store/backup.aspx.

UK Data Service. "Research Data Lifecycle." Last modified 2019. https://www.ukdataservice.ac.uk/manage-data/lifecycle.

UK Data Service. "Version Control and Authenticity." Last modified 2021. https://www.ukdataservice.ac.uk/manage-data/format/versioning.aspx.

White, Ethan P., Elita Baldridge, Zachary T. Brym, Kenneth J. Locey, Daniel J. McGlinn, and Sarah R. Supp. "Nine Simple Ways to Make It Easier to (Re)Use Your Data." *Ideas in Ecology* 6, no 2 (2013). https://ojs.library.queensu.ca/index.php/IEE/article/view/4608.

Whitmire, Amanda. "Data Management throughout the Research Lifecycle." Figure. Figshare. Last modified August 20, 2013. https://doi.org/10.6084/m9.figshare.774628.v2.

Woodbury, David. "Metadata Tutorial." Vimeo, 2009. https://vimeo.com/3161893.

White, Ethan P., Elita Baldridge, Zachary T. Brym, Kenneth J. Locey, Daniel J. McGlinn, and Sarah R. Supp. "Nine Simple Ways to Make It Easier to (Re)Use Your Data." *Ideas in Ecology and Evolution* 6, no. 2 (August 30, 2013).

Assessment: Research data life cycle diagram, metadata standard explanation, data management plan.
Draw a diagram of the research data life cycle and list at least three steps that might occur during each stage. Choose and explain a metadata standard.
Write a data management plan for their survey data.

Week 14

Topic: Advanced tools. Broad overview of SQL, Python, R, MySQL, Hadoop, Tableau, and others
Readings:
Salehian, Iman. "Tableau Public." In *Introduction to Digital Humanities: Concepts, Methods, and Tutorials for Students and Instructors*, by Johanna Drucker with David Kim, Iman Salehian, and Anthony Bushong, 89–95. Last modified March 12, 2020. https://web.archive.org/web/20200312112540/http://dh101.humanities.ucla.edu/?page_id=163.

Lerner, K. Lee, and Brenda Wilmoth Lerner,

eds. "SQL." In *Computer Sciences*, 2nd ed., 2:238–40. Software and Hardware. Detroit, MI: Macmillan Reference USA, 2013.

Pierson, Lillian, and Jake Porway. "Computing for Data Science." Part 4 in *Data Science for Dummies,* 2nd ed. Somerset, NJ: John Wiley & Sons, 2017.

Pierson, Lillian, and Jake Porway. "Talking About Tableau Public." In *Data Science for Dummies*. Somerset, NJ: John Wiley & Sons, 2017.

Sheikh, Nauman. "Big Data, Hadoop, and Cloud Computing." Chapter 11 in *Implementing Analytics: A Blueprint for Design, Development, and Adoption*. San Francisco: Elsevier Science & Technology, 2013.

"What Is MySQL?" In *MySQL 8.0 Reference Manual*. MySQL, 2021. https://dev.mysql.com/doc/refman/8.0/en/what-is-mysql.html.

Assessment: Advanced data tool overview. Choose and explain the use of an advanced data tool.

Week 15
Topic: Additional topics. Big data and its implications for the future
Readings:
Meyer, Robinson. "The Cambridge Analytica Scandal, in Three Paragraphs." *Atlantic*, March 20, 2018. https://www.theatlantic.com/technology/archive/2018/03/the-cambridge-analytica-scandal-in-three-paragraphs/556046/.

Pence, Harry E. "What Is Big Data and Why Is It Important?" *Journal of Educational Technology Systems* 43, no. 2 (December 1, 2014): 159–71.

https://doi.org/10.2190/ET.43.2.d.

Assessment: Data presentation (Final)

Data Presentation—Due Week 15
1. Create a video presentation based on the data set you have been working with all semester. This is the final project for the course, so be sure to incorporate all of the previous concepts, including but not limited to
    a. What the research question is and how your data set answers it
    b. The elements of your data collection plan
    c. How you analyzed your data
    d. Telling a story to a specific audience
    e. Choosing appropriate visualization types and following best practices for data visualization design
    f. Your plans for data storage, back-up, security, and reuse
2. You may use any tool you like to create the video, but it must include visual aids (with at least one data visualization), text, verbal explanation, and relevant citations. If you are not sure what tools to use, I recommend creating slides in PowerPoint and then recording yourself presenting your slides using the VidGrid recorder (https://app.vidgrid.com).
3. Upload your video here, or provide a link to it.

REVIEWS/ASSESSMENT STRATEGY
Each week of the course includes one or two mini-assessments as described above, such as a discussion board, video creation assignment, reflection paper, or quiz. However, the three major assessments for the course all focus on methods of using data to communicate effectively. Students administer a survey on a topic of their choice and then share the results as a memo in week 5, a data visualization in week 10, and a video presentation in week 15, adding background information related to their topic along the way.

ADAPTING THE RECIPE
After having offered the course twice, I would suggest making some modifications.
- The major assignment to conduct a survey could be expanded to give students additional data collection options, such as focus groups or interviews.
- The final video presentations students submitted would have benefitted from a time limit to encourage concision.
- One of the shorter assessments was to create instructional videos explaining specific features of Excel, which were submitted directly to the instructor but would be more beneficial if collected in a library where other students could see them.
- Regarding course materials, students had difficulty identifying biases in empirical research articles and might benefit from more in-depth coverage and examples.
- They also submitted data visualizations that used color to convey meaning without providing additional data labels, suggesting that at least one reading about how to create accessible visualizations should be added.

Student feedback on the course has generally been positive. In course evaluations, one student said that the course "was well organized, balanced, clearly communicated and provided ample opportunity to tailor learning to real-life situations," and another student said that it "allowed for flexibility on deadlines while encouraging a manageable pace." The primary criticism of the course was that the content level was too low for students who had taken other data-focused courses first or were further along in their careers.

ADDITIONAL RESOURCES

Syllabus available at https://drive.google.com/file/d/17Ad-BexYUTUEoI7pmaB10xUf6L-GLaMPB/view?usp=sharing.

Baking a Data Layer Cake
Scaffolding Data Skills through Video Vignettes

Shannon Sheridan, Research Librarian, Pacific Northwest National Laboratory, shannon.sheridan@pnnl.gov

NUTRITION INFORMATION

This research data service began when the University of Wyoming moved to online instruction in mid-March for the remainder of the spring 2020 semester. A faculty member approached me with a request to support data skills in his undergraduate education class. Students were creating surveys via Google Forms, then visualizing and analyzing the results as a final project. These students had limited experience working with data. The instructor searched for adequate online videos to embed in his course shell but found that most were either far too long or poorly made. To provide support for the students, I made short video vignettes demonstrating three skills: creating a Google Forms survey, extracting data from the survey, and creating basic data visualizations in Excel. Much like baking a layer cake, comprised of frosting, cake, and toppings, each of these skills could technically stand on its own. When combined, however, the result was a delicious data layer cake where students were able apply the skills to their final project. All the videos were under 5 minutes. Since the students did not need all the information or skills at once, the instructor was able to scaffold the videos in the relevant online modules throughout the remaining weeks of the course.

TARGET AUDIENCE AND NUMBER SERVED

Since this recipe provides a "take-out" option, it serves any number of students who need to learn discrete data skills.

LEARNING OUTCOMES

While learning objectives vary depending on the video subject, they all focus on data literacy. In general, students will determine the appropriate scope of the application of a data skill and how to use these data skills in their coursework. The student learning outcomes (SLOs) for the library's portion of the final project were

- Create a survey in Google Forms
- Extract survey data to an analyzable format
- Demonstrate data visualization skills with Excel
- Communicate findings based on data via visualizations

COOKING TIME

Chefs will need 2–3 hours prep time for each 3–5 minutes final product (video).

DIETARY GUIDELINES

With instruction moving online and most classes using some form of an online learning management system, short video vignettes provide an opportunity to incorporate real-life data skills in a class, bolstering students' data literacy within a discipline. Data vignettes present an excellent opportunity to scaffold data skills throughout a class. It easily integrates with classes, while allowing data practitioners new modes of data education delivery. Students can practice one component, like focusing on creating the perfect creamy icing, without having to worry about constructing a whole layer cake at once. These short videos can easily be embedded throughout online courses to augment other class materials, while also having the potential to be useful in teaching face-to-face, such as to practice skills before a lesson or have an easy way to review key skills.

INGREDIENTS

- Faculty member interested in teaching their students data skills.
- Learning management system. Our university uses Canvas, through which the instructor was able to easily share the videos in relevant modules.
- Video editing software. I used Camtasia to record and edit the videos, then uploaded them to VidGrid for captioning and to facilitate embedding. Other options are Panopto or QuickTime.
- A good microphone/headset.
- Instruction skills.

- Data topic. In this case, it is extracting and visualizing survey data.

PREPARATION

1. Consult with the instructor on the topics they wish to cover.
2. Plan out what skills are needed to achieve the desired data learning outcome.
3. Decide on skills featured in each video.
4. Determine number of videos needed.

INSTRUCTIONS

1. Write up a brief outline, followed by a fleshed-out script. This will result in the cleanest sounding audio. One of my videos for the SLO "Demonstrate data visualization skills with Excel" has a script of about 500 words. The goal of these videos is to give students the knowledge they need to perform a discrete skill that can be scaffolded. (See the OSF project in the Additional Resources section for copies of the final scripts.) When brainstorming what topics are suitable for this format, keep in mind that these are meant to be short videos. These videos are a shortcut to getting new skills in the hands of students quickly. Talking with the professor will help you prioritize what to put in your script. While many of the details depend on the instructor and course goals, there are a few general things to keep in mind when writing. Unless the goal is to teach theory, avoid theory and only narrate the how of each topic. When creating my scripts, I choose to focus on just the information students need to complete the assignment. In the video "Demonstrate Data Visualization Skills with Excel," I don't explain why a student might select different chart types. Instead, I focus on what buttons to click to change them. Speaking with the instructor confirmed that chart choice was a topic with which the students should be familiar. Finally, make sure to read your script aloud once or twice before recording it. What looks fine in print might sound stilted aloud—people rarely write exactly how they speak. It also gives you a new way of listening to the instructions to see if the flow of information makes sense.
2. If interested in reuse, make the videos discipline-agnostic. I created a fake data set for the Excel video. This way other students or teachers could more easily learn or reuse the videos. This can be particularly helpful if you're the only data cook at your institution!
3. Record audio, then screen grab your content. This could be PowerPoint, a live action demonstration of using a tool, or some combination. Some of my videos were screen capture only, and others used a combination of live demonstrations with PowerPoint slides, as shown in figure 1.

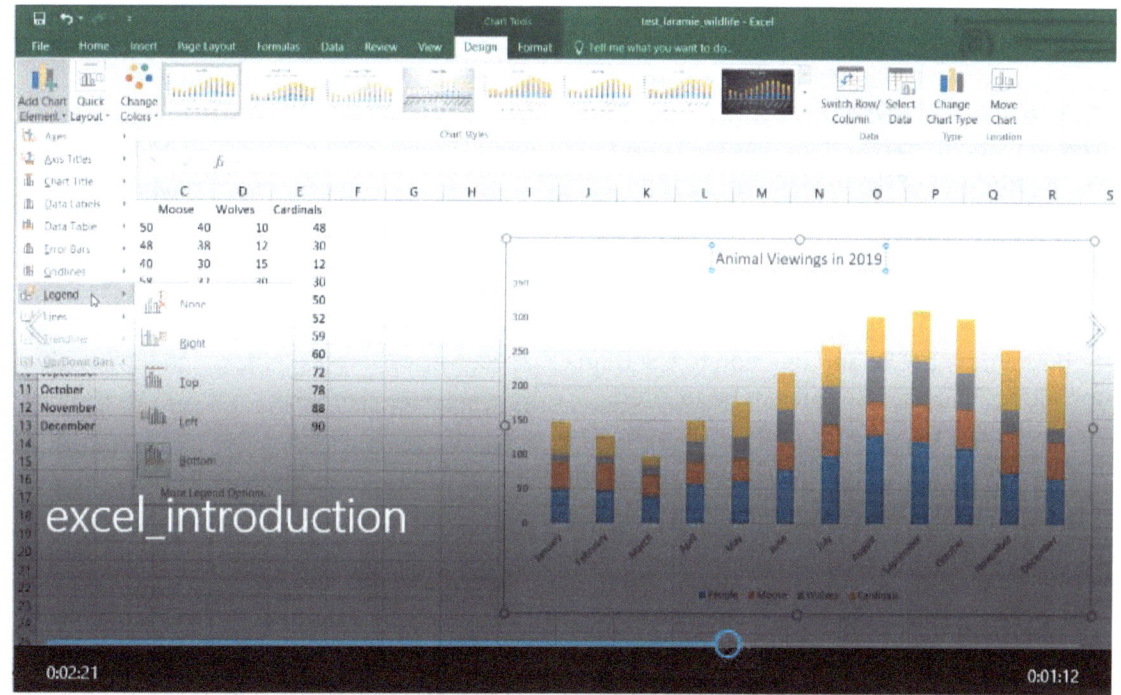

FIGURE 1
Screen shot of live capture used in Excel presentation.

4. Edit the video. Make sure it is no longer than 5 minutes. A link to an OSF project in the Additional Resources section leads to copies of the final videos.
5. Make sure captioning is available for accessibility.
6. Share the videos with the instructor. They will embed them throughout the course in the relevant online modules. The University of Wyoming uses Canvas, so the instructor was able to take the embed codes I sent him and place them in the appropriate spots throughout the class. The three short videos were spread over the last few weeks of class, as students created their surveys, pulled the data, and visualized the results.

REVIEWS/ASSESSMENT STRATEGY

The success of these videos was measured by the number of video views and a feedback survey created by the instructor. A link to the survey can be found in the *Additional Resources* section. This survey asks students about their general opinion of data visualization and their use of and reaction to the training video. This assessment can be tailored depending on your video topic and SLOs.

ALLERGY WARNING

In these videos, I taught students how to whip the egg whites into stiff peaks. I did not explain the chemical process behind it. For this reason, there is not much theory in the videos. The videos were supplemented with deeper explanations from the instructor, such as why a student would want to create a regression line on a scatterplot.

ADAPTING THE RECIPE

This cake can be made in any flavor combination imaginable. These vignettes can be placed in online course shells even if a class is meeting in person. Practitioners and instructors would be able to scaffold skills throughout a course or multi-week project (instead of a one-shot session at the beginning of the project) and have easy access to them whenever they are needed. The data topics also have the potential to cover a variety of skills, scaffolded or not. Other potential video vignettes could show students how to download R and RStudio before an in-class workshop or show graduate students how to register for an ORCID.

ADDITIONAL RESOURCES

Note: While the first instance of these vignettes was focused on Excel, since then additional videos were created for other coding and data topics. These videos and scripts are also available together on the OSF site or individually at the links below.

Schroer, Joseph. "Data Visualization Survey." 2019. https://forms.gle/azxnZchZkcYLsTtj9.

Sheridan, Shannon. "Baking a Data Layer Cake: Additional Resources." Last updated July 27, 2021. osf.io/gu89w.

Full set of videos: https://app.vidgrid.com/content/a7zyw6lNmvhd.

Full set of scripts and videos available for download: https://osf.io/gu89w/.

"How to… Create a Questionnaire in Google Forms." Video, 3:09. https://use.vg/Je9Xvs

"How to… Retrieve Answers from Google Forms." Video, 3:12. https://use.vg/zlFWs6.

"How to… Create Visualizations in Excel." Video, 3:33. https://use.vg/uImKjy.

"Downloading R and RStudio." Video, 3:41. https://use.vg/6O1bL4.

"Creating a Project and Working Directory." Video, 4:43. https://use.vg/2kU3va.

"Packages." Video, 3:00. https://use.vg/ivrYyJ.

"Getting a Dataset in R." Video, 3:07. https://use.vg/qUMBrP.

"Introduction to ggplot2." Video, 26:23. https://use.vg/kl79qZ.

"ggplot2: How to Make a Bar Chart." Video, 4:33. https://use.vg/Lptrqt.

"ggplot2: How to Make a Scatterplot." Video, 2:59. https://use.vg/KuuNS4.

"ggplot2: Intro to Aesthetics." Video, 3:30. https://use.vg/C5C65S.

Building Data Literacy through Scaffolded Workshops
Experiences and Challenges

Jiebei Luo, Financial & Economic Data Analysis Librarian, New York University, jl14614@nyu.edu; Yaqing (Allison) Xu, Data and Visualization Librarian, Thomas P. O'Neill, Jr. Library, Boston College, allison.xu@bc.edu

INTRODUCTION

For academic librarians conducting a one-shot workshop, the conflict between time limitations and the amount of information to be delivered is a challenge. In order to address this problem and also expand attendance at our workshops, Data Services in the Boston College Digital Scholarship department designed a series of workshops focused on different stages of the research data life cycle held across the span of one semester.

One recent example of our pedagogical approach is from the spring 2020 semester (as shown in figure 1). In response to the 2020 census, we selected the census as the central theme of the scaffolded data workshops for that semester. Accordingly, the following series of workshops on the topic were developed based on working with research data in social sciences.
1. Data/Statistics Sources in Social Sciences
2. Analyzing Census Data in Excel
3. Demographic Data Visualization in Tableau (using census data)
4. Data Management in Social Sciences

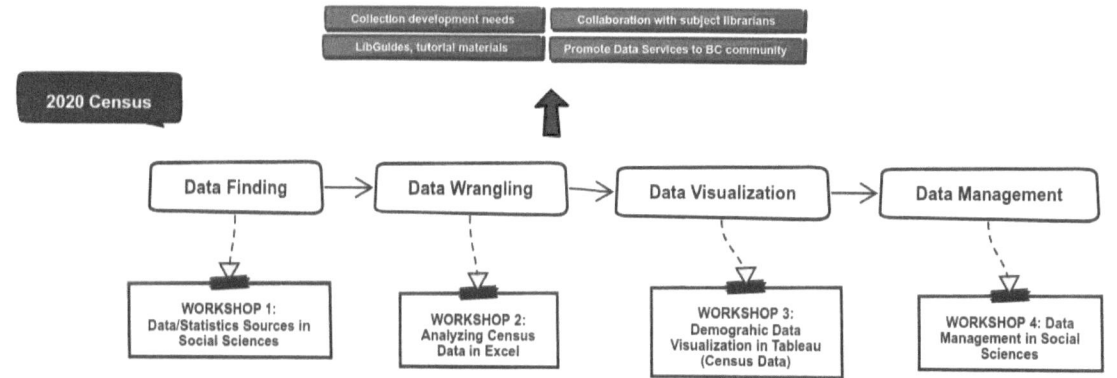

FIGURE 1
Sample data workshop flow design

TARGET AUDIENCE AND NUMBER SERVED

Workshops are suitable for all college members, including undergraduate and graduate students, as well as researchers and librarians who are interested in learning about data resources and improving data literacy skills. In particular, undergraduates and entry-level graduate students will find the workshops useful when they start working on research projects with a data component, such as a senior thesis or an introduction to research methodology course. A cap of 15 people for one workshop with one helper allows the instructor to monitor the group and provide timely troubleshooting assistance.

LEARNING OUTCOMES

By attending the workshops, participants will
- evaluate data and statistics critically, and access data sources effectively
- differentiate research data life cycles and determine data needs in different stages
- apply data processing skills to accomplish a specific analytical purpose
- visualize and present data effectively and efficiently
- evaluate data management needs, and make plans for data management

COOKING TIME

A few weeks within one academic semester; each workshop is 90 minutes long.

DIETARY GUIDELINES

This workshop series provides an alternative to the more common one-shot library instruction sessions. Because the workshops

are designed in a scaffolded fashion, attendees are able to acquire knowledge and skills for working with data that match several frames within ACRL's *Framework for Information Literacy for Higher Education*:
- *Information Creation as a Process:* The workshops provide students with opportunities to explore the dynamic nature of working with research data, requiring ongoing attention to different elements in the research data life cycle.
- *Research as Inquiry:* Meaningful data analysis and visualization come from a comprehensive understanding of the subject. In particular, effective visualization requires researchers to formulate their questions and deepen their understanding as their research progresses.

INGREDIENTS
- Laptop, internet access
- Software installation: Excel, Tableau Desktop (student version)

PREPARATION
- *Preparation for the workshop series:* Select a trending topic that is of interest to researchers across different disciplines and that illustrates skills related to different stages of the research data life cycle. Digital scholarship librarians lead in designing, creating, and conducting the workshops, but they should partner with subject liaisons on the data acquisition session as well as promotion activities.
- *Preparation for workshop 1, data finding:* Consult various data sources beforehand and select a short list of sources to demonstrate in the workshop. Using the 2020 census as an example, the instructor may (1) begin the session by providing background information such as the history of the U.S. Census (https://www.census.gov/history/) and the evolution of its questions and sampling methodology; (2) introduce the Census Bureau as the government agency that conducts and publishes census surveys, explaining in particular the concept of data versus statistics and illustrating with corresponding sources from the census; (3) prepare data sources tailored to local areas after going through the country-level data (e.g., The data for our census sessions came from Mass.gov and data sources published by the city of Boston); and (4) consult with subject liaisons and identify library guides published by peer libraries to further improve the coverage and depth of the data sources to be introduced in this workshop.
- *Preparation for workshop 2, data wrangling:* Excel is used as the data wrangling tool for this session. Prepare step-by-step tutorials for how to use Find and Replace for quick data cleaning, Conditional Formatting for a range of data, Sort and Filter for easy slicing and reorganizing, PivotTable, and VLOOKUP.
- *Preparation for workshop 3, data visualization:* Prepare a cleaned and merged data set (inherited from the data wrangling session) for participants who missed workshops 1 and 2. Provide the installation instructions for Tableau in advance, and ask all participants to have the software installed before coming to the workshop. Create a Tableau workbook template that covers various visualization types using the selected data set. Post everything to Google Drive, GitHub, or other university cloud storage so that workshop participants can download them. Create a step-by-step tutorial (https://bcds.gitbook.io/learn/tutorials/data-visualization/untitled-1/using-tableau-to-visualize-covid-19-data) to show how to connect data to Tableau and create visualizations with Tableau; this tutorial serves as a guide to help students follow the workshop materials and replicate the demonstrated visualizations during and after the workshop.
- *Preparation for workshop 4 data management:* Equip yourself with knowledge of data management needs specific to the topic of that semester. For instance, survey data involving human subjects will require IRB review and typically de-identification of participant data. Output data in different disciplines may require software-independent formats for long-term preservation and different repositories.

INSTRUCTIONS
Workshop 1: Data Finding
The first workshop focuses on data finding and covers basic knowledge of data and statistics using the census as an example.
- Following the introduction of the history of the US census, start the data-finding journey by searching a specific demo-

graphic variable in the census; for example, US population characteristics under Explore Census Data (https://data.census.gov/cedsci/), which can reveal more granular levels, such as by gender, by age group, or by adding more variables such as education, employment, or income.
- Introduce IPUMS (https://www.ipums.org) as the platform to extract census microdata as the example of raw data.
- Perform searches in ICPSR and Dataverse to demonstrate locating data sets published as the permanent preservation versions in disciplinary-focused data repositories.
- Introduce a few databases based from the library's subscriptions. In our example, we promote DataPlanet, Statista, and PolicyMap.

Workshop 2: Data Wrangling

Daily consultations with patrons have revealed and confirmed a significant demand for basic data manipulation and analysis skills using tools such as Excel.
- Utilize the American Community Survey 5-Year Data (https://www.census.gov/data/developers/data-sets/acs-5year.html) to demonstrate basic skills such as Filter and Sort to subset data such as a specific race group, and apply conditional formatting to highlight the data that are below a certain threshold of education attainment.
- Illustrate how to calculate the average income of each age group with aggregate data ranging from county to state level. Use VLOOKUP for merging external data, such as data downloaded from PolicyMap, with the working data set.
- Leave time for attendees to try out the functions during the session and share the resulting spreadsheet after the session.

Workshop 3: Data Visualization

By inheriting the processed data from the data wrangling session, which contains a clean data set including census geographic locations (e.g., county and state, population, education, race, income, and employment), the instructor can use Tableau to perform visualization tasks based on the following workflow.
- To get started in Tableau, explain how to import a processed data set; outline major structures within Tableau, such as Sheet, Dashboard, and Story; and demonstrate visualization tasks in an individual Sheet.
- When working in an individual Sheet, demonstrate the use of rows and columns, qualitative and quantitative variables, various chart types, and formatting.
- Use Dashboard or Story to present the analytical results in an integrated fashion.
- Introduce different chart types during the workshop (e.g., start by creating a basic chart such as a bar chart in Tableau, and then walk through more advanced charts such as maps and time lines that help visualize the data in an interactive fashion).

Workshop 4: Data Management

A data management session wraps up the series. Issues in managing research data, such as data format, identifying a data repository, and planning for long-term storage and access, are discipline-specific and usually more relevant to faculty and graduate students.
- Use the 2010 decennial census (https://www.census.gov/history/www/through_the_decades/index_of_questions/2010.html) questionnaire as an example of conducting a survey involving human subjects.
- By navigating a DMPTool (https://dmptool.org) template, explain data management concepts, including best practices for data sharing, metadata, data repository selection, and long-term preservation.
- Encourage participants to create an account in the DMPTool and start creating a data management plan for a mock decennial survey using the template provided in the DMPTool.

CHEF'S NOTES

We would like to take this opportunity to document several challenges during our exploration of this innovative workshop format. For instance, our approach to workshop development requires one to continuously acquire in-depth knowledge of new data sources and also keep up with emerging tools for data handling. In addition, identifying a trending topic that can attract a wide audience on campus while providing enough flexibility to support a series of workshops centered around it is another challenging task.

Data Literacy Appetizers
LibGuide Data Instruction Modules for Undergraduates

Beth Hillemann, Research and Instruction Librarian for the Social Sciences, Macalester College, hillemann@macalester.edu;
Aaron Albertson, Patron Experience Supervisor, Hennepin County Library-Plymouth Branch, alberaa@gmail.com

NUTRITION INFORMATION

At Macalester College, a small, private liberal arts college in St. Paul, Minnesota, we created LibGuide-based data literacy modules to teach our undergraduate students basic skills and knowledge for finding, using, and managing data. We went into the kitchen in an experimental mood, tinkering with different ingredients and cooking times for our recipe. We came up with modules that may be combined with other dishes to form a complete meal in class or served cafeteria-style with students picking and choosing the bits that work best for their needs. The topics for the seven modules are What Is Research Data?; Planning for Your Data Use; Finding and Collecting Data for Your Research; Keeping Your Data Organized; Intellectual Property and Ethics; Storage, Backup, and Security; Documentation.

TARGET AUDIENCE AND NUMBER SERVED

This recipe is designed to feed undergraduate students with little to no data literacy or data management knowledge. It is highly flexible in order to appeal to the tastes of both individual students and classroom faculty or librarians. The modules may be consumed in any order and may be combined with other recipes (e.g., assignments, lectures, videos) as desired. While some of the modules include a case study about research in the field of political science, they have been designed to be suitable for any student beginning with data-based research in any field. The modules were not designed for a specific course, but rather to fill a need we saw across disciplines.

LEARNING OUTCOMES

Each module has its own learning outcomes.
- Module 1: Students will be able to define different types of research data and demonstrate how they fit in with the scholarly research process.
- Module 2: Students will be able to explain what data management is and be able to recognize and use best practices for managing their own projects.
- Module 3: Students will be able to implement strategies for finding and gathering data.
- Module 4: Students will be able to list strategies for keeping data files organized.
- Module 5: Students will be able to articulate a basic understanding of intellectual property as it pertains to data. Students will be able to demonstrate ethical practices in the use of data for their own research.
- Module 6: Students will be able to explain data storage practices and security measures.
- Module 7: Students will be able to explain documentation practices for data and analysis.

TIME ALLOWANCE

Cooking and eating times are highly variable, depending on local conditions. Each module is designed to provide a quick overview with strategies that a student may apply to their own project. Students may decide to devour all modules in an hour or two or snack on them over several weeks. Each module should take students 10–20 minutes to complete, although the more complex the research project, the longer it might take to apply the principles learned. Indeed, a student might benefit from circling back to various modules as their project progresses.

DIETARY GUIDELINES

Using LibGuides gave us a set framework to fit into our website and the flexibility to be easily edited by multiple staff. Our goal was to meet dietary standards for an introduction to data literacy and management for undergraduate students. Specifically, by completing all of the modules, students will have acquired and, if working with a project, demonstrated data literacy skills in the following frames from ACRL's *Framework for Information Literacy for Higher Education*:

- Searching as Strategic Exploration: understanding different types of data, who is likely to produce them, how one might access them, and the ethical considerations for using data collected by others.
- Authority Is Constructed and Contextual: understanding data provenance and the ethical considerations for reusing and remixing data.
- Information Creation as a Process: understanding the data life cycle in the context of a research project.
- Information Has Value: understanding for what purposes data are collected, secured, used, and shared.

INGREDIENTS

Our data appetizers use Springshare's LibGuides and are publicly available at https://libguides.macalester.edu/data1. They may be used as they are, or copied and remixed as needed, according to a CC-BY-NC 4.0 International License. The modules are the crucial starter ingredient. Simply add undergraduate students with a data assignment or project, and then your own flavoring and seasoning for variation.

INSTRUCTIONS

We encourage chefs to take as much of our content as they wish if they want to implement data instruction modules at their own institutions. All of our content may be reused, mixed with other content, or edited to suit your local needs. We chose to create the modules in LibGuides because it is a familiar tool to us and it fits well with our website. The modules could easily be adapted to other platforms. They were designed to be a broad overview of data literacy and data management concepts for undergraduate students. They do not go into any great depth on any particular issue. See figure 1 for an example. Below are the addresses for each module. Individual modules start with an overview page and end with a "Let's Review" page that contain a brief quiz on the content covered in the module:

1. What Is Research Data? https://libguides.macalester.edu/data1
2. Planning for Your Data Use: https://libguides.macalester.edu/data2
3. Finding and Collecting Data for Your Research: https://libguides.macalester.edu/data3
4. Keeping Your Data Organized: https://libguides.macalester.edu/data4
5. Intellectual Property and Ethics: https://libguides.macalester.edu/data5
6. Storage, Backup, and Security: https://libguides.macalester.edu/data6
7. Documentation: https://libguides.macalester.edu/data7

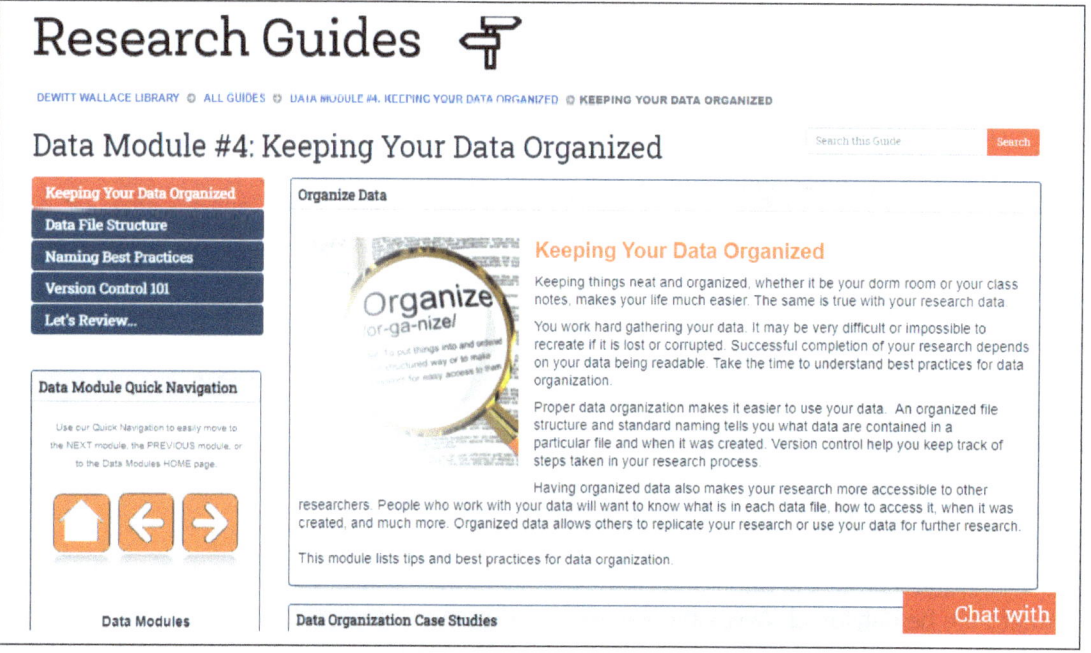

FIGURE 1
Example module on a Research Guide (https://libguides.macalester.edu/data4)

CHEF'S NOTES

The modules may be consumed in any amount and in any order. We recommend a guided class assignment in conjunction with individual or group data projects. The appetizers may be eaten all at once or spread out throughout a semester.

The recipe is designed to work at the individual level, with little to no guidance needed from an instructor. However, pairing it with a class-based assignment might be most appetizing for students.

Clearly, there is much more that could be said about intellectual property, data collection methods, or any other of the topics we touched on. Additionally, we did not address some topics at all, such as data analysis and visualization. Another issue to consider when adapting or adopting these modules is that we work in the social sciences, and therefore our knowledge in this area drove much of our content, especially the case study portions. If you do create new modules or adapt our content we encourage you to share it in the spirit of a community potluck. We look forward to your new creative dishes!

REVIEWS/ASSESSMENT STRATEGY

Students assess their own learning by taking brief quizzes found in the "Let's Review" page included at the end of each module. Web traffic data shows that the original recipe modules are heavily used. We plan on assessing the effectiveness of the modules in conjunction with a class at Macalester in the future.

ADDITIONAL RESOURCES

All of our data modules use Springshare's LibGuides (https://springshare.com/libguides/), though other content management systems could be used.

RECIPE INSPIRATION

Our thanks to Macalester librarians Ron Joslin and Louann Terveer for their inspiration, dedication, and help with the creation of the data modules.

Data as Curation
Framing Data Creation as a Critical Practice through Collections-Based Research Inquiry

Gesina A. Phillips, Digital Scholarship Librarian, Digital Scholarship Services, University Library System, University of Pittsburgh, GAP64@pitt.edu; Tyrica Terry Kapral, Humanities Data Librarian, Digital Scholarship Services, University Library System, University of Pittsburgh, tyt3@pitt.edu; Matthew J. Lavin, Assistant Professor of Humanities Analytics, Data Analytics, Denison University, lavinm@denison.edu; Aaron Brenner, Associate University Librarian for Digital Scholarship, Creation, and Technology, University Library System, University of Pittsburgh, abrenner@pitt.edu

NUTRITION INFORMATION
This lesson is based on the work of the project "From Collection Records to Data Layers: A Critical Experiment in Collaborative Practice" (CaD@Pitt). This project focuses on increasing the visibility and discoverability of library collections, making library collections data accessible for computational use, and enabling scholars to extend and enrich collections data with critical, research-driven layers of additional data. This lesson is a distillation of several topics from the modular lesson framework that we designed as a scaffolded introduction to working with library collections data.

This lesson will fit well in classes that already engage with data or visualization, and it can bring a new dimension to classes engaging with digital library collections or with a content area that is well served by a particular library collection.

TARGET AUDIENCE AND NUMBER SERVED
This lesson was designed with an undergraduate audience in mind. It works best in a humanities classroom with fewer than 20 students or even in a slightly larger discussion-based or seminar-style class.

LEARNING OBJECTIVES
1. Understand the possibilities and shortcomings of data-driven humanities inquiry
2. Appreciate data work as interpretive and complicated
3. Investigate collections objects in order to better understand the content and context of the selected research collection
4. Better understand connections between data work and visualization

COOKING TIME
Two 75-minute sessions with homework

DIETARY GUIDELINES

Framework Applications
This lesson applies the following frames from ACRL's *Framework for Information Literacy for Higher Education*:
- Information Creation as a Process
  - Create data to encounter interpretive challenges of the process
  - Critique data creation methods
  - Consider the data's future life and reuse
- Authority Is Constructed and Contextual
  - Conceptualize themselves as creators and bring their own experiences or expertise to the process of creating data
  - Engage in the data creation process to experience and (re)conceptualize data creation as a series of choices based on interpretation
- Research as Inquiry
  - Formulate research questions that can be answered with data
  - Encounter foundational questions about using data for inquiry
- Information Has Value
  - Connect experiences creating data with various downstream uses (e.g., analysis, visualization)
  - Experience the labor required to create high-quality data

INGREDIENTS
1. Item records from one or more library

collections (e.g., a run of a periodical, a group of archival images that share a similar theme, a digital collection from a common source, a selection of texts that share some characteristic); this set of item records will serve as your *collection*.

» Desirable features for a collection include aptness of collection content to class themes; robustness of metadata; potential for surfacing issues of ethics, bias, or reductiveness; and quality of digitization that allows learners to deeply investigate each item.

» Consider what data to add or what existing data you might enhance if you were to build on the existing metadata. For example, you might find that the genre "archival document" is used in various item records when a more specific term might be useful. Could your class add more specific genre descriptors? Would a controlled vocabulary, such as "information artifacts" terms from the *Art and Architecture Thesaurus* (https://www.getty.edu/research/tools/vocabularies/aat), help guide consistent data collection?

» Consider common features of collection items (e.g., the structure of documents, availability of temporal and geographical locations, use of imagery) that may be used as data elements. Then, consider the amount of variation in those features across the collection, and how they might support visualizations of similarity and difference.

Assigned to:	Item Title	Item URL	(Data collection field 1)	(Data collection field 2)
Student Name	Example Item	https://library.edu/example-item		

FIGURE 1
Data collection spreadsheet. https://drive.google.com/file/d/13IMqzAWcFCViYCfJNWXKL3vVg11SWjAT/view

2. Data Design Plan worksheet. The Data Design Plan (see appendix) walks learners through the process of creating a research question, proposing methods of answering it using data and visualization, and planning to collect the data in a thoughtful and systematic fashion.

3. Data collection spreadsheet (figure 1)
   » The spreadsheet may be a single, shared file with multiuser input (e.g., Google Sheets) or a template that is distributed separately to each learner. A single shared file will ease the creation of a data set that can be used for visualizations.
   » The instructor should pre-populate this shared file with "base" data about items; the nature and amount of these base data may depend on what is feasible to collect before the lesson. It is necessary to include at least an item title and a link to the digital record page for each item.
   » The number of items in the spreadsheet depends on the size of the collection, the size of the class, and the number of items needed for a robust visualization.
   » Create a column or columns for data that learners will collect. Think about whether controlled vocabularies are needed, whether drop-downs or data validation are useful, or whether additional instructions are needed for consistent data creation.

4. *Optional*: Collaborative notepad (e.g., Padlet)
   » A shared notetaking space will allow learners to document their decision-making processes and questions as they occur.

PREPARATION
- Create your collection (see Ingredients for instructions).
- Review the Data Design Plan worksheet, updating to suit your class, and prepare it for distribution to learners.
- Prepare the spreadsheet for data collection (see Ingredients for instructions).
- Choose discussion questions (see examples in Instructions).
- Decide how to arrive at a shared plan for data collection homework. Will you choose a plan from those presented by the students, or create your own data design plan and ask learners to enact it? In either case, data must be gathered from the collections objects themselves

(rather than requiring outside research) and must support visualization.

INSTRUCTIONS

These instructions are for a series of two lessons embedded in a class. Each session is imagined to be 75 minutes at a minimum, with a necessary break in between.

Presession Assignment (Optional)

Learners explore data visualization resources to get familiar with principles and types of data visualization (see Additional Resources below for suggested resources).

Lesson 1

1. (10 minutes) Introduce learners to the chosen collection. Have them take 3–5 minutes to explore and get familiar. Discuss the interesting characteristics of the collection, noting the existing metadata.

    Example discussion questions:

    » *For periodicals:* What details about the periodical can you glean from the first issue (how and why it was created, what did it publish, how did it imagine its audience, etc.)?

    » *For an archival collection:* What details about the collection can you gather from the finding aid? What topics are discussed most often? What is the date coverage of the collection? What kinds of documents are included?

    » *For a digital collection:* What details about the collection can you gather from the digital collection summary? What additional details can you gather by clicking on some of the items and viewing them? What topics are discussed most often? What is the date coverage of the collection? What kinds of documents are included?

    » *General questions:* Is there any important information you want to know that you cannot immediately find? If so, what?

    » *For library metadata:* What information isn't captured in library metadata? Why might that be?

2. (10 minutes) Introduce a few basic visualization types.

    Example visualization types (see Additional Resources for supporting resources):

    » Display counts or other numerical value for categorical data: bar graph

    » Distribution of one continuous variable: histogram or box plot (depending on sample size)

    » Relationship of two continuous variables: scatter plot

    » Change over time: line plot

    » Comparison of distributions of a continuous variable faceted by category: multiple box plots

    » Geographical density: heat map or choropleth

3. (10 minutes) Generate questions about the collection that could be investigated by adding new data and creating visualizations. Introduce the data collection spreadsheet to get learners thinking about rows and columns and what can be counted, classified, or measured. Make sure to problematize/complicate aspects of counting and measurement.

    Example general discussion questions:

    » How can we begin to think about the creators of these materials? Are there multiple items created by the same person?

    » What were the imagined audiences of these documents? Were they intended for a specific person (as with a letter) or for wide consumption?

    » What types of organizations or institutions are represented in the data?

    » Are there any features of these collection items that might be difficult or problematic to represent as data in a spreadsheet?

    Example discussion questions for periodicals:

    » What does the cover of a periodical tell you about the content?

    » Might a community emerge around this publication? Who might be a part of that community?

    » How is the editorial voice of the periodical established? What kind of larger narrative is established through issue themes, and so on?

    » What kind of art appears alongside written content? Photographs, illustrations, and so on? What are the subjects?

4. (15 minutes) Introduce the Data Design Plan worksheet. Learners complete the worksheet in pairs or small groups.

5. (20 minutes) Groups present their plans. Through class discussion, focus on help-

ing each group craft a strong proposal, encouraging constructive questioning and collaborative solution finding.

Discussion questions for groups and their peers, organized by topics that the instructor might like a particular group to consider more fully:

» *Utility of research question and clarity of purpose:* State the purpose of the inquiry in your proposal. How would you describe the significance or importance of your research question? Who is most likely to benefit from or value this inquiry?

» *Appropriateness of data to answer the proposed question:* To what degree does the data collection proposed relate to the research question? How well would these data answer or address that question? Can you identify any questions of bias, important nuance, or representativeness of specific people or populations?

» *Feasibility:* How difficult or easy do you think it would be to collect the data that you propose? What practical concerns can you identify? (For example, the amount of time it would take; how much effort, training, and special equipment you would need; and whether the data you're looking for can be found in the collection items themselves.)

» *Scalability:* Would the plan still work if you had 1,000 items to evaluate? Thirty? Why or why not?

» *Subjectivity:* If three people, on their own, collected data as you propose, how similar or different would their results would be? Does the plan ask the data collector to make judgment calls?

» *Data ethics:* In what ways, if any, do the proposed data reduce the agency of individuals being represented? How would you feel if you were represented in this material and then as data following this plan? Are there issues of historic mistreatment of communities that might make this data layer problematic? Are there steps you might take to avoid harm or the appearance of disrespect?

6. (10 minutes) Decide on plans to implement. Have learners test data entry process.

7. *Homework:* Learners use the instructor-prepared data collection spreadsheet to collect data from assigned collection items. If using a collaborative notetaking document, encourage learners to note strategies and challenges as they go along. The homework due date should give the instructor sufficient time to prepare visualizations for class discussion in Lesson 2.

Lesson 2

1. (5 minutes) Learners explore the completed data set.
2. (15 minutes) Reflect on the data collection process, either as a class or in small groups. Draw from the shared notetaking document (if used) to prompt discussion.

Example reflection questions:

» How did it feel to make choices about the relevancy of data? Did you struggle to decide what was important enough to include? Did you leave anything out or reduce nuance when making those choices?

» Were there occasions where the data collection plan made it easy to make a decision? Did you feel like you had a good idea of how to implement it?

» Where did the data collection plan fail? Were there occasions where you weren't sure how to make a decision?

» Did you learn anything new about the collection or about the specific items that you looked at?

3. (30 minutes) The instructor, using the Data Design Plan as a guide, demonstrates the creation of one or more data visualizations. The instructor may prepare the visualization ahead of the class session and then walk through the most relevant aspects of the process with the class.

4. (25 minutes) Reflect on visualizations and the full experience of approaching collections material through a data lens.

Example reflection questions:

» What insights do the visualizations provide into our shared research question?

» Are there multiple ways that you could interpret some of these visualizations? Are there ways in which they could be confusing or misleading?

» Do the visualizations raise any new questions or areas of interest about this collection?

» If we were to collect more data to

supplement what we worked on together, what new data points might you suggest?

» What parts of the research question might be better investigated using another type of research, such as close reading?

» How does it feel to think about collections items as a source of data?

REVIEWS/ASSESSMENT STRATEGY

This lesson is assessed based on the discussion with and between learners. Lesson 1 asks learners to demonstrate engagement with collections materials and an understanding of the constructed nature of data. It also asks them to do some hands-on work as they create a data collection plan and begin to create data. This hands-on learning continues through the homework assignment, which also results in the creation of a deliverable that can be assessed based on the quality and appropriateness of the data collected. Lesson 2 asks learners to reflect on the challenges they encountered, potential uses of the data they created, and the ways in which they needed to adjust their strategy while creating data. They will also be asked to reconsider their research questions through the lens of visualization and to reflect on this use of their data during class discussion.

ADAPTING THE RECIPE

- Given more time or sessions, you could incorporate instructor or peer review of the Data Design Plans. The full CaD@Pitt learning modules include an adaptable "Critique a Layer" section: https://cadatpitt.github.io/modules.

- You might collaboratively design the data collection schema by deciding on new columns and data entry standards as a class. In most contexts this approach is not recommended due to concerns about time and cognitive load for learners.

- Given more time for plan definition and review, you might ask each group to enact their own Data Design Plan rather than choosing one shared class plan. In this case, the instructor would need to decide how to select among the plans for the guided visualization exercise.

- Given more time or a class that is more focused on data visualization, students may create their own visualizations during the second session.

ADDITIONAL RESOURCES
Lesson 1

Data Visualizations Types

Data Visualisation Catalogue. Accessed March 30, 2021. https://datavizcatalogue.com/.

- "Treemap," https://datavizcatalogue.com/methods/treemap.html.
- "Heatmap (Matrix)," https://datavizcatalogue.com/methods/heatmap.html.
- "Choropleth Map," https://datavizcatalogue.com/methods/choropleth.html.
- "Bubble Chart," https://datavizcatalogue.com/methods/bubble_chart.html. See also "Bubble Plot," From Data to Viz, https://www.data-to-viz.com/graph/bubble.html.

Organizing Data

Wickham, Hadley and Garrett Grolemund. "Tidy Data." Chapter 12 in *R for Data Science*. 2017. https://r4ds.had.co.nz/tidy-data.html.

Lesson 2

Data Visualization Tools

Python/Seaborn

Luvsandorj, Zolzaya. "6 Simple Tips for Prettier and Customised Plots in Seaborn (Python)." Towards Data Science. October 5, 2020. https://towardsdatascience.com/6-simple-tips-for-prettier-and-customised-plots-in-seaborn-python-22f02ecc2393.

Nurhidayat, Rizky Maulana. "Python Data Visualization with Matplotlib—Part 1." Towards Data Science. Last modified November 4, 2020. https://towardsdatascience.com/visualizations-with-matplotlib-part-1-c9651008b6b8.

Nurhidayat, Rizky Maulana. "Python Data Visualization with Matplotlib—Part 2." Towards Data Science. Last modified November 4, 2020. https://towardsdatascience.com/python-data-visualization-with-matplotlib-part-2-66f1307d42fb.

R/ggplot

Wickham, Hadley and Garrett Grolemund. "Data Visualisation." Chapter 3 in *R for Data Science*. Accessed March 30, 2021. https://r4ds.had.co.nz/data-visualisation.html.

Tableau

Tableau Desktop and Web Authoring Help. "Dashboards." Accessed March 30, 2021. https://help.tableau.com/current/pro/desktop/en-us/dashboards.htm.

Data Visualization Design

Chiasson, Tina, and Dyanna Gregory. *Data + Design: A Simple Introduction to Preparing and Visualizing Information*. 2014. https://trinachi.github.io/data-design-builds/titlepage01.html.

Creative Bloq Staff. "How to Design Better Data Visualizations." April 24, 2014. https://www.creativebloq.com/how-design-better-data-visualisations-8134175.

Custer, Charlie. "11 Design Tips for Data Visualization." *Dataquest* blog. October 25, 2018. https://www.dataquest.io/blog/design-tips-for-data-viz/.

UK Home Office. "Designing for Accessibility." Poster. Accessed March 30, 2021. https://github.com/UKHomeOffice/posters/blob/master/accessibility/dos-donts/posters_en-UK/accessibility-posters-set.pdf.

APPENDIX: DATA DESIGN PLAN TEMPLATE

https://github.com/CaDatPitt/instructional-modules/blob/master/Data-Design-Plan-2021.pdf

1. Inquiry: *Think about the themes, subjects, individuals, material qualities, or other features of the library collection that spark your curiosity.*

  a. What feature(s) would you like to investigate about this collection?
  b. What is a research question that you could ask about this feature of the collection?
  c. How do you imagine you could answer this question using only information gathered from the objects in the collection (no outside research)?

2. Visualization: *Think about how you could collect data so that it might be visualized in order to provide some insight into the question you formulated above.*

  a. What kind of data would you collect: categorical, quantitative, temporal, geographical, etc.?
  b. What type of chart, graph, or other visualization might you create using that data?

3. Data Point(s): What specific data would you need to collect to investigate your question using visualization? Does any information already exist in the collection metadata that you could use?

4. Data Elements: Think about each bit of data you propose to collect as a column in a spreadsheet, where the rows are individual items from the collection. Each column will have a name and rules for how you enter the value.

What would you name the new column(s) in your dataset? This name will be your column header. What's an example value for the data you would enter in that field?

To answer these questions, fill out the following fields. Copy, paste, and fill out the fields for each column you create after the first.

Example:
Column Name: Publication Date
Data Value Example: 1990-12-01

5. Challenges: Do you anticipate any challenges in collecting the data you need? If so, what? *Think in particular here about the difficulty of collecting the data and/or potential areas of sensitivity such as how the data represents individuals or groups of people.*

Quantitative Data Skills for Undergraduates
A Seminar Series for Social Science Students

Whitney Kramer, Industrial and Labor Relations Research and Data Librarian, Catherwood Library, Cornell University, wbk39@cornell.edu; Amelia Kallaher, Biomedical Data Manager, Sage Bionetworks, amelia.kallaher@sagebase.org

NUTRITION INFORMATION

This recipe is for a series of six 60-minute social science data-focused lessons that provide a broad overview of the social science research life cycle with respect to working with quantitative data. This seminar teaches undergraduate researchers to locate and evaluate quantitative data for a senior honors thesis or other undergraduate research project. It helps fill in the blanks for many undergraduates, who are often taught statistical analysis skills but not the basics of the research and research data management life cycles, including how to find the data that they need to analyze. Teaching these much-needed skills can help fill a void in both undergraduate knowledge and instructional offerings. This recipe can easily be adapted for librarians who would like to teach data literacy concepts, regardless of discipline, through a combination of synchronous and asynchronous methods.

TARGET AUDIENCE AND NUMBER SERVED

Our target audience for this seminar was upper-level undergraduate students who were completing a senior thesis or other in-depth research project. This can be modified based on your instructional needs.

This seminar was originally designed for 10–20 undergraduate participants, but can accommodate over 50 students through asynchronous learning. Teaching the seminar virtually eliminates physical space restrictions and enables the librarian to open up the seminar and associated online content to any interested students.

LEARNING OUTCOMES

Over the course of the six-week seminar, students will

- articulate the research process with emphasis on how to choose a research topic
- recognize sources for locating statistics, spatial data, and other quantitative data formats
- demonstrate the importance of data organization for collaborative group projects
- examine how data can be shared, reused, and replicated
- identify the elements of a data citation
- interpret and create a codebook

COOKING TIME

45–60 minutes per session, plus additional prep and planning time for each session based on the needs of the students, librarians, and optional guest speakers or panelists.

DIETARY GUIDELINES

This curriculum aligns with all six frames of ACRL's *Framework for Information Literacy for Higher Education*, especially Information Creation as a Process (e.g., collaborating around organizing data and how to create codebooks), Information Has Value (e.g., locating and properly citing data sets), and Research as Inquiry (e.g., understanding the iterative process of research and the research data life cycle). This recipe provides instructions for replicating an instructional seminar on data literacy in the social sciences, as well as more generalized instructions for utilizing Zoom and Canvas to construct a librarian-led seminar geared toward whatever data literacy topic you are hoping to address in your instructional session.

INGREDIENTS

- Librarians
- Zoom or other video software (or a space for in-person instruction)
- Access to Canvas or another learning management system
- Guest speakers or panelists (optional)

PREPARATION

1. *Determine the needs of your constituents:* Identify the research methods that are required for undergraduate honors

theses. For example, one-shot instruction sessions may not sufficiently support undergraduates who are looking for librarian assistance for their social science data needs. Creating a longer seminar to address multiple concepts such as how to successfully find and organize quantitative research data can provide more support than a single session. Librarians who wish to teach a similar seminar at their own institution should begin by considering the data literacy needs of their campus constituencies. Would any of the library's typical one-shot sessions benefit from being expanded into a longer seminar? Are you getting similar questions from different student groups that suggest there might be some kind of instructional gap that a library seminar could fill?

2. *Determine your target audience:* Are there particular student groups whose data needs are underserved by their school's curriculum or the library's typical instructional offerings? For example, your target audience of undergraduate students undertaking a senior thesis or other research project may not be enrolled in a for-credit research methods class. We want to target students who are starting from scratch and have not otherwise been exposed to a broad array of social science data topics.

3. *Create course content:* Once you have determined the data literacy needs of your constituents, curate the content of your seminar based on the needs of the groups on your campus. For example, we chose to focus on the theoretical elements of locating and working with social science data within the context of the research life cycle. These are not typically covered in statistics classes or in workshops offered by other groups on campus. Students are introduced to core social science research concepts and receive in-depth instruction on various elements of finding and organizing social science data sets for their research. (See Instructions for sample seminar topics.)

4. *Conduct outreach:* Targeted outreach works best for this type of instruction. Look at the programs or student populations that would benefit the most from the specific content of your seminar series, and focus on reaching out to these specific groups instead of performing blanket outreach. For example, try advertising directly to undergraduate honors programs. You may find that most of your synchronous live engagement will come from students in one specific honors program. They may be using the seminar to fulfill a research requirement for their program in a virtual space.

INSTRUCTIONS

Each session of the seminar consists of a 20-to-30-minute presentation from a librarian on that week's topic and a 20-to-30-minute presentation by a guest speaker or panel, with time for questions at the end. For virtual instruction, we found that it was helpful to have one librarian teaching the class session and one providing technological assistance to participants and monitoring the chat for questions.

For each session, the general outline follows:
1. Welcome students to the seminar and outline the content of that day's lesson.
2. Review contents of previous sessions (if necessary).
3. Teach the lesson through a lecture and accompanying slide deck (or similar).
4. Any optional interactive elements of your presentation can be included here.
5. Introduce guest speakers and have them present (if applicable).
6. Question-and-answer period (students may unmute themselves or interact via the chat).
7. Wrap up—introduce next week's seminar and remind students who to contact if they have any questions.

Sample Topics

We focused on the following six topics for our seminar series, which was loosely structured around both the research life cycle and the research data management life cycle:

1. *Choosing Your Research Topic: How to Plan for a Successful Research Project:* This session focused on the research life cycle concepts of developing a research question and beginning to plan a research project. Because we didn't know how much research experience our attendees had, we wanted to start off with an overview of the research process. Our guest speakers for this session were a panel of three graduate students in the social sciences who spoke about their own research and paths to pursuing a PhD and took questions from our primarily undergraduate audience.

2. *Telling a Story with Data: How to Locate Data for Your Research:* This session focused on locating social science data to help support (or refute) a research question. In this session, we reviewed the different ways one can structure one's research and the different types of social science data, along with common sources for social science data. We also discussed data visualization and introduced students to Tableau. Our guest speaker for this session was the GIS and geospatial data librarian, who presented an overview of geospatial data and showed where students can locate social science–focused geospatial data.
3. *Collaborating with Data: Tools for Data-Focused Group Projects:* This session focused on working with data in groups and the concepts and tools needed for a successful collaboration. We discussed how to use collaboration tools that our students have access to, to collaborate with other researchers, particularly with respect to research data. Our guest speaker was the Research Data Management Services Group (RDMSG) coordinator, who discussed best practices for collaboration and workflow management and provided an introduction to the Open Science Framework (OSF).
4. *Organizing Your Data: How to Organize, Share, and Replicate Your Research:* We introduced the research data management life cycle, with a focus on data replication and validation and why this is important in social science research. While undergraduates often work with cleaned and packaged data, it is important for them to understand what it takes to get data to that point. Our guest speaker was a research associate from the Cornell Center for Social Sciences, who runs the center's Results Replication service.
5. *Citing Data: How to Cite Data for Your Research Project:* This session delved into data citation. Students are not commonly taught how to cite data, so we devoted a session to the concept. Our guest speaker was the data editor for the American Economic Association, who is also a professor at our institution.
6. Codebooks: How to Find, Read, and Write a Codebook: In our final session, we discussed codebooks. Whether students use already existing data or are performing original research, knowing how to read and write a codebook is an important part of the research process. We discussed best practices for codebook creation. Our guest speaker was the librarian from the Cornell Center for Social Sciences, who discussed her experiences with creating codebooks and led students in a simple exercise of generating a codebook from a provided sample data set.

Post-session Follow-up

After a live session is complete, post the recording of the instruction session and accompanying materials to Canvas or your learning management system so that students who attended the live session can review materials and students who are attending asynchronously have access to the recording and materials as soon as possible. If you are teaching the seminar in person, it is a best practice to post materials as soon as possible for students to access while it is still fresh in their minds.

REVIEWS/ASSESSMENT STRATEGY

For virtual instruction, it can be challenging to assess the seminars only by attendance metrics. There can often be a significant difference between the number of registrants and the amount of actual engagement with the live webinars and the asynchronous materials on Canvas. We found that students engaged more with the materials on Canvas than during the live webinars. However, conducting targeted outreach may encourage students to attend the live seminar to engage with the material and ask questions in real time.

ADAPTING THE RECIPE

The structure of this series can be adapted depending on the library instruction needs of your institution. The combination of live instruction (whether over Zoom or in person) with the asynchronous components in the accompanying Canvas course can be utilized for instructional scenarios that are not suited for a one-shot instruction session or a for-credit course. We highlighted guest speakers or panelists at each session of our workshop instead of hands-on activities. If you do not include guest speakers, you can add in hands-on activities (such as using a particular data tool and software package) or allow more time for discussion. Keep in mind that students who don't attend the live lectures may miss out on participating in the activities. You may want to consider recording separate demos or linking to self-paced tutorials for students to view the content asynchronously.

ADDITIONAL INFORMATION

For additional information on the original seminar, including course outlines, please contact the authors at wbk39@cornell.edu or amelia.kallaher@sagebase.org.